THE (not so) LITTLE BOOK *of* ASTRONOMICAL CURIOSITIES

and Explanations of Celestial Phenomena

Robert J. Douglas

Octant Books
Mill Valley, California

octantbooks@gmail.com

Library of Congress Control Number: 2022909076

Publisher's Cataloging-in-Publication Data
provided by Five Rainbows Cataloging Services

Names: Douglas, Robert J., 1937- author.
Title: The (not so) little book of astronomical curiosities and explanations of celestial phenomena / Robert J. Douglas.
Description: Mill Valley, CA : Octant Books, 2022. Includes bibliographical references and index.
Identifiers: LCCN 2022909076 (print) | ISBN 978-0-578-37708-7 (paperback)
Subjects: LCSH: Astronomy--Popular works. | Solar system--Popular works. | Universe--Popular works. | Planets--Popular works. | BISAC: SCIENCE / Space Science / Astronomy.
Classification: LCC QB44.3 .D68 2022 (print) | LCC QB44.3 (ebook) | DDC 520--dc23.

To my wife, Fran Marsh

PREFACE

This book is intended for amateur astronomers and more generally for anyone with an interest in astronomy. I hope it will further pique the reader's interest in the solar system and beyond. Fifty-nine eclectic topics fall into two general types, with some overlap. Many of the topics selected are ones that I think might contain a surprise and will intrigue you—"Wow" topics for short. Other topics provide explanations for various often-asked questions about celestial phenomena. Some of these latter topics involve mathematical arguments. A few of these arguments are included within the topics themselves, but most are in appendices at the end of the book and are intended for those of a more mathematical bent. Those who prefer to avoid mathematical intricacies can concentrate on the main text which I believe they will find quite interesting.

While reading the book one is struck by the number of scientific disciplines—including chemistry, optics, geology, and physics (even a little quantum mechanics)—that are involved in discussing astronomy. I do not assume the reader has a background in any of these areas. My goal is to carefully provide sufficient information to appreciate their application to astronomy.

Because there are 249 diagrams and images plus five tables accompanying the text, I have chosen to use higher-quality ink and paper. While these upgrades increase the cost of the book, they are essential for fully enjoying the subject matter.

Robert J. Douglas, Ph.D.
Professor Emeritus of Mathematics and Computer Science
San Francisco State University

About the author: Since childhood Robert Douglas has been interested in astronomy and science fiction. He is an active night-sky observer using several telescopes he has acquired over the years.

He received his Ph.D. in mathematics in 1968 from the University of Washington, Seattle. After spending an International Combinatorial Year in the Department of Statistics at UNC Chapel Hill, he joined the mathematics faculty at San Francisco State University. Additionally, he has been a member of the computer science faculty since the department's inception in 1983. He remained at SFSU (with brief time off for sabbaticals at Harvard and the University of Washington and to study computer science at MIT on an NSF grant) until his retirement as Professor of Mathematics and Computer Science in 2000.

He is a member of the San Francisco Amateur Astronomers, the Rose City Astronomers (Portland, Oregon), the Tucson Amateur Astronomy Association, the Planetary Society, and the American Mathematical Society.

He has traveled to Australia nine times for its night skies, a journey he hopes all amateur observational astronomers and astrophotographers can make at least once in a lifetime.

Acknowledgments: My thanks to Alan Agrawal, M.D.; Steve Gottlieb, M.A. (Mathematics), a *Sky & Telescope* Contributing Editor; and Akarsh Simha, Ph.D. (Physics) for their many helpful suggestions, and to my wife, Fran Marsh, for her editorial and graphic assistance.

On the cover: The Homunculus of the star Eta Carinae
NASA/HST/J. Morse/K. Davidson   http://chandra.harvard.edu/photo/1999/0099/

CONTENTS

...strange things may be generally accounted for
if their cause be fairly searched out.
—Jane Austen, Northanger Abbey c 1798

TOPIC 1. Miranda is perhaps the most bizarre moon in the solar system. It also has the highest known cliff in the solar system.

When the physicist Isidor Rabi learned of the discovery of the sub-atomic particle the muon he said, "Who ordered that?" The muon's existence had come as a complete surprise to physicists. Likewise, astronomers were stunned when the first images of Uranus' moon Miranda were returned from Voyager 2 in January 1986. The surface was a hodgepodge of dissimilar terrains. The regions seemed sewn together like Frankenstein.

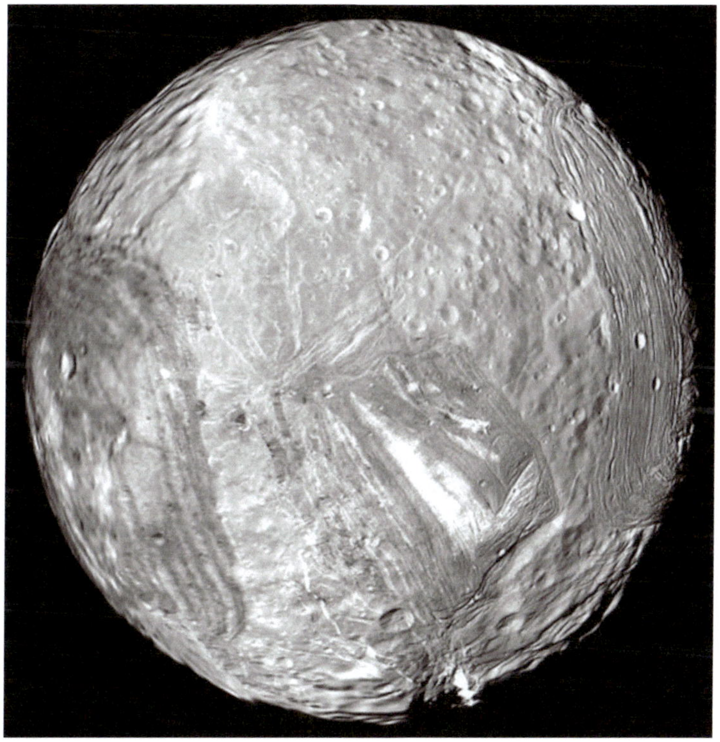

Figure 1. Miranda imaged by NASA's Voyager 2. January 24, 1986. Credit: NASA/JPL-Caltech

Of the five moons of Uranus known prior to the Voyager 2 mission, Miranda is the smallest and the closest to Uranus. Its mean diameter is 293 miles (472 kilometers), and it orbits Uranus in 34 hours at a mean distance of about 80,642 miles (129,780 kilometers).

Voyager 2 was a flyby mission and only imaged the southern hemisphere. Unusual terrain lying in the northern hemisphere will have to wait for another mission. If Miranda has a rival for its peculiarity, it would be Io, Jupiter's volcanic moon (see Topic 7).

From the surface of Miranda, Uranus would be a cyan (bluish-green) ball roughly 22.5 degrees in diameter, or about 45 times the angular diameter of our Moon as seen from Earth. That would indeed be an awesome sight. See *Appendix 1.1* for the calculation. Uranus' cyan color is caused by methane absorbing red light.

Two conspicuous regions in Figure 1 are the "Chevron," a bright, V-shaped feature in a darker, rectangular region, and the oblong, grooved terrain, called coronae, in the upper right, bringing to mind ancient Rome's Circus Maximus chariot racetrack. Regarding the rectangular region, it is unusual to see straight lines in natural terrain.

Several theories[2,3,4] have been put forward to explain the strange appearance of Miranda, one of which is that Miranda in the past had a more eccentric orbit which resulted in tidal heating due to Uranus' gravity (similar to that on Io due to Jupiter's gravity), causing upwelling of softer material.

Now for something that future ecotours in the solar system will advertise: The ultimate experience for those who love to climb cliffs is on Miranda, namely *Verona Rupes*—the highest cliff in the solar system. See Figures 2 and 3. The cliff is currently estimated to be an astonishing 12 miles (20 kilometers) high.[1,2] If you fell or jumped off of it, it would take about 12 minutes to reach the bottom. The long fall time is due to the very low acceleration of gravity on the moon, namely 0.079 meters/second^2 (compared to the Earth's 9.807 meters/second^2). See *Appendix 1.2* for the calculation. If you were climbing it and fell (perhaps not too near the top) you might even survive. But you might want to take an inflatable airbag with you. If you jumped off from the top of the cliff you would be hitting the bottom at about 125 mph (202 kilometers per hour). See *Appendix 1.2* for the calculation.

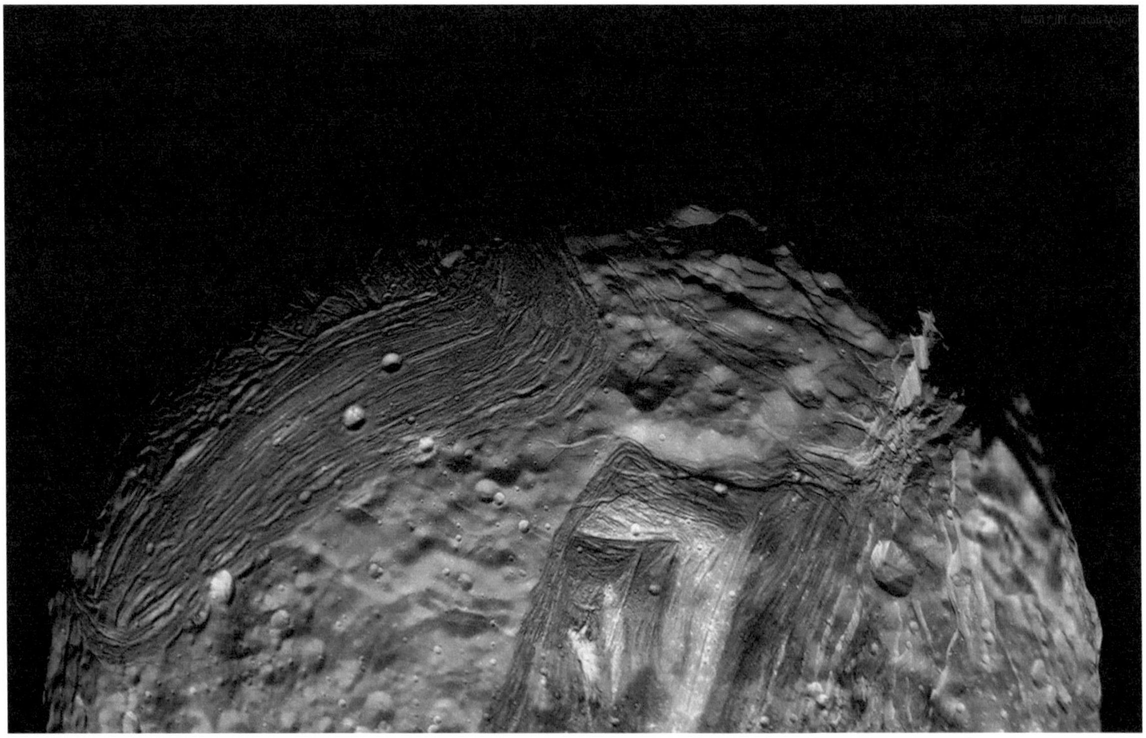

Figure 2. The Circus Maximus region (on the right in Figure 1) is prominent here in the upper left. Verona Rupes (cliff) is on display in the upper right edge and is pointed to by the Chevron. Credit: Voyager 2 (NASA)

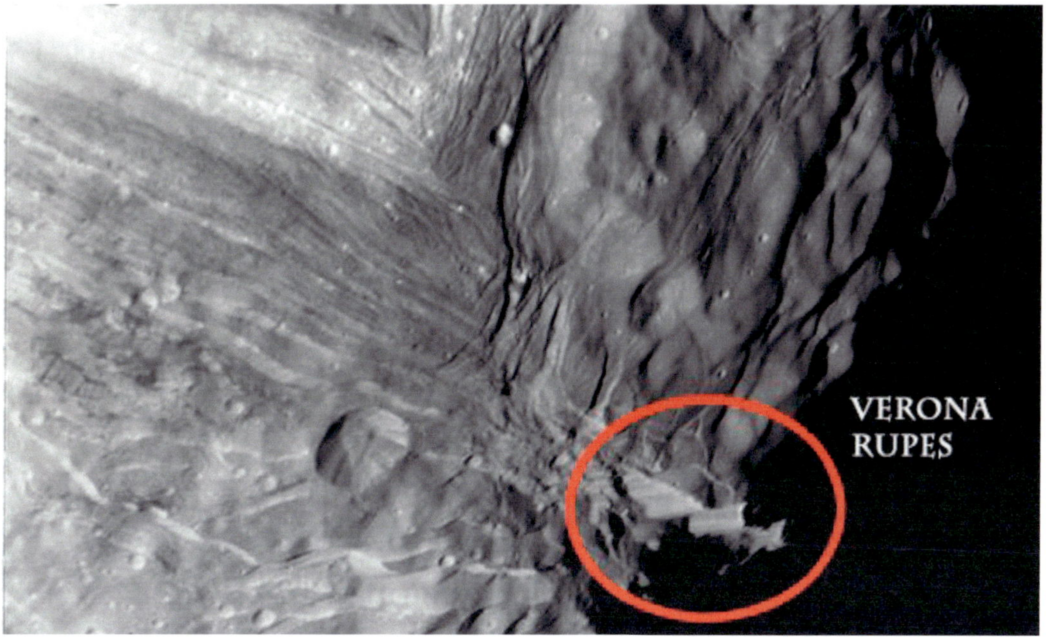

Figure 3. A closer view of Verona Rupes. Credit: Voyager 2 (NASA)

Figure 4. Base jumping off the tallest cliff in the Solar System. Credit: Erik Wernquist, www.erik wernquist.com/wanderers. Wernquist also produced visuals for a beautiful, short film Wanderers*, written and narrated by Carl Sagan—from his book "Pale Blue Dot," courtesy Ann Druyan.*

To give you some perspective on *Verona Rupes'* height, the highest cliff on Earth is on Mount Thor in northern Canada. The mountain itself is 5,495 feet (1,675 meters) high—a little over a mile. Its cliff drops 4,101 feet (1,250 meters) and overhangs at an angle of 105 degrees—steeper than vertical.

Figure 5. Mount Thor on Baffin Island in northern Canada. Its cliff is the largest drop on Earth. Credit: bwallpaperhd.com

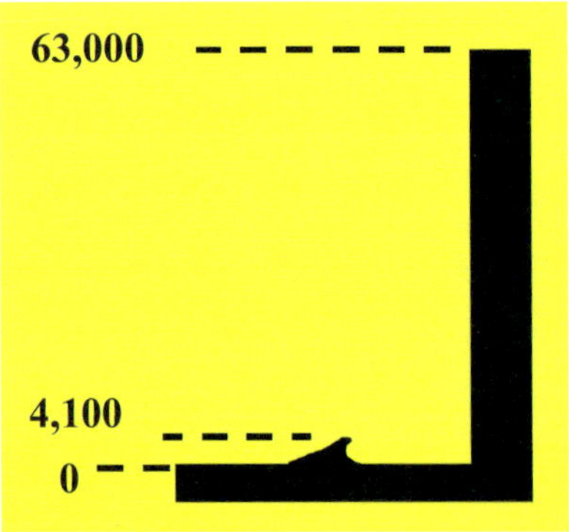

Figure 6. A height comparison. Mt. Thor's cliff on the left at 4,100 feet vs Verona Rupes on the right at 63,000 feet.

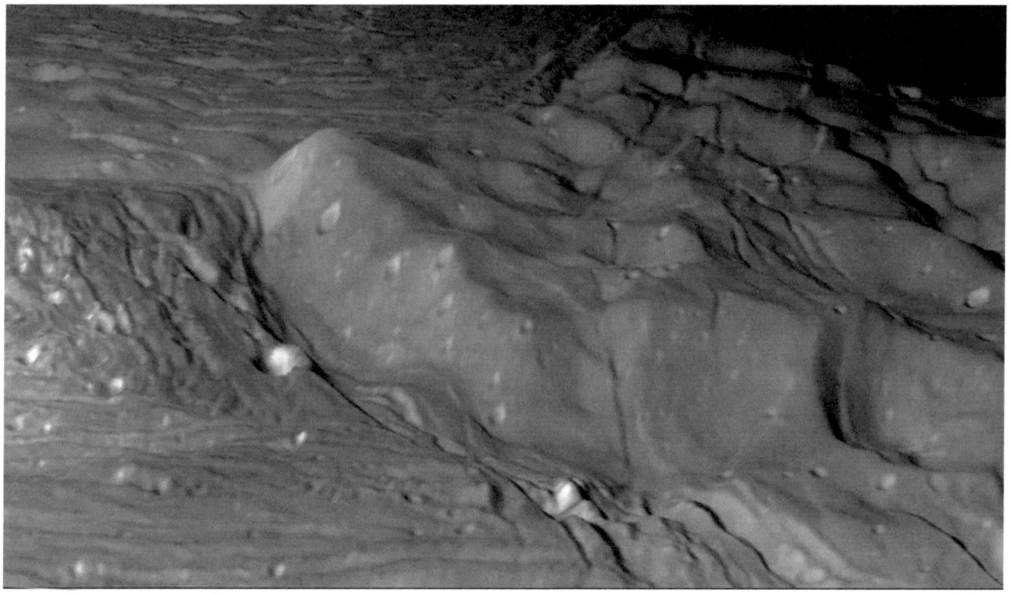

Figure 7. A closeup of the terrain just above the Chevron in Figure 2. It shows cliffs (not Verona Rupes). In the upper left is the edge of the Circus Maximus. Credit: Voyager 2 (NASA)

References

1. NASA Voyager 2 (November 27, 2016). *Verona Rupes: Tallest known cliff in the solar system.* apod.nasa.gov. https://apod.nasa.gov/apod/ap161127.html

2. Choi, Charles Q. (October 3, 2014). *Bizarre shape of Uranus' "Frankenstein" moon explained.* space.com. https://www.space.com/27334-uranus-frankenstein-moon-miranda.html

3. Schenk, Paul M., Gwynn, David, & Tutor, James (n.d.). *The solar system in 3D: Coronae Miranda, Uranus.* Lunar and Planetary Institute, Universities Space Research Association. https://www.lpi.usra.edu/publications/slidesets/3dsolarsystem/slide_37.html

4. Geological Society of America. (September 18, 2014). Miranda: An icy moon deformed by tidal heating. *ScienceDaily*. www.sciencedaily.com/releases/2014/09/140918162222.htm

It is like the two are staring at each other
never breaking their gaze,
as they waltz around the solar system...

—*Maggie Aderin-Pocock*
The Book of the Moon: A Guide to Our Closest Neighbor

TOPIC 2. Pluto's largest moon is Charon. How would Charon appear from Pluto's surface?

Everyone knows that our Moon is tidally locked (also called gravitationally locked) to the Earth, that is, the Moon rotates once on its axis for each revolution of the Moon about the Earth. So, on Earth we can see only one side of the Moon. Indeed, many moons in our solar system are tidally locked to the planet they revolve around.

In the case of the dwarf planet Pluto we have an unusual situation. Pluto's largest moon Charon (half the diameter of Pluto itself) is tidally locked to Pluto, but Pluto is also tidally locked to Charon. So if you were on the "far" side of Pluto, you would never see Charon. Getting in your Pluto car and driving to the other side of Pluto would result in quite a shock when in the sky there would appear this enormous moon. How large would Charon be to an observer on Pluto? The answer is 3.76 degrees, i.e., about seven times the diameter of our Moon. See *Appendix 2.1* for the mathematical details. And Pluto would be 7.18 degrees, i.e., about 14 (Earth) Moon diameters in size as seen from Charon.

Figure 1. Pluto and Charon as viewed by NewHorizons, July 11, 2015.

It is interesting to note that it is not the case that a moon revolves around its primary. Strictly speaking both the moon and its primary revolve about a point called the *barycenter* (or center of gravity) which is on a line between the centers of the two objects. Usually the primary (e.g., the Earth) is so much more massive than its moon that the barycenter lies inside the primary. In the case of Pluto and Charon the barycenter lies outside the surface of Pluto. Pluto and Charon have been called double planets.

Some statistics for Pluto and Charon: Pluto is 1,477 miles in diameter and Charon is 750 miles. The distance between their centers is 12,160 miles.

Change is inevitable. Change is constant.

—Benjamin Disraeli

TOPIC 3. The sky has changed dramatically over the time of human civilization. For most of human history Polaris was not special—it was not the pole star until relatively recently. During the Roman empire there was no pole star. The ancient Greeks could see the Southern Cross from the Mediterranean. Not now. And for the Egyptians at the time of the first Pharaohs the sky was different still.

When the pyramids were built around 2550 BCE, the star Thuban in the constellation Draco appeared very near true north. True north or the *north celestial pole* (NCP) refers to the northern point on the celestial sphere to which the Earth's axis points. At its closest approach in 2830 BCE Thuban was less than 10 arc minutes (one arc minute = 1/60th of a degree) from the NCP. At the present time Polaris is about 40 arc minutes from the NCP and will be closest just after 2100 when it will be about 27 arc minutes away—not as close as Thuban got. Keep in mind that the Moon and Sun are each 30 arc minutes (a half a degree) in diameter. At the time of the ancient Greeks (~350 BCE) the NCP lay halfway between Thuban and Polaris—about 10 degrees from each. During the entire history of the Roman empire there was no pole star. It wasn't until about the 12th century CE that Polaris might reasonably be considered the Pole Star.

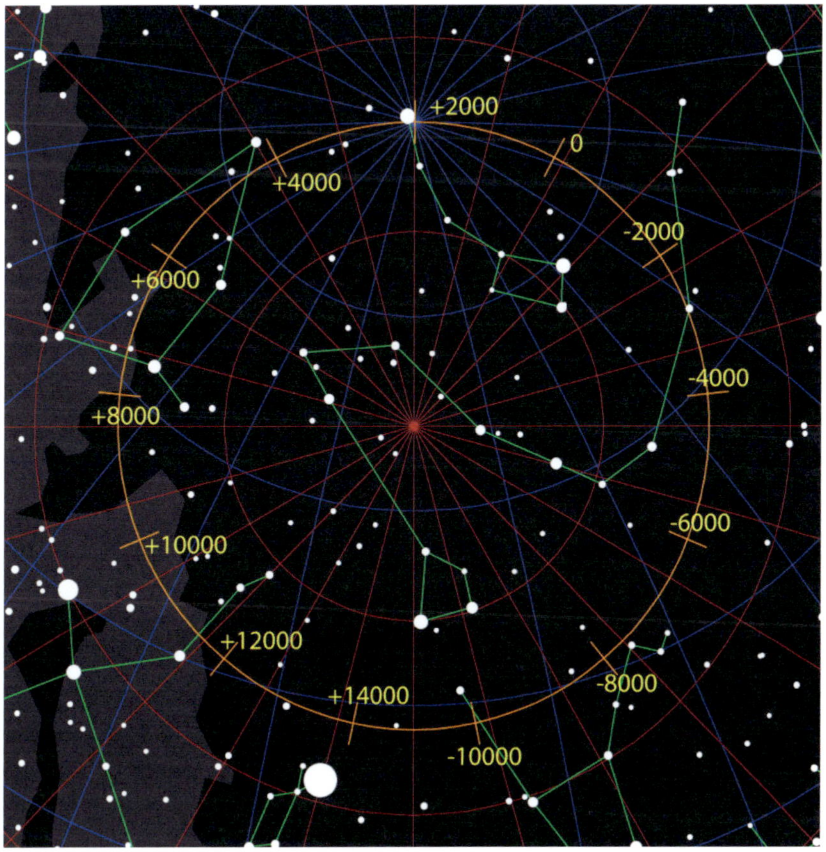

Figure 1. The circle traces the location of the NCP for one complete wobble of the Earth's axis. Credit: Tau'olunga

Figure 1 shows the circle that the NCP traces over about 26,000 years. Polaris is at 12 o'clock. Thuban is the star on the circle CW from Polaris that lies between the years -2000 and -4000. You can see that other stars near the circle and CCW from Polaris will become the pole star in the future. But other than the first-magnitude star Vega (at about 6:30 o'clock), Polaris is the brightest. In about 14000 CE, Vega will be within five degrees of the NCP, not anywhere near as close as Polaris comes. Interestingly, at the same time, Canopus (the second-brightest star in sky) will be within eight degrees of the south celestial pole (SCP). That is because Vega and Canopus are virtually antipodes of each other (opposite each other on the celestial sphere).[2] In about 2800 BCE, the first-magnitude star Achernar was about eight degrees from the SCP, at the same time that Thuban was within about 10 arc minutes of the NCP. Currently the SCP is in the constellation Octans and is without a pole star.

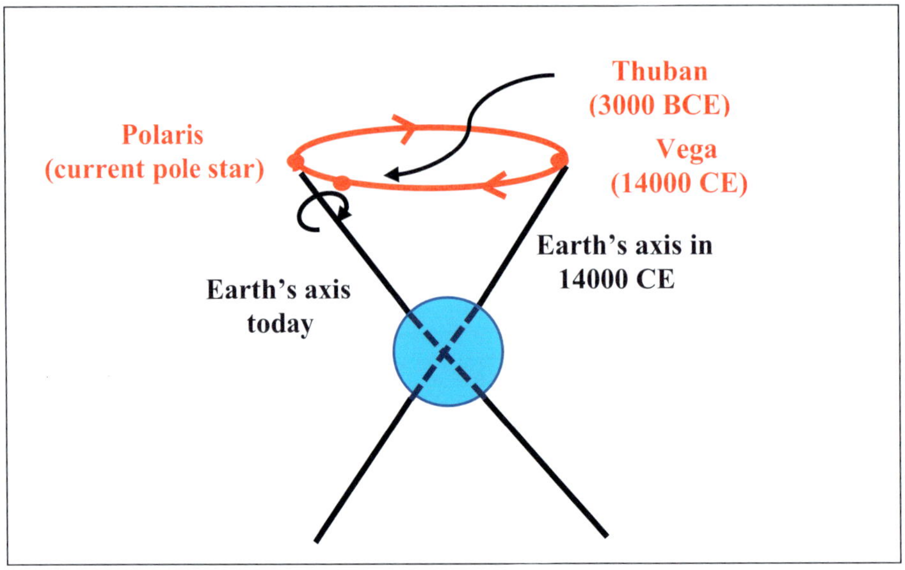

Figure 2. The Earth's axis spins like a top or a gyroscope completing one circuit in 26,000 years.

Why is the NCP (and the SCP) changing in a circular cycle? The reason is that the Earth is spinning like a top. So, in addition to the Earth rotating about its axis, its axis rotates as well, tracing a large circle on the celestial sphere. But unlike the rapid rotation of the axis of a top, the Earth's axis makes one circle against the background stars about every 26,000 years. The Earthly top precesses (wobbles) slowly (at least compared to a human life span). The cause of this wobble is the combined gravitational attraction of the Sun and Moon on the Earth's equatorial bulge. This bulge is from centrifugal force caused by the Earth rotating about its axis every 24 hours. See *Appendix 3.1* for more details. The wobble of the Earth's axis produces a related effect called the "*precession of the equinoxes*." By this we mean a shifting or rotation of the equinoxes against the background stars (the stars in the zodiac).

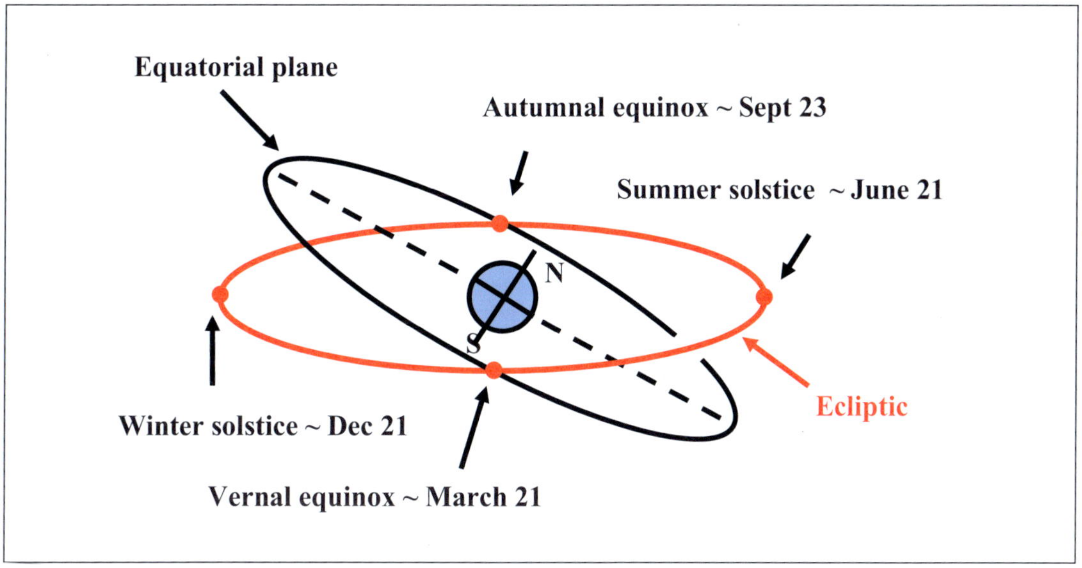

Figure 3. The ecliptic (red ellipse) defines the plane of the of the Earth's orbit around the Sun. It is the projection of the orbital plane of the Earth onto the celestial sphere (the background stars). The black ellipse is the projection of the Earth's equator onto the celestial sphere (and is currently inclined 23.5 degrees to the ecliptic plane). From the point of view of an observer on Earth, the Sun travels in one year around the red ellipse.

First we need some terminology. The *celestial equator* is the projection of the Earth's equator onto the celestial sphere (an imaginary sphere centered on the Earth and containing the stars as seen from Earth). The Earth revolves about the Sun, and the plane defined by the Earth's path intersects the celestial sphere in a circle called the *ecliptic*. The ecliptic is the path that the Sun takes against the background stars over the course of one year as seen from Earth. The inclination of the plane determined by the ecliptic with the plane determined by the celestial equator is currently about 23.5 degrees. The two points where the ecliptic and celestial circles intersect are called *equinoxes*, and when the Sun is at either of these points the length of the day and night are (almost) the same everywhere on Earth. One of these points is called the *autumnal equinox* (~September 23) and the other is the *vernal equinox* (~March 21).

Now you might think that if the axis of the Earth wobbles 360 degrees in about 26,000 years, that the ecliptic (one large circle) would swivel around the celestial equator (a second large circle) with the two equinoxes rotating 360 degrees around the celestial equator in 26,000 years. And you would be correct. This is called the *precession of the equinoxes.* At the time of the ancient Greeks the point on the celestial sphere for the vernal equinox was in the constellation Aries. It was there from 1865 BCE to 68 BCE. It is now in Pisces until 2597, when it moves into Aquarius (the dawning of the age of Aquarius).

In addition to the equinoxes we also have two other important points: the solstices. As seen from Earth, when the Sun is at the farthest northern point on the ecliptic from the celestial equator, we are at the *summer solstice;* when the Sun is at the farthest southern point on the ecliptic from the celestial equator, we are at the *winter solstice.* (Of course, the words summer and winter are only applicable from the vantage point of someone in the northern hemisphere—the reverse holds in the southern hemisphere.) At the time of the ancient Greeks the point on the celestial sphere for the summer solstice was in the constellation Cancer and that for the winter solstice was in Capricorn. Hence the circles of latitude for the summer solstice and winter solstice are respectively called the Tropic of Cancer and the Tropic of Capricorn. At solstices it is at these circles of latitude where the Sun is found directly overhead. But the solstice points have moved due to precession of the equinoxes so that now these points are actually in the constellations Gemini and Aquarius. Should we thus change these names to the Tropic of Gemini and the Tropic of Aquarius? Well, the old names have stuck despite this change.

The above discussion sounds technical, but *the precession of the equinoxes actually has changed what parts of the night sky are visible at any given latitude.* For example, at the time of the ancient Greeks the Southern Cross was visible from the Mediterranean, but is not now. That should be clear as the NCP has shifted position in the night sky, so all other points on the celestial sphere must have also changed position in the night sky.

It is also easy to see this with another argument. Consider a person at a (northern) latitude looking at a star a little above his/her southern horizon. See Figure 4a. Then consider the extreme case of the orientation of Earth 13,000 years later (so the Earth's axis has rotated 180 degrees around the circle traced by the NCP). The same person (much older now) again tries to see the same star. It is clear from Figure 4b that now the star has sunk below the southern horizon.

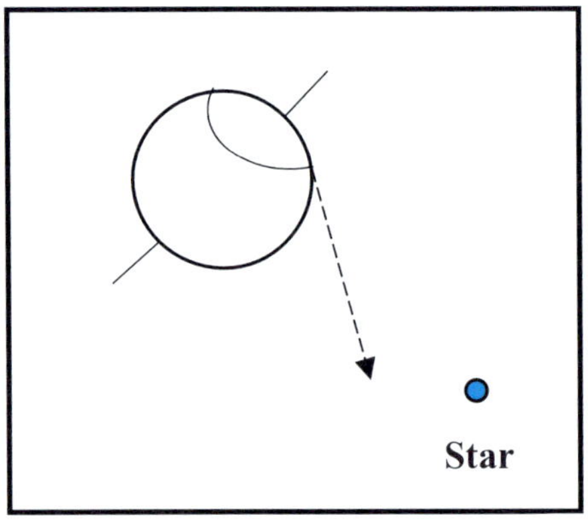

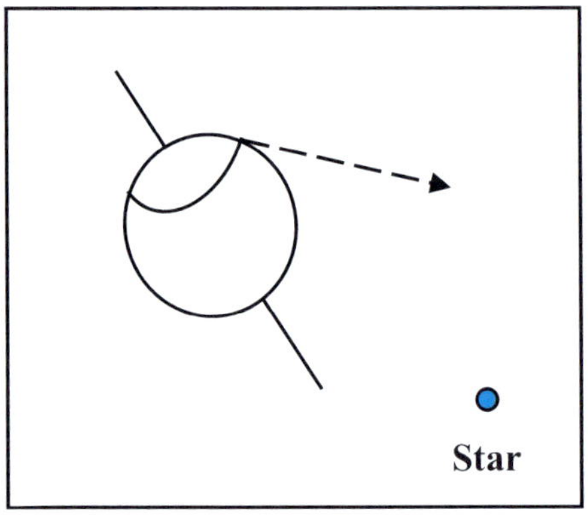

Figure 4a (left). The dashed line is tangent to the Earth's surface at the location of a person, defining the person's southern horizon. A southern star is visible. Figure 4b (right). Thirteen thousand years later the Earth's axis has rotated 180 degrees around the circle traced by the NCP. The same star is now below the southern horizon.

One might very well ask if the angle of about 23.5 degrees between the plane of the celestial equator and the ecliptic plane (the angle of obliquity) changes over time. Indeed it does. That will also affect where the NCP and the Tropic of Cancer (and Capricorn) are against the fixed stars. This effect is due to the combined gravitational attraction of the other planets on the Earth causing the orbital plane of the Earth to precess (relative to distant stars) in a period of approximately 71,000 years. So here it is *the ecliptic plane* that is changing. This is in contrast with the 26,000-year wobble in which the plane of the celestial equator changes. Both effects combine to cause the obliquity to oscillate in an approximately 41,000-year period. But the change is only about ±1.3 degrees, the average obliquity being 23.3 degrees.

If it were not for the Earth having a large moon, this change would be much greater and would lead to large climate fluctuations occurring over any given spot on the Earth. Such a scenario would have adversely affected the ability of life to arise and to sustain itself. Think of a tropical area having a sheet of ice over it in a relatively short (geologically speaking) time period. Living creatures would have to adapt quickly or perish.

While 1.3 degrees is a small angular deviation in the sky, there is a rather dramatic effect on land. Currently the Tropic of Cancer is close to its average position, but is changing at about its fastest rate, namely about 14 meters (approximately 45.5 feet) a year. In 10 years that is 1.5 football fields. In 1908 the Japanese colonial government in Taiwan erected a concrete landmark at the spot where the Tropic of Cancer crossed a newly completed railroad line. Since then the Taiwanese government has made the area a park. But now the Tropic of Cancer has moved entirely out of the park and the administration cannot acquire more land.[1]

I give another illustration to show the effect of the precession of the equinoxes. If someone is asked what a year is, the response might be that it is the time for the Earth to make one orbit around the Sun. But that year is the orbital period of the Earth around the Sun, called a *sidereal year*, which is the time it takes for the Sun to move from one location with respect to the background *"fixed" stars* back to the same location.

However, that is not what a year is with respect to our calendars. We use a calendar that is based on the *seasons*, not on the fixed stars. We use a *solar year*, also called the *synodic year or tropical year*, which is the time the Sun takes to return to the same position in the cycle of seasons, as seen from Earth. For example, it is the time from vernal equinox to vernal equinox, or from summer solstice to summer solstice. That is a more useful year for agricultural purposes—of interest to all peoples in the history of the world. One sidereal year is 365 days, 6 hours, 9 minutes, 10 seconds. One solar year is 365 days, 5 hours, 48 minutes, 45 seconds. So a sidereal year is about 20 minutes longer than one solar year.

Suppose the Mayans constructed a temple so that the Sun's rays on summer solstice at noon would pass through an opening in the temple and light up an inside passageway. Further suppose they presumed that in each succeeding year this event would repeat. If the year-long period is defined as a solar year, then the precession of the equinoxes is already taken into consideration. No problem. But if by a year one means a sidereal year, then after just one year the Sun would be seen in the passageway not at noon, but 20 minutes before noon. And after three years, the event would occur one hour too early. Using the sidereal year doesn't take into consideration the precession of the equinoxes, with disastrous consequences.

Finally, let us instead assume that the passageway was desired to be constructed so that the star Sirius could be seen from it when Sirius, year after year, reached its highest point in the sky (i.e., was on the meridian). From our earlier discussion we see that this is not possible as the positions of the stars in the whole sky are shifted due to precession of the equinoxes. From some latitudes Sirius would be visible for many years during the 26,000-year precession, and for many other years, from those same latitudes, it would never be visible.

References

1. Rubincam, David, Chao, B. Fong, & Bills, Bruce (June 1998). The incredible shrinking tropics. *Sky & Telescope,* p 36-38.
2. Berman, Bob (March 2021). Voyage to the antipodes. *Astronomy Magazine,* p 12. Here Berman points out that Vega and Canopus are antipodal to each other and the consequence that they will simultaneously be (almost) pole stars.

Timothy Leary and I kept the same hours. He believed, as I do, that 'After midnight, all things are possible.'

—Hunter S. Thompson

TOPIC 4. During a meteor shower, why does one see more meteors after midnight than before midnight?

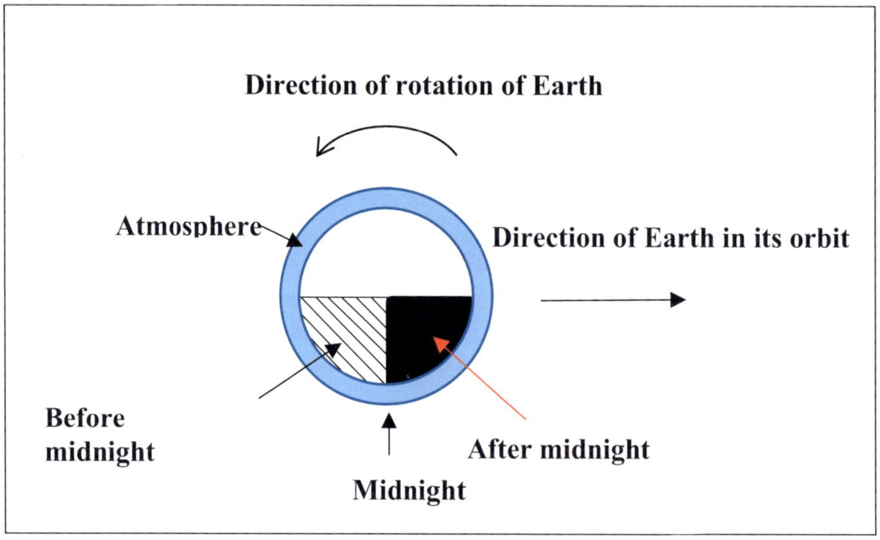

Figure 1. View of the Earth looking down on the North Pole. Imagine that the entire Earth is embedded in a stream of meteoroids (not shown).

Think of the Earth as a car moving rapidly left to right through air filled with rain drops. The front windshield will hit many more drops than the rear windshield (regardless of the direction the rain drops are moving). The front windshield corresponds to that half of the Earth (the right half in Figure 1) plowing into the meteoroid particles, and the rear windshield corresponds to the other half of the Earth. Since we can only see the meteoroids at night and since the Earth in Figure 1 rotates CCW, it is thus best to be viewing *after* midnight. (If the Earth rotated CW, which it does not, then the best views would be before midnight.)

In Figure 1, keep in mind that the meteoroid stream extends well above and below the surface of the paper. So even when on the night side before midnight, some particles will be seen burning up in the Earth's upper atmosphere when looking toward the horizon. See [1] for a good discussion of meteor showers.

References

1. Lynch, David K. (2008). What causes meteor showers?—How to observe meteor showers. https://geology.com/articles/meteor-shower.shtml

A cosmic quotation mark?
No, it's just another moon of Saturn

—Jason Major, space blogger[5]

TOPIC 5. Saturn's third largest moon, Iapetus, is strange: It is walnut-shaped with one side much darker than the other.

Darker and lighter hemispheres.
Iapetus is Saturn's third-largest moon with a mean diameter about 1,471 kilometers (914 miles). When Giovanni Domenico Cassini discovered Iapetus on October 25, 1671, he noticed an oddity. He first observed the moon when it would have been on the western side of Saturn. Then he attempted to see it when it should have been on the eastern side—its period around Saturn is 79 days. This failed. In 1672 he repeated his observation of Iapetus on the western side, but again failed to see it on the eastern side.

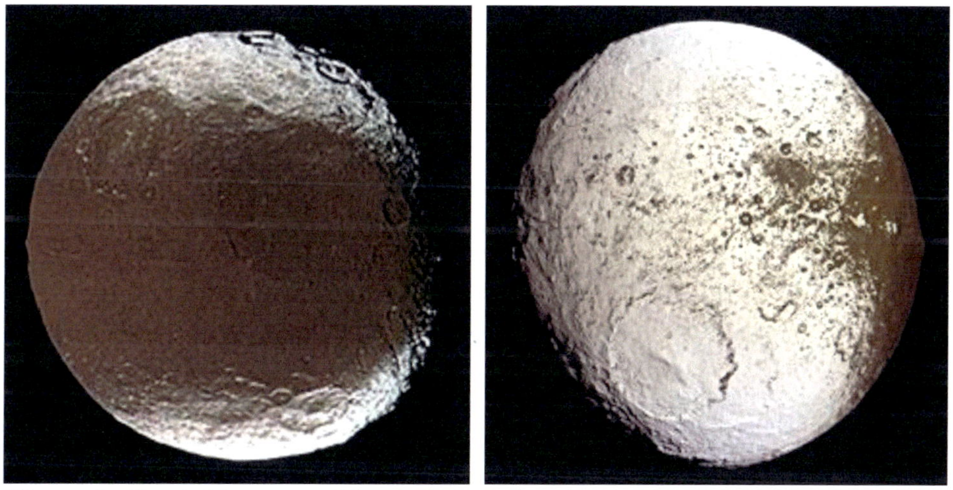

Figure 1. Cassini images of Saturn's two-faced moon Iapetus.
Credit: NASA/JPL/Space Science Institute

In 1705, with an improved telescope, he managed to see the moon on Saturn's eastern side. But it was much fainter. The brighter side has magnitude about 10.2; the darker side has a magnitude about 11.9, too faint for 17th century telescopes. Cassini correctly deduced what the problem was: Iapetus' leading hemisphere in its orbit around Saturn must be darker than the trailing hemisphere—and Iapetus was tidally locked to Saturn, so it always presented the same face to its planet. Thus, as seen from Earth, Iapetus always displays its bright hemisphere when it is on the western side of Saturn (at the top in Figure 2) and its darker hemisphere when on the eastern side (at the bottom in Figure 2).

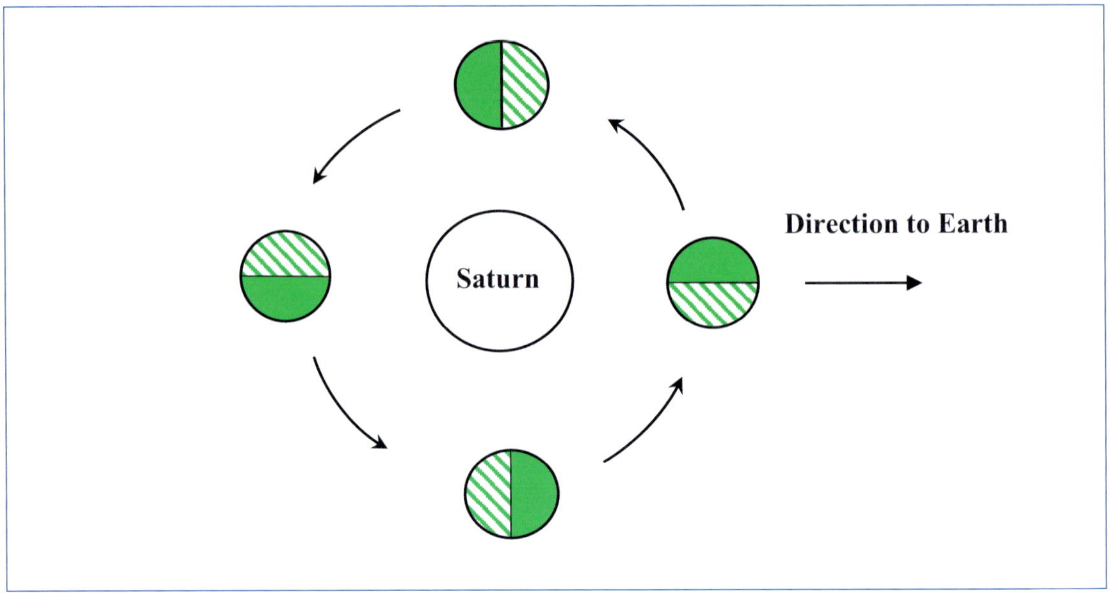

Figure 2. Iapetus orbiting Saturn with its leading hemisphere solid green and its trailing hemisphere striped green. In the diagram we are looking down from above Iapetus' orbital plane. The diagram is accurate as Iapetus revolves once on its axis for every orbit it makes around Saturn.

But why the striking difference in albedo (the proportion of the incident light or radiation that is reflected by a surface) between the two hemispheres? A leading theory is that material ejected from the moon Phoebe gets deposited onto the leading hemisphere of Iapetus, darkening it. It is similar to the case of a car moving in the rain. More rain drops fall on the car's front windshield than on its rear windshield. So it is natural that Iapetus' leading hemisphere would pick up any particles that the moon moves through. And since the moon always has the same face toward Saturn, its leading hemisphere is always the same physical face of the moon and keeps accumulating material.

The Phoebe ring is part of the explanation.
Of the nine largest moons of Saturn, Phoebe is the farthest from Saturn, with Iapetus being the next moon in. The irregular-shaped Phoebe, with a mean diameter of 213 kilometers (132 miles), is the second-smallest of the inner nine moons and largest of the outer ones. It revolves in retrograde motion around Saturn, indicating it was not one of Saturn's original moons, but was captured. The orbital plane of Phoebe is inclined only slightly (about five degrees) to the orbital plane of Saturn.

In October 2009 the Spitzer Space Telescope discovered a large, tenuous "ring" visible in the infrared around Saturn. The ring is in Saturn's orbital plane and consists of material ejected from Phoebe by micrometeorites. Material that, like Phoebe itself, orbits Saturn in retrograde motion. So Iapetus' leading hemisphere is moving into material that is streaming toward it (i.e., opposite to the direction Iapetus is moving).

Because Saturn's axial tilt (the angle between Saturn's rotational axis and a perpendicular to Saturn's orbital plane) is 27 degrees, the angle between Saturn's equatorial plane (which is Saturn's ring plane) and its orbital plane is also 27 degrees. So the Phoebe ring is tilted 27 degrees to the ring plane of Saturn. The ring's thickness is many times the diameter of Saturn.

Phoebe, embedded in its ring, moves through it with an up-and-down motion which occurs as Phoebe shifts above and below Saturn's orbital plane by about five degrees in each direction. Iapetus itself has an orbit inclined about 15 degrees to Saturn's ring plane. Material from the Phoebe ring drifts inward toward Iapetus. Figure 3 below depicts the situation. The discovery of the Phoebe ring supports the theory that material from Phoebe is "painting" one face of Iapetus.

Figure 3. The Phoebe ring is tilted 27 degrees to Saturn's ring plane; Iapetus' orbit is tilted 15 degrees to Saturn's ring plane. Phoebe moves up and down while traveling through its ring. Credit: NASA/JPL-Caltech

A thermal feed-back loop: another part of the explanation.

It is further thought[1] that Iapetus' dark region, which absorbs more heat from the Sun than the brighter hemisphere, causes water ice on the dark side to sublimate and find its way to the cooler hemisphere. There it condenses maintaining that hemisphere's brightness and the darker side's darkness. This exacerbating thermal feed-back loop leaves little surface ice remaining on the dark side. The dark region is now called *Cassini Regio* after its discoverer. So the dark area is due to two factors: dark material from Phoebe painting one face of Iapetus and ice vaporizing, with the water vapor (under a weak gravitational force of the moon) migrating and condensing onto the opposite face. (Iapetus, with a density only 1.2 times that of liquid water, consists mainly of ice as opposed to rock.)

By the way, a good mnemonic for remembering the names of the nine largest moons of Saturn in their order from Saturn is MET DR THIP: Mimas, Enceladus, Tethys, Dione, Rhea, Titan, Hyperion, Iapetus, and Phoebe. The first seven of these moons orbit in essentially the same plane as Saturn's rings. So the thin rings would not be easy to discern from them. But Iapetus' orbit, inclined about 15 degrees to the ring plane, gives a non-edge-on view of the rings.

The Walnut Shape.

The Cassini spacecraft mission to Saturn was launched in 1997, entered orbit around Saturn in 2004, and was deliberately flown into Saturn (after passing through its ring plane) in 2017 to avoid its crashing into and contaminating any of Saturn's potentially habitable moons. The spacecraft passed close to several of Saturn's moons giving us spectacular images. In 2004 and 2007 Cassini flew close to Iapetus, with 2007 being the closest pass. Below are three images of Iapetus showing that it has a ridge along its equator which extends for about 30 percent of the way around the moon. This ridge and the moon's oblateness (its equatorial diameter is greater than its polar diameter) gives it the appearance of a walnut. Iapetus is equatorially flattened by 4.5 percent as opposed to only 0.3 percent for the Earth.[4]

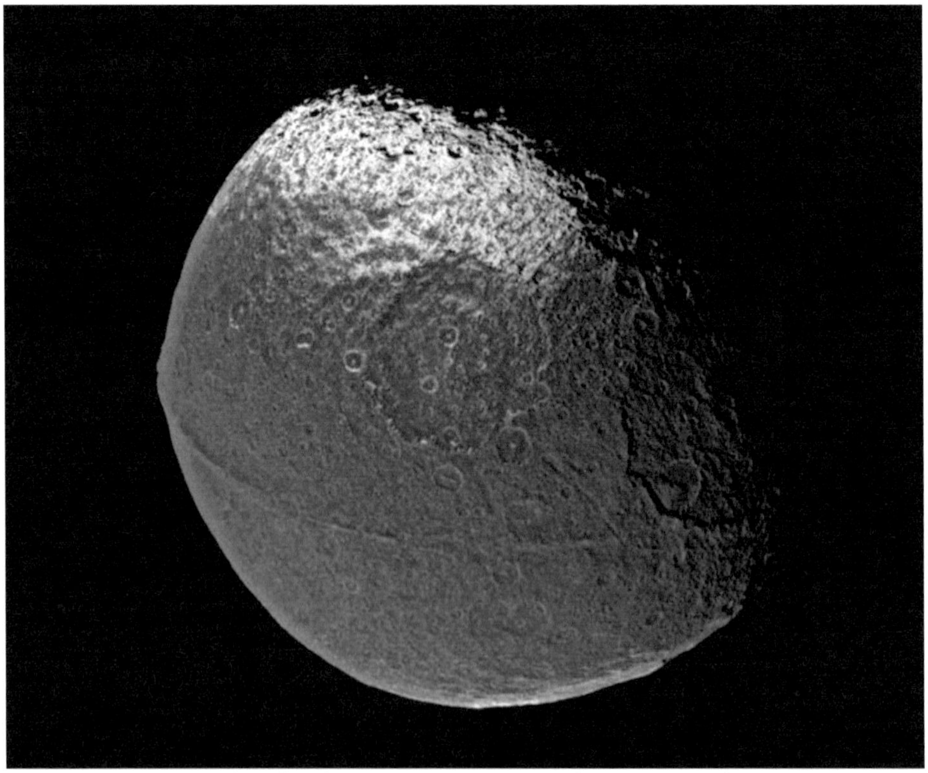

Figure 4. Image taken by the NASA Cassini probe in December 2004.
Credit: © NASA/JPL/Space Science Institute

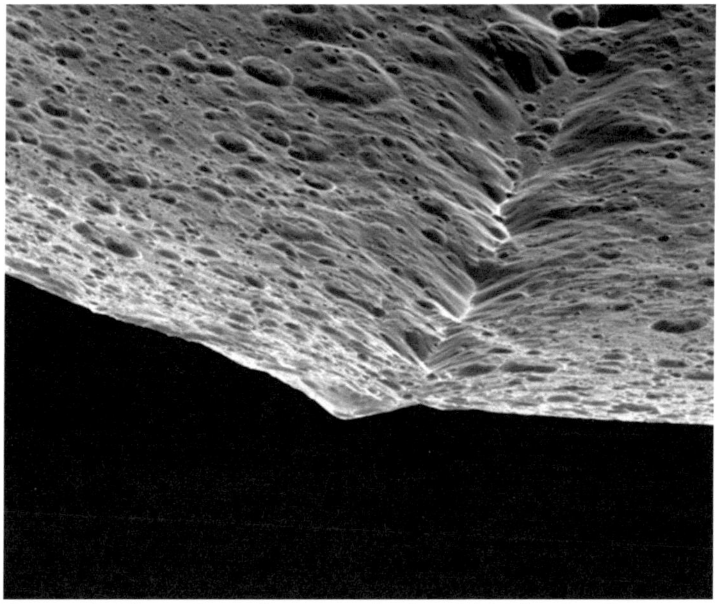

Figure 5. The ridge, photographed in 2004 by the Cassini spacecraft. Credit: © NASA/JPL/Space Science Institute

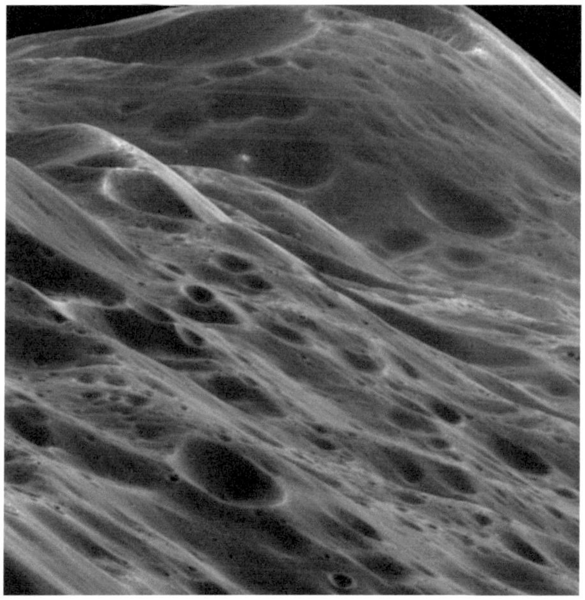

Figure 6. Close-up view showing mountainous terrain that reaches about 10 kilometers (six miles) high along the equatorial ridge of Iapetus. The image was taken on September 10, 2007, when the spacecraft was about 1,227 kilometers (762 miles) from the surface. Credit: NASA/JPL/Space Science Institute

The ridge is about 1,300 kilometers (808 miles) long[2] and is in the Cassini Regio (the dark region of the moon). It has peaks up to 20 kilometers (12.4 miles) high.[2,4] Compare that to Mt. Everest which is 5.5 miles above sea level. The ridge system includes sets of isolated peaks, ridge segments more than 200 kilometers long, and some sections with three nearly parallel ridges.[2] While Iapetus' bright face doesn't have a ridge similar to the dark face, a series of isolated peaks 10 kilometers (six miles) high runs along the equator.[3] These mountains, discovered by Voyager 1 and 2, are referred to as the Voyager Mountains. Since the ridge is heavily cratered, it must be ancient.

Scientists debate how the ridge might have formed. One idea is that Iapetus at an earlier time was rotating much faster than today and material bulged up. Another idea[4] is that a ring of material around the moon rained down onto the moon's equatorial region. And this ring may have formed from a moon circling Iapetus (a moon with a moon) that broke up when this moon approached too close to Iapetus (reaching Iapetus' Roche limit). The Roche limit is the closest distance from the center of a body that a satellite can approach without being pulled apart by the primary body's gravitational field.

References

1. Reddy, Francis (March 2018). Saturn's small worlds. *Astronomy Magazine*, p 32. This idea was first proposed in 2010 by John Spencer at the Southwest Research Institute in Boulder, Colorado, and Tilmann Denk at the Free University of Berlin.
2. Porco, C. C., et al. (February 25, 2005). Cassini imaging science: Initial results on Phoebe and Iapetus. *Science, 307*, p 1237–1242.
3. Cassini solstice mission, NASA/JPL/Space Science Institute. https://web.archive.org/web/20110610025306/http://saturn.jpl.nasa.gov/photos/imagedetails/index.cfm?imageId=2760
4. Dombard, A. J., Cheng, A. F., McKinnon, W. B., & Kay, J. P. (March 7, 2012). Delayed formation of the equatorial ridge on Iapetus from a subsatellite created in a giant impact. *Journal of Geophysical Research, Planets, an AGU Journal, 117*(E3). https://doi.org/10.1029/2011JE004010
5. NASA/JPL/Space Science Institute. Composite by Jason Major. https://lightsinthedark.com/2013/09/03/a-cosmic-quotation-mark-no-its-just-another-moon-of-saturn/

A clever person solves a problem.
A wise person avoids it.

—Albert Einstein

TOPIC 6. Two moons of Saturn are in virtually identical orbits. You might think they would collide. They don't. What is going on?

The two small moons Epimetheus and Janus, respectively about 120 and 180 kilometers (75 and 112 miles) in diameter, orbit Saturn in the same direction. Their orbits only differ by about 50 kilometers (31 miles) in distance from Saturn, less than the radius of either body. Since the moon that is slightly closer to Saturn travels at a slightly faster speed than the outer moon, one might think that when it catches up with the outer orbiting moon it would collide with it.

But something else happens. One might humorously call it a "dance" or a "passing protocol." The inner moon gravitationally attracts the outer moon, slightly slowing it down. Simultaneously the outer moon gravitationally attracts the inner moon, speeding it up. Since the total angular momentum of the two moons is a constant, the velocity changes cause the inner moon to move a little farther away from Saturn and the outer moon to move a little closer to Saturn. In effect the two moons switch orbital places due to an exchange of angular momentum. This "dance" takes about 100 (Earth) days, and the two moons keep a healthy minimum distance (of about 15,000 kilometers or 9,320 miles) between them during this period. Then, after about four years, they repeat this performance in reverse. Because Janus is about four times as massive as Epimetheus, its radial movement is about one-fourth the radial movement of Epimetheus. The twins' co-orbital situation is the only one known in the solar system. See Figure 2.

Figure 1. Epimetheus (lower left) and Janus (right) as seen by the Cassini spacecraft on March 20, 2006, two months after swapping orbits. Credit: NASA/JPL/Space Science Institute

In Figure 1, Janus is actually farther away (by about 40,000 kilometers) from the Cassini spacecraft than Epimetheus. They seem closer due to foreshortening—i.e., because we are viewing the two moons with a small angle between them, the distance between them appears foreshortened. If we viewed them when they both were in a plane perpendicular to our line of sight, the distance between them would not be foreshortened.

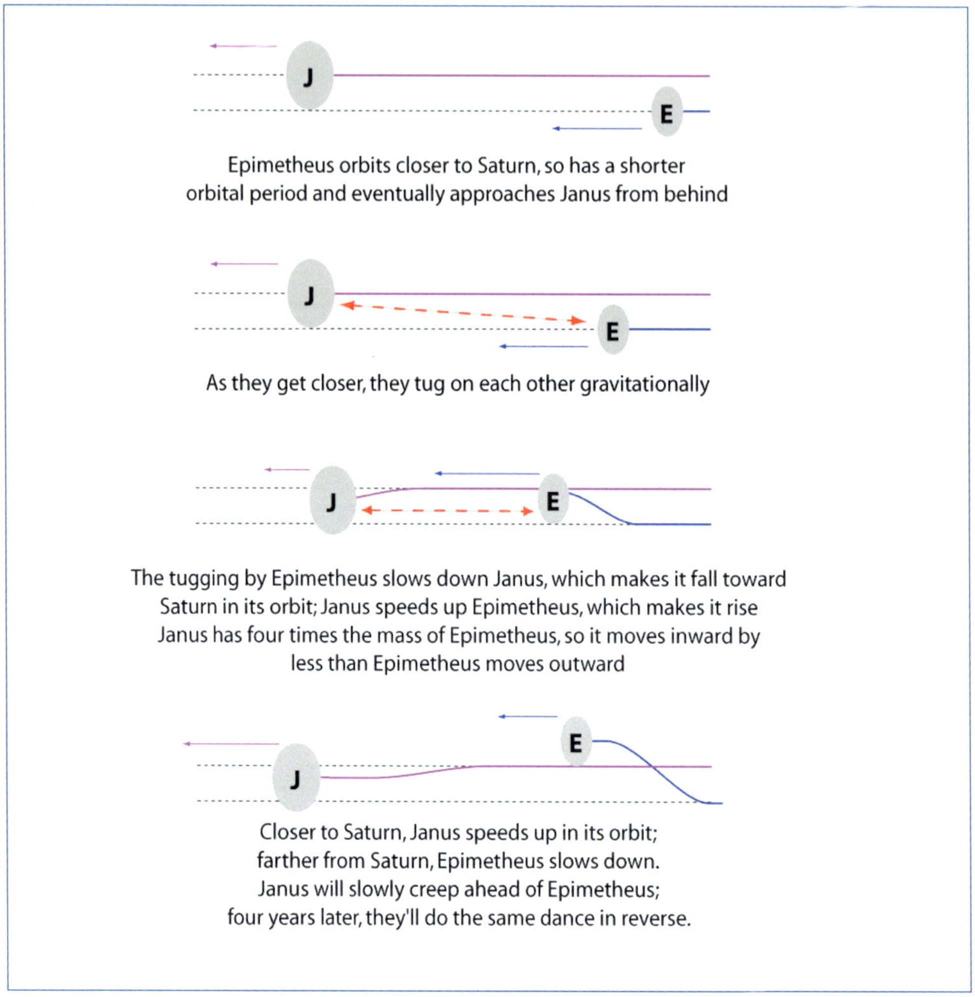

Figure 2. In this depiction, Saturn is located below each of the four sections. Credit: Emily Lakdawalla, the Planetary Society blog[1]

References

1. Lakdawalla, Emily (February 7, 2006). The orbital dance of Epimetheus and Janus, Planetary Society blog. https://www.planetary.org/articles/janus-epimetheus-swap

Hurricanes, tornadoes,
magnetic fields around planets and stars,
and whole galaxies themselves
are all toroidal energy systems.

—Cosmometery.net

TOPIC 7. The Jupiter-Io plasma torus and flux tube: An unusual, interconnected magnetic and electrical phenomenon in our solar system.

Io, the closest of the four Galilean moons of Jupiter and slightly larger than our Moon, has more than 400 volcanoes on its surface. Over 100 are erupting an any given time, constantly reforming its hellish-looking surface.

Figure 1. Galileo orbiter's true-color image of Io. The dark spot just left of the center is the erupting volcano Prometheus. Whitish plains on either side of it are coated with volcanically deposited sulfur dioxide frost (SO_2), whereas the yellower regions contain a higher proportion of sulfur. Credit: Galileo Orbiter

Figure 2. Eruption at Tvashtar Catena, February 22, 2000. Credit: Galileo Orbiter

The heat that produces volcanic eruptions on Earth is primordial heat left over from when Earth first coalesced in addition to heat from radioactive elements. For Io, heat is generated from the gravitational pull of three other bodies: Jupiter and its moons Europa and Ganymede. Forces of these three bodies combine to knead or flex the material constituting Io. The gravitational stretching and shrinking generates great heat from internal friction, resulting in the many volcanoes. The exact composition of the outflowing low-viscosity lava is not yet known.[2] It could be basalt or ultramafic lava (igneous rock with a low silica content and high Mg and Fe mineral content).[2] But the crust of Io contains a considerable amount of sulfur and sulfur dioxide (SO_2). So some of the sulfur and sulfur dioxide ejected by the eruptions returns to coat the surface. Molecules that don't return to the surface are broken apart into neutral and ionized oxygen and sulfur. The neutral atoms are then ionized by ultraviolet radiation and electron impacts. The total mass loss from Io is thought to be about one ton per second[1] which produces a large mass of ionized sodium atoms (and other ions) and electrons surrounding Io.

There is another player in this drama. The very large and strong magnetic field of Jupiter sweeps down over Io. This magnetic field is rotating around Jupiter in sync with Jupiter's rotation, which is one rotation of 360 degrees in 9.8 hours. Io itself makes one orbit around Jupiter at the much slower rate of 42 hours. So Jupiter's rapidly rotating and powerful magnetic field accelerates the slower-moving sodium ions away from Io and distributes them in a large *plasma torus* (donut-shaped ring) that circles Jupiter with its center in the equatorial plane of Jupiter. Io's orbit is slightly inclined to this equatorial plane, but lies completely inside the torus.

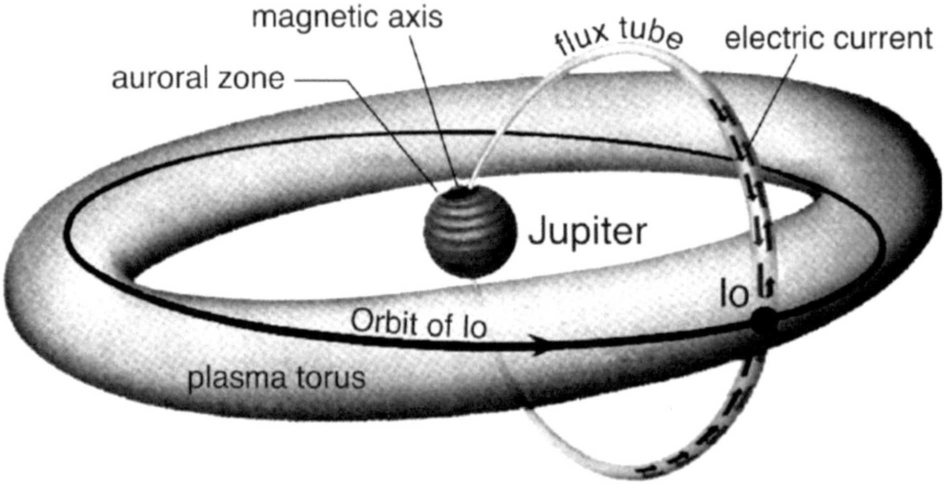

Figure 3. Jupiter's plasma torus and flux tube. Credit: See reference [1]

In addition, all of the free electrons from the ionized sodium atoms surrounding Io flow along the magnetic field lines of Jupiter's magnetic field. This produces a *flux tube* from Jupiter's north magnetic pole to Io and from Jupiter's south magnetic pole to Io. Electrons spiraling inside these tubes create a powerful electric current circling from each pole to Io and back to that pole. Magnetic field observations from the Voyager spacecraft give an estimated electric current in Io's flux tube of 2.8 million amps.[1] That would make one powerful toaster.

The electrons in the flux tubes strike Jupiter's upper atmosphere and produce auroras around the magnetic poles. These auroras are permanent, unlike the transient auroras on Earth. Furthermore, the electrons are spiraling (hence accelerating) at near light speed in the flux tubes and thus produce radio waves. These Io-dependent radio waves are more intense and predictable that the non-Io-dependent waves produced by particles spiraling along field lines in Jupiter's magnetosphere. Radio waves were first detected emanating from Jupiter in 1955, and Jupiter is now known to be the strongest radio source in our solar system after the Sun.

Figure 4. Aurora on Jupiter. An image composite from two different Hubble observations. Credit: NASA, ESA, and J. Nichols (University of Leicester)

The auroras were photographed during a series of Hubble Space Telescope Imaging Spectrograph far-ultraviolet-light observations taking place as NASA's Juno spacecraft approached and entered into orbit around Jupiter. The full-color disk of Jupiter in this image was separately photographed at a different time by Hubble's Outer Planet Atmospheres Legacy (OPAL) program, a long-term Hubble project that annually captures global maps of the outer planets.

Radiation belts (radiation not in the sense of radioactivity, but in the sense of charged particles) occur as clouds or zones of electrons and ions (protons, oxygen, and sulfur) in places along Jupiter's magnetic field lines. The protons emanate from Jupiter's ionosphere and the solar wind, though the solar wind plays a minor role. Jupiter's radiation belts are millions of times more intense than the Earth's Van Allen belts, which were discovered by Explorer 1 in 1958. Most of the ions and electrons in Earth's belts are from the solar wind, with some from the Earth's atmosphere. Earth's ions are protons and helium ions, not the heavier ions of oxygen and sulfur found in Jupiter's belts.

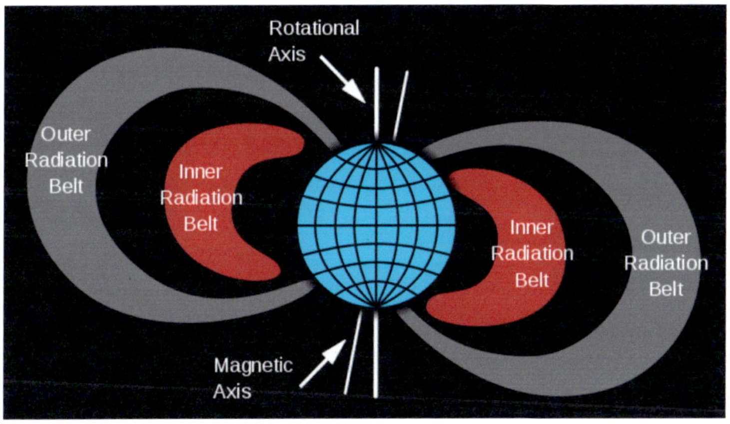

Figure 5. Earth's two Van Allen Radiation Belts. Credit: Wikimedia Commons

References

1. Saur, J., Neubauer, F., Connerney, J., Zarka, P., & Kivelson, M. (2004). Plasma Interaction of Io with its Plasma Torus. In F. Bagenal, T. Dowling, & W. McKinnon (Eds.), *Jupiter: The Planet, Satellites and Magnetosphere* (Chapter 22, p 538). Cambridge University Press. In this source, the diagram in Figure 3 is labeled "Sketch of the Io plasma torus and the general geometrical setup after Audouze et al." https://lasp.colorado.edu/home/mop/files/2015/08/jupiter_ch22-1.pdf

2. Davies, Ashley Gerard (July 2020). Io—The Volcanic Rosetta Stone. *Sky & Telescope,* p 14-21.

There is major interest [in Ceres] at this stage in quantifying the habitability potential of the deep brine reservoir....

—Julie Castillo, planetary scientist
NASA/JPL August 2020

TOPIC 8. Ceres' version of "The Lighthouse at Alexandria" and Lone Mountain on Ceres.

The dwarf planet Ceres was discovered on January 1, 1801, by Giuseppe Piazzi at the Palermo Astronomical Observatory in Sicily. In his telescope it didn't appear as a disk, only as a moving, star-like object. Soon similar objects were discovered, and they all were given the name "asteroids" (Greek: asteroid, star-like). The asteroid belt lies between the orbits of Mars and Jupiter and consists of protoplanetary bodies that either never coalesced into planets or are the remains of a few planetesimal-sized objects that collided and broke up. Ceres and Vesta are the two largest, with diameters of 952 kilometers (592 miles) and about 530 kilometers (329 miles) respectively. Ceres, unlike the other asteroids, was large enough to form a sphere and is now classified as a dwarf planet (a promotion), making Vesta the largest asteroid. Comprising about 25 percent of the mass of the entire asteroid belt, Ceres lies just inside the "snow line" (the distance from the Sun where liquid water condenses into solid ice grains) which is about three AU. (One AU or astronomical unit is the average distance of the Earth from the Sun.) Ceres is the smallest of the dwarf planets and the only dwarf planet whose orbit is entirely inside Neptune's orbit—its orbit is in fact entirely between Mars and Jupiter. Pluto and Eris are currently the two largest dwarf planets known, with Pluto's diameter being slightly greater than that of Eris, but with Eris being the more massive of the two. While Vesta is mainly rocky with a substantial nickel and iron core, Ceres' density is low, suggesting substantial water ice below its surface. Ceres is believed not to have formed in the asteroid belt, but farther out in the solar system and thus is composed of lighter material. It has migrated inward to its present position.

It is interesting to note that all the inner planets from Mercury to Mars have transits across the face of the Sun that could be seen from Ceres. The Earth transited in 1814 and will again in 2081.[14] Tickets for the latter event will be expensive.

NASA's Dawn spacecraft, powered by an ion engine, launched in September 2007 and orbited Vesta from July 2011 to September 2012. In March 2015 Dawn started orbiting Ceres as close as 35 kilometers (22 miles). Its mission ended when it ran out of fuel in November 2018.

Figure 1. The Dwarf Planet Ceres. Credit: NASA/JPL-Caltech

Figure 2. Ceres' Occator Crater. Credit: NASA/JPL-Caltech

The white spot.[1, 2, 3, 4, 5, 6, 7, 8, 9]

As Dawn approached Ceres, it detected one very obvious and strange feature on the surface—a very bright, white spot—like a headlight beaming straight on. Only the largest telescopes on Earth could barely detect any features on the surface of Ceres, but this mysterious white area was somewhat visible on one hemisphere. As Dawn got closer, it was seen that the white area was inside a prominent crater, which now has the name *Occator.*

An outcropping of ice was an obvious possibility, but the idea that the white area was ice was problematic, as any ice on the surface would sublimate to water vapor over time. Sublimation is the process where a solid material changes to a gas without first becoming a liquid. For temperatures and pressures on the surface of the Earth, dry ice (carbon dioxide) is the classic example of a substance that sublimates. One idea was that the ice might be replenished by cryovolcanoes spewing water brine that would freeze. But ultimately Dawn has revealed that the white material consists of salts.

What type of salts? Earlier it was suggested that the salt was hydrated Epsom salt ($MgSO_4.6H_2O$), but it has been determined it is sodium carbonate ($NaCO_3$), and this has been found in great abundance. Also found are the ammonia-bearing salts ammonium chloride (NH_4Cl) and/or ammonium bicarbonate (NH_4HCO_3). The average surface of Ceres, which has low reflectance, has been found to contain clays containing ammonia, some calcium and magnesium carbonates, and magnesium phyllosilicates. (Phyllosilicates, or sheet silicates, are interconnected six-member rings of SiO_4^{-4} tetrahedra that extend outward in infinite sheets. Clay is hydrated aluminum phyllosilicate.) The above-mentioned salts, especially the sodium carbonate, as well as a lower amount of clay with ammonia, produce a white appearance contrasting with the generally dark surface material.

Most of 92 bright spots examined spectroscopically in Palomba et al.[8] show features similar to the low reflectance of the average Ceres surface, with other bright spots at variance from this. Palomba et al. propose a single evolutionary path from very bright to less bright. The bright spot in Occator crater, which has been named *Cerealia Facula* (Latin: facula, little torch or bright spot), is Ceres' version of "The Lighthouse at Alexandria."

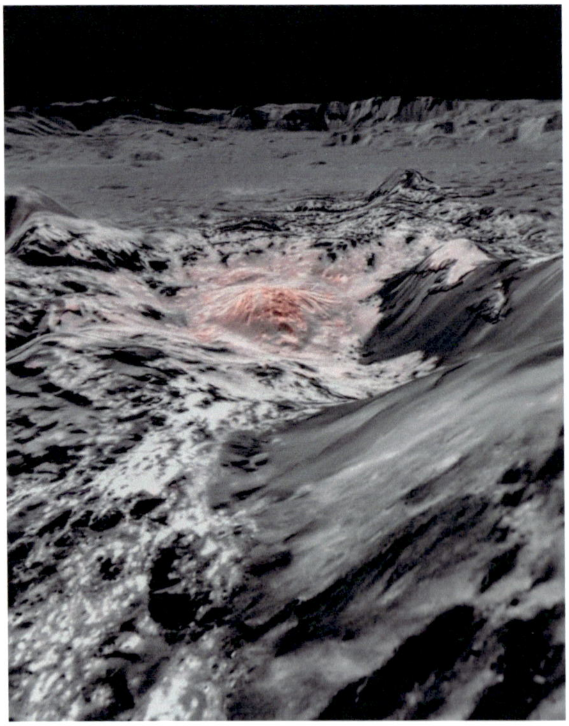

Figure 3. Ceres' bright spots in Occator Crater. This mosaic created by images from NASA's Dawn spacecraft uses false color to highlight salty water that was pushed up from a reservoir beneath the dwarf planet Ceres' crust.
Credit: NASA/JPL-Caltech/UCLA/MPS/DLR/IDA

Figure 4. This simulated perspective view shows Occator Crater. It measures 57 miles (92 kilometers) across and 2.5 miles (4 kilometers) deep and contains the brightest area on Ceres. Credit: NASA/JPL-Caltech/UCLA/MPS/DLR/IDA

Dawn detected over 300 white spots,[6] virtually all inside craters. It is believed that when impacts blasted out these craters, brine rose to the surface, froze, and then sublimated, leaving only the salts on the surface. Some of the brine that rose up was from the impact forcing up pre-existing frozen brine. In other cases, pre-existing liquid reservoirs below the surface were activated by the heat from the impact.[15] Scientists now believe the water ice in Occator Crater originates from a reservoir 40 kilometers deep and hundreds of kilometers wide.[16]

Lone Mountain...and its accompanying crater. [10,11,12,13,17]

In 2015 the Dawn spacecraft also discovered a major peak on Ceres almost half the height of Mt. Everest on Earth, but on a much smaller world. Named *Ahuna Mons*, the mountain tops out at four kilometers (13,000 feet) high. Its base is elliptical, 21 x 13 kilometers (13 x 8 miles), and the slopes are 30 to 40 degrees. Considerable amounts of sodium carbonate have been detected on its slopes (the same material found in abundance at Cerealia Facula). Lone Mountain is about 10 degrees south of Ceres' equator and is now known (from gravity measurements obtained by NASA's Dawn spacecraft mission to Ceres) to be a cryovolcano (cold volcano). Instead of erupting lava (molten rock) a cryovolcano erupts a mixture of water, salts, and mud. The mixture then oozes out and freezes.

One obvious question is why only one mountain is found on the entire surface of Ceres? Well, over 30 more have been found. But they all have aspect ratios (height divided by width of base) much smaller than that of Ahuna Mons. The surface had to be scrutinized to find these low mounds. It is thought that all of these mound-mountains have undergone *viscous relaxation.* Just as the weight of a glacier on Earth liquifies its bottom, allowing the glacier to slowly move, the weight of the mound-mountains would cause a subsidence in the material underneath, resulting in slumping. Of course ice on Ceres is much stiffer at the very low temperatures there, and it could take millions of years for a mountain to be reduced to a low, broad mound. Ahuna Mons' location near Ceres' equator where temperatures are higher than at the poles suggests the mountain is no more than 240 million years old, fairly young geologically speaking.

Ahuna Mons lies right next to a crater whose depth and breadth are about the same as the height and breath of the mountain. One might ask whether a meteor strike (with an angle of inclination almost vertical) could have heated the underground material initiating cryovolcanism? But then one would have to explain why one doesn't see a tall mountain next to any of the numerous other craters on Ceres' surface. We all would like to be around when robotic- or human-inhabited research stations strategically located in the solar system can give us answers to all (or at least many) of our questions. Humans are a curious lot.

Figure 5. Ahuna Mons. Credit: NASA/JPL-Caltech/UCLA/MPS/ DLR/IDA

Figure 6. A simulated view of Ahuna Mons made with bonus stereo pictures (no vertical exaggeration) which Dawn acquired from an altitude of 240 miles (385 kilometers). The bright streaks on the steep slopes are sodium carbonate deposited during eruptions. Credit: NASA/JPL-Caltech/UCLA/MPS/DLR/IDA

Figure 7. Ahuna Mons. This photo was assembled from images taken by the Dawn spacecraft. Credit: JPL-Caltech/NASA, UCLA, MPS, DLR, IDA

Figure 8. In one of the last views that Dawn transmitted, Ahuna Mons is silhouetted against the blackness of space.
Credit: NASA/JPL-Caltech/UCLA/MPS/DLR/IDA

References

1. Landau, Elizabeth, & Greicius, Tony (June 29, 2016). Recent hydrothermal activity may explain Ceres' brightest area. NASA. https://www.nasa.gov/feature/jpl/recent-hydrothermal-activity-may-explain-ceres-brightest-area

2. Landau, Elizabeth, & McCartney, Gretchen (July 24, 2018). What looks like Ceres on Earth? NASA. https://www.jpl.nasa.gov/news/what-looks-like-ceres-on-earth

3. Lewin, Sarah (June 29, 2016). Mistaken identity: Ceres' mysterious bright spots aren't Epsom salt after all. space.com. https://www.space.com/33302-ceres-bright-spots-new-composition.html

4. De Sanctis, M. C., et al. (June 29, 2016). Bright carbonate deposits as evidence of aqueous alteration on (1) Ceres. *Nature, 536*(7614), p 54–57. https://www.nature.com/articles/nature18290
5. NASA. (February 16, 2017). Dawn discovers evidence for organic material on Ceres (Update). Phys.org. https://phys.org/news/2017-02-scientists-geology-ceres.html
6. Taylor Redd, Nola (December 14, 2017). Dwarf planet Ceres' bright spots suggest an ancient ocean. space.com. https://www.space.com/39094-dwarf-planet-ceres-bright-spots-geologic-activity.html
7. Mckinnon, Mika (October 2, 2017). The mysterious bright spots on Ceres may have a common origin. *NewScientist*, https://www.newscientist.com/article/2149150-the-mysterious-bright-spots-on-ceres-may-have-a-common-origin/ This article references [8].
8. Palomba, E., et al. (March 1, 2019). Compositional differences among bright spots on the Ceres surface. *Icarus, 320*, p 202-212. https://www.sciencedirect.com/science/article/abs/pii/S0019103517303135
9. Vu, Tuan H., Hodyss, Robert, Johnson, Paul V., & Choukroun, Mathieu (July 2017). Preferential formation of sodium salts from frozen sodium-ammonium-chloride-carbonate brines—implications for Ceres' bright spots. *Planetary and Space Science*, *141*, p 73–77. https://www.sciencedirect.com/science/article/abs/pii/S0032063316304214
10. University of Arizona. (September 17, 2018). Ceres takes life an ice volcano at a time. https://phys.org/news/2018-09-ceres-life-ice-volcano.html
11. Sundermier, Ali (September 2, 2016). NASA just found an ice volcano on Ceres that's half the size of Everest. sciencealert.com. https://www.sciencealert.com/nasa-just-found-an-ice-volcano-on-ceres-that-is-half-the-size-of-everest
12. Ruesch, O., et al. (September 2, 2016). Cryovolcanism on Ceres. *Science, 353*(6303). https://zenodo.org/record/1231277#.YSmNcd9lCbg
13. Choi, Charles Q. (June 10, 2019). A weird mud plume may have built the highest peak on dwarf planet Ceres. space.com. https://www.space.com/ceres-highest-mountain-mud-plume.html
14. http://www.scientificlib.com/en/Astronomy/SolarSystem/Ceres.html
15. Scully, J. E. C., Schenk, P. M., et al. (August 10, 2020). The varied sources of faculae-forming brines in Ceres' Occator crater emplaced via hydrothermal brine effusion. *Nature Communications*, *11*(3680). https://www.nature.com/articles/s41467-020-15973-8
16. NASA/JPL. (August 10, 2020). Mystery solved: Bright areas on Ceres come from salty water below. https://www.jpl.nasa.gov/news/mystery-solved-bright-areas-on-ceres-come-from-salty-water-below
17. NASA. (June 12, 2019). Scientists solve mystery of Ceres' lonely mountain, NASA solar system exploration. solarsystem.nasa.gov. https://solarsystem.nasa.gov/news/967/scientists-solve-mystery-of-ceres-lonely-mountain/

It is represented to us as having been
the most extraordinary sight that has ever been seen
even by the oldest of our inhabitants.

—Pittsburgh Weekly Gazette, November 18, 1833

TOPIC 9. The Leonids and the Great Meteor Storm of 1833.

On the night of November 12-13, 1833, perhaps the greatest meteor shower ever mentioned in history occurred. Meteor showers that remotely approach the number of meteors per minute that fell on this date are referred to as meteor *storms*. People rushed outside. Many were frightened, and some thought it was the beginning of the end of the world. It was this event that is said to mark the beginning of the science of meteors. Abraham Lincoln commented on it years later.[1] That night Lincoln happened to be traveling on a train with a deacon of the Presbyterian church. The deacon rapped on Lincoln's door and said, "Arise, Abraham, the day of judgment has come!" Springing from his bed and peering from the train's window, Lincoln saw the stars falling in great showers. Years later, as President, he was asked if his confidence in the permanency of the Union was not beginning to be shaken. Lincoln loved to make a point with a story, and, in answering, he gave his account of the meteor storm. He ended by concluding, "The world did not come to an end then, nor will the Union now."

The Figure 1 engraving below was created over 50 years after the event and was presented in a biblical context, not an astronomical one. The Figure 2 and 3 engravings, however, were created contemporaneously with the event.

Many observers noted that the meteors "were like the ribs of a gigantic umbrella."[3,4] Figures 2 and 3 both show the "dark cloud" around the radiant point, mentioned below.

The "snake" in Figure 2 is likely not an exaggeration as evidenced by what Professor Denison Olmsted (also mentioned below) wrote from Yale on November 13, 1833:

> One ball that shot off in the north-west direction, and exploded near the star Capella, left, just behind the place of explosion, a phosphorescent train of peculiar beauty. This line was at first nearly straight but it shortly began to contract in length, and dilate in breadth, and to assume the figure of a serpent folding itself up, until it appeared like a small luminous cloud of vapour.

Figure 1. Woodcut print of the November 13, 1833, meteor storm. It was originally published in the Seventh-day Adventist book "Bible Readings for the Home Circle," Review and Herald Publishing Company, April 1888. Joseph Waggoner, a Seventh-day Adventist minister, saw the event when he was about 13. Fifty years later, around 1887, he described the scene to the Swiss artist Karl Jauslin who depicted it in a painting. The engraving of it by Adoph Völlmy is shown on the left[3] *and is the most famous of several depictions of this event.*

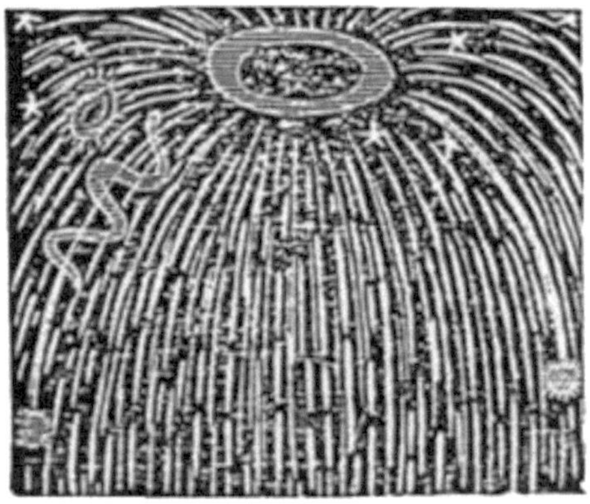

Figure 2. This woodcut appeared in the November 20, 1833, issue of The Old Countryman*—a New York weekly journal. It was created under the direction of Henry J. Pickering, one of the journal's editors, who witnessed the event. It was then reproduced in other newspapers and journals such as* Mechanics' Magazine *(New York) November 1833, p 287.* [3]

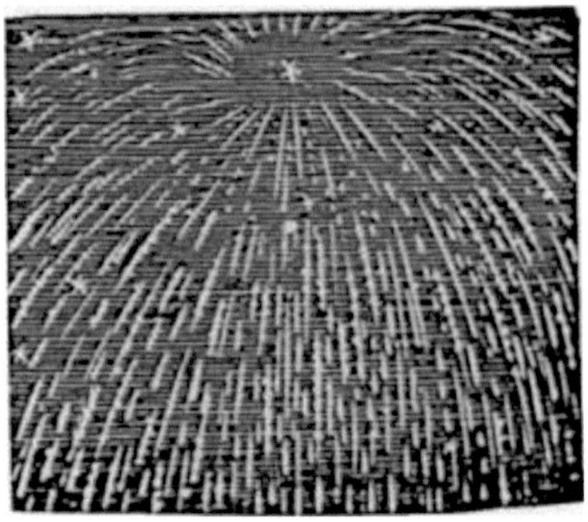

Figure 3. This woodcut appeared in the November 1833 edition of the Mechanics' Magazine *(New York) p 287.*[3]

The event occurred over North America. People started noticing an unusual number of meteors as early as 10 p.m. on the 12th, not realizing what was about to occur between 2 a.m. until dawn on the 13th. (See Topic 4 for why it is after midnight when most meteors are seen during a shower.) Some people described a dark area or cloud in the sky from which the meteors originated where they seemed to be moving slower. This area actually contained the *radiant point* of the storm. It was in the sickle of the constellation Leo (hence the current name "Leonids") and moved westward, because of the eastward rotation of the Earth, as it got later. Other meteors, scattering as they encountered the atmosphere, were seen to move faster and to move laterally across the sky in arcs. The lack of lateral movement near the radiant point made the area look darker. The fact that there seemed to be a "dark cloud" testifies as to how brightly lit the rest of the sky must have been by the meteors that were not streaming almost directly toward an observer. Modern estimates put the peak rate at 100,000 meteors per hour[2] or about 1,700 a minute. There were fireballs rivaling the Moon in brightness. Many left a trail of visible smoke, twisted by winds in the upper atmosphere, and often many such trails could be seen simultaneously. A smoke trail could linger for up to 20 minutes.[2] One particularly bright fireball was referred to as "the serpent," as it left a snake-like trail of smoke.[2]

Prior to the 1833 event, no comprehensive scientific study of the nature of meteor showers had been published. Some thought that shooting stars were a type of atmospheric phenomenon, possibly involving electrical activity. Some ancient Greeks believed meteors were associated with strong winds.[5] The fact that stones were sometimes found on the ground afterward didn't support the atmospheric idea. The eminent French mathematician and scientist Pierre-Simon de Laplace suggested in 1802 that meteors were debris blasted from volcanoes erupting on the Moon. Some thought that the phenomena took place mainly between the clouds and the ground. Though none of these ideas were correct, some progress was made. In the 18th century Edmond Halley suggested that meteors were of a cosmic origin. Alexander von Humboldt and his companion Bonpland witnessed and reported on the 1799 major meteor shower when they were in Cumana, Venezuela. Humboldt mentioned that this event was seen as far away as Germany and Greenland, so it was not a local phenomenon. He also noticed that the meteors seemed to emanate from a point in the sky, now known as the "radiant point."

The first person to publish a comprehensive study on meteors was Denison Olmsted (1791-1859), a professor of mathematics and natural philosophy at Yale University. He witnessed the 1833 meteor storm and was the critical figure in the birth of this new scientific study. His two-part article[6,8] included observations and opinions of others, and he gave his explanations of the phenomenon.[8] One of his conclusions[9] was that

> ...the radiant point, which indicated the position of the source from which the meteors emanated, ...was independent of the Earth's rotation, and consequently at a great distance from it, and beyond the limits of the atmosphere.

Another conclusion[10] was that

> the meteors fell towards the Earth in straight lines and in directions, which, within considerable distances, were nearly parallel with each other.

He also deduced [11]

> that the Meteors of Nov. 13th, consisted of portions of the extreme parts of a nebulous body, which revolves around the Sun in an orbit interior to that of the Earth, but little inclined to the plane of the ecliptic, having its aphelion near to the Earth's path, and having a periodic time of 182 days, nearly.

He used the term "comet" for the nebulous body (the source of the meteors), but was mistaken about the orbit of the comet being between the Earth and the Sun.

Olmsted acknowledged[12] that Alexander C. Twining, a civil engineer at West Point, had pursued nearly the same track of investigation and arrived at some results very similar to his. (In regard to Twining, see [7, 13].)

It did not go unnoticed that the meteor storms of 1799 (mentioned above) and 1883 both occurred on November 12. Could they be a recurring of the same phenomenon? In the 1860s another Yale professor, Hubert A. Newton, found historical records of 13 major meteor showers between the years 902 CE and 1833 with a cyclical period of 33.25 years. He predicted another would occur in 1867. And it did. So there was a cloud/trail of meteoric debris that cyclically was producing meteor showers at the Earth. Work then proceeded on determining the actual orbit of the trail (which would give the orbit of any comet responsible for the debris). This effort involved John Couch Adams in England, Giovanni Schiaparelli in Italy (the same person who famously "discovered" Martian canali), and Urbain Le Verrier in France. (Adams and Le Verrier are best known for their calculations leading to the discovery of Neptune.) By 1866 the debris trail's orbit had been determined. And instead of the orbit being between the Sun and Earth, as thought by Olmstead and Twining, it extended out as far as the orbit of Uranus. Astronomers then noticed that the orbit of the debris trail was almost identical to that of Comet 55P/Temple-Tuttle, which was independently discovered by Wilhelm Tempel on December 19, 1865, and by Horace Parnell Tuttle on January 6, 1866. It was now clear that the debris trail was material ejected from the comet. The meteoroid swarm (i.e., the trail) can be deflected by the giant planets Jupiter and Saturn, and was by Saturn in 1870 and by Jupiter in 1898, which caused the expected 1899 event to be a big disappointment.[2] But on November 17, 1966, another great meteor storm was visible from North America. The last (at the time of the writing of this book) major Leonid shower in North America was in 1998.

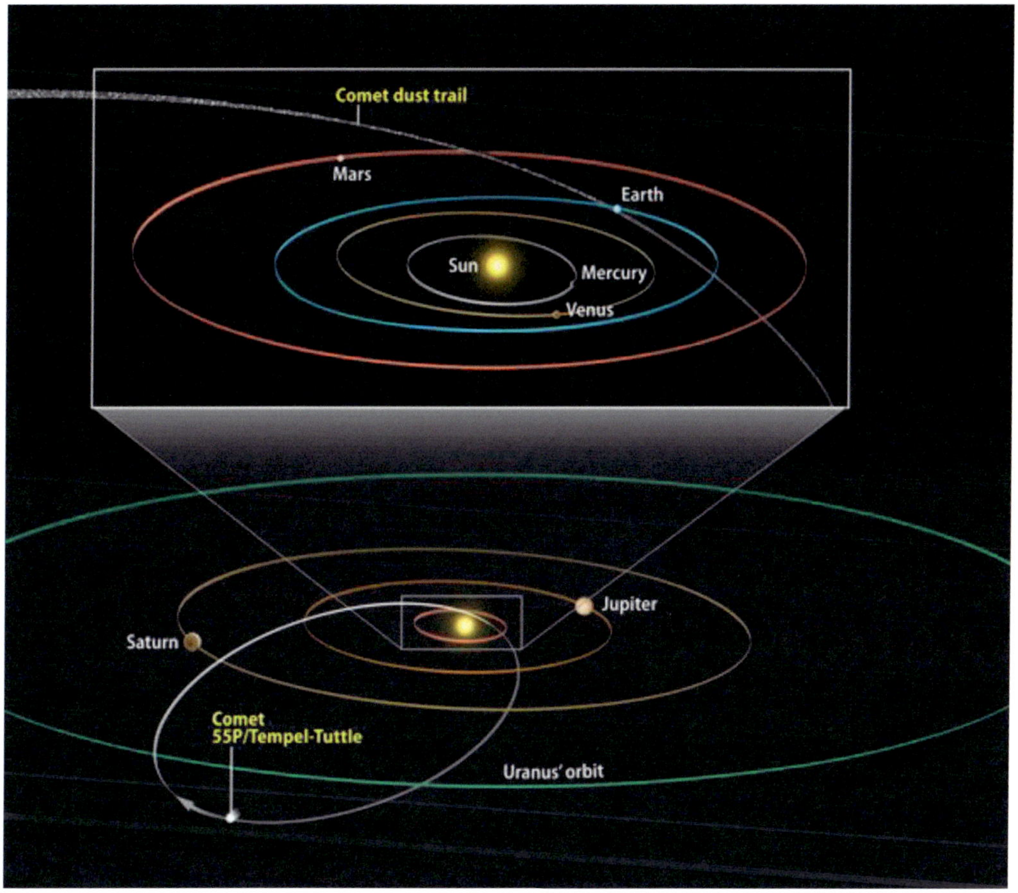

Figure 4. Comet 55P/Tempel-Tuttle orbit and dust tail. Credit:Roen Kelly, Astronomy Magazine, December 2011, p 55

References

1. Olsen, Donald W., & Jasinski, Laurie E. (November 1999). Abe Lincoln and the Leonids. *Sky & Telescope,* p 34–35.

2. Sandereson, Richard (November 1998). The night of raining fire. *Sky & Telescope,* p 30-36. This interesting article includes many of the salient facts about the Leonids.

3. Hughes, David W. (January 1995). The world's most famous meteor shower picture. *Earth, Moon, and Planets, 68*(1-3), p 311-322. https://doi.org/10.1007/BF00671522

4. Roche, John (Ed). (1990). *Physicists look back, studies in the history of physics.* p 270-272. Adam Hilger. ISBN 0-85274-001-8. The section of interest in this edited book is written by David Hughes, who is the author of reference [3].

5. Gheorghe, A. D., & McBeath, A. (2007). Meteor beliefs project: Meteors as prognosticators of strong winds in Classical thought. *WGN, Journal of the International Meteor Organization, 35*(4), p 92-94. http://adsabs.harvard.edu/full/2007JIMO...35...92G

6. Olmsted, D. (1834). Observations on the meteors of 13 Nov. 1833. *American Journal of Science and Arts, 25,* p 363-411.

7. ibid. p 405.

8. Olmsted, D. (1834). Observations on the meteors of 13 Nov. 1833. *American Journal of Science and Arts, 26,* p 132-174.

9. ibid. p 140.

10. ibid. p 147.

11. ibid. p 172.

12. ibid. p 174.

13. Twining, A.C. (1834). Investigations respecting the meteors of Nov. 13, 1833. Remarks upon Prof. Olmsted's theory respecting the cause. *American Journal of Science and Arts, 26,* p 320-350.

...larger than the planet Mercury
and more massive than Pluto
...in significant ways it resembles a planet
more than it does a typical moon.

—William B. Hubbard, planetary scientist

TOPIC 10. Titan.

Fly like an eagle.

On Earth we have many sports. In the future if humans have a base on Titan we might invent a new sport that could only be played there, namely flying and gliding by attaching wings (perhaps made of titanium) to our arms. Titan has a thick atmosphere whose surface pressure is about 1.6 times that on Earth. The pressure would be even greater but for Titan's low gravity. Titan's atmosphere is about 10 times as deep as Earth's, and with its low surface gravity the age-old dream of flying somewhat like a bird is probably possible. Titan's surface gravity is only about 14 percent of Earth's and lower than on our Moon, which is approximately 17 percent of Earth's surface gravity. You might wonder how this could be, given that Titan is the second-largest moon in the solar system, having a diameter of 3,200 miles (only slightly smaller than 3,273 miles, the diameter of the largest moon, Ganymede). Our Moon's diameter is only 2,159 miles (3,475 kilometers). But the density of icy Titan is much less than that of our rocky Moon.

Another factor contributing to Titan's lower surface gravity is that someone on Titan's surface is 1,600 miles from its center, while someone on our Moon's surface is only 1,080 miles from the Moon's center. From Newton's law, the force of gravity is inversely proportional to the square of the distance between the *centers* of the objects. Newton showed that the mass of a spherical object (where at any distance from the center of the object the density is roughly uniform) can be considered to be entirely concentrated at the object's center. So being closer to the center translates into a stronger gravitational force.

This fact also comes into play when talking about a black hole. The gravitational force on an object, say one AU from a black hole, is the same as the gravitational force of a star of the same mass as the black hole on that object. But the event horizon of a black hole is very small (about 30 kilometers or 19 miles from the center of a 10-solar-mass black hole) in comparison to the radius of a star (our Sun has a radius of 700,000 kilometers or 435,000 miles). So an object could get much closer to a black hole than to the star and yet not be inside it. It is when this situation occurs that the power of the black hole's gravity really comes into play—giving it its scary "Jack the Ripper" reputation.

Titan is the only moon in the solar system that has an atmosphere to speak of, and its thick atmosphere blocks radiation from reaching the surface.[3] It is a very cold -179° C (-290° F). Compare that with the record cold temperature on Earth of -89° C (-128° F) at Vostok Station in Antarctica in 1983. Because of the external air pressure, a more rigid spacesuit would not be necessary.[1,4] A thick, insulated, well-heated spacesuit made of looser fabric that allowed oxygen to flow through it might be substituted. Of course an oxygen mask would be needed.

The Cassini Mission with the Huygens Probe.

At wavelengths visible to humans the surface of Titan cannot be seen. An atmospheric haze of hydrocarbon molecules such as methane obscures the surface. The atmosphere is about 98 percent nitrogen (Earth's atmosphere is 78 percent nitrogen), about two percent methane (CH_4),[7] a slight amount of hydrogen (H_2), and some other trace gases. In 1980, Voyager 1's visible-light view of Titan revealed only an orange ball.

Twenty-four years later, the surface was seen with radar aboard the Cassini spacecraft orbiting Titan. Cassini entered orbit on July 1, 2004, and released the Huygens probe on January 14, 2005. Parachuting through the atmosphere for 2 1/2 hours, the probe landed on the surface, staying "alive" there for 72 minutes. Huygens' findings, including images taken during descent, were astounding. Much has also been learned from radar images of the surface taken from the orbiter. Cassini ended its mission on September 15, 2017, after passing through the rings of Saturn, with a fatal dive into Saturn itself. This maneuver was to prevent a possible future crash into one of Saturn's moons, contaminating it with Earth bugs—for example a crash into a moon such as Europa or Enceladus with the potential for life in a subsurface ocean.

Figure 1. A composite image of Saturn's moon Titan taken by the Cassini spacecraft. Credit: NASA

An abbreviated version of NASA's description of the image follows:

> This composite image shows an infrared view of Titan from NASA's Cassini spacecraft, acquired on November 13, 2015. The spacecraft's visual and infrared mapping spectrometer (VIMS) instrument made these observations, in which blue represents wavelengths centered at 1.3 microns, green represents 2.0 microns, and red represents 5.0 microns. A view at visible wavelengths (centered around 0.5 microns) would show only Titan's hazy, orange-colored

atmosphere. The near-infrared wavelengths in this image allow Cassini's vision to penetrate the haze and reveal the moon's surface.

During this Titan flyby, the spacecraft's closest-approach altitude was 6,200 miles (10,000 kilometers).

The view looks toward terrain that is mostly on the Saturn-facing hemisphere of Titan. The scene features the parallel, dark, dune-filled regions named Fensal (to the north) and Aztlan (to the south), which form the shape of a sideways letter "H."

Several places on the image show the surface at higher resolution than elsewhere. These areas, called subframes, show more detail because they were acquired near closest approach. They have finer resolution, but cover smaller areas than data obtained when Cassini was farther away from Titan.

Near the limb at left, above center, is the best VIMS view so far (September 2017) of Menrva, Titan's largest confirmed impact crater. Similarly detailed subframes show other regions: eastern Xanadu, the basin Hotei Regio, and channels within bright terrains east of Xanadu.

Among the spectacular features discovered by Cassini are lakes and seas, not of liquid water, but liquid methane and ethane (C_2H_6), vast dunes, and eroded mountains and river beds created from liquid methane rain or springs of liquid methane. (UV radiation should interact with methane in the atmosphere to produce ethane.)

The dunes (some up to a couple of hundred meters high[11]) are not made of silicate sands (e.g., in the form of quartz), but of some other molecules, quite likely hydrocarbons and nitriles (see below). Dunes cover at least 12 percent and possibly as much as 20 percent of Titan's surface, in comparison to four percent on the Earth.[8]

In an area the size of Australia, called *Xanadu*, are mountains and river networks which wind through hills and valleys. The mountains, consisting of ice as hard as granite, are mostly between 2,000 and 7,000 feet high,[14] comparable to the Appalachians in the eastern United States.

On Titan methane can exist near its *triple point*, the temperature and pressure at which the solid, liquid, and gas of a substance coexist in thermodynamic equilibrium. Earth's counterpart is water. Titan has clouds of methane. (And as on Earth, clouds consist of liquid droplets and ice crystals.) Also, in its atmosphere are various other hydrocarbons (molecules with hydrogen atoms attached to carbon atoms) and nitriles (molecules with a carbon atom triple bonded to a nitrogen atom). Some examples are propane (C_3H_8), hydrogen cyanide (HCN), benzene (C_6H_6), ethane (C_2H_6), and acetylene (C_2H_2).[7] Titan's lithosphere (bedrock) is mainly water ice.

Based on data taken by Cassini between 2006 and 2011, Titan's surface deforms by more than 10 meters (33 feet) over one nearly 16-day orbit around Saturn.[9,10] Though Saturn's gravity produces this tidal deformation, these "tides" are not Earth-like. Tidal budges on our planet (180 degrees apart) circle the Earth since the Earth is not tidally locked to (does not keep one face toward) the Moon. But Titan is tidally locked to Saturn. The situation is similar to that discussed in Topic 45, "Surface flexing and strange cycloidal loops on the surface of Europa." One spot on Titan's surface, directly facing Saturn, and the moon's antipodal point are the only places where tidal bulges occur. And no tides would occur at all if Titan's orbit around Saturn were a circle. But it is an ellipse. During Titan's elliptical orbit, the changing distance from Saturn causes a changing gravitational force from Saturn on

Titan. The large, 10-meter flex would not be possible if Titan were completely solid throughout. (A solid Titan would produce only about one meter of flex.) It is thought that the increased bulging is due to a liquid ocean under Titan's surface.[9] Imagine yourself spending time at one of these two locations and experiencing the up-and-down motion.

Titan's Seas and Lakes.
Cassini has detected more than 300 areas of liquid, from depressions of about 90 acres (the lower limit of detection) to the size of the three seas, called *Mare*, namely *Kraken Mare, Ligeia Mare*, and *Punga Mare.*[2] The largest sea, Kraken Mare, has five times the surface area of Lake Superior, which makes it slightly larger than the Caspian Sea. The total liquid area in the northern hemisphere is 35 times that in the southern hemisphere.[5] But large-scale depressions in the southern hemisphere look like they once contained liquid. What is going on?

It is thought[2,4,5] that this is due to Saturn's elliptical orbit around the Sun. Saturn is about 12 percent farther from the Sun at aphelion than at perihelion. Saturn's orbital period is 29.5 years, and its axial tilt with respect to its orbital plane is about 27 degrees. Titan's orbital plane around Saturn is inclined to Saturn's equator by only 0.384 degrees, close to zero. So Titan's axis is close to being parallel to Saturn's axis. We thus have Titan's axis also being tilted about 27 degrees to the orbital plane of Saturn and at the same time being almost perpendicular to Titan's orbital plane around Saturn. The former fact means that there are seasons on Titan, and the latter fact means that Titan's axial precession is irrelevant with respect to its seasons. Whatever the axial precession period, Titan's celestial north pole would trace out only a tiny circle in the firmament, unlike the situation for the Earth. When Saturn is closest to the Sun, Titan is tilted so that it is summer in Titan's southern hemisphere. When Saturn is farthest from the Sun, Titan's summer is in its northern hemisphere. Saturn travels slower in its orbit when it is farthest from the Sun. Consequently, while Titan's summers in the southern hemisphere are shorter than those in its northern hemisphere, they are hotter. The thought is that these hotter summers in Titan's southern hemisphere cause (over a period of many orbital periods of Saturn) evaporation of liquid methane and ethane in its south, with the gases moving to the north where they condense forming the northern lakes/seas.

But the position of Titan's summers/winters on Saturn's orbit varies over a time period with a dominant frequency of about 45,000 years.[5] And about 32,000 years ago the situation was the reverse of today.[4] Back then the transport of gases would have been southward and Titan's southern hemisphere would have featured the lakes and seas. The main driver[5] in the 45,000-year period is the apsidal (or perihelion) precession rate of Saturn, which is 19.5[6] arc seconds per (Earth) year. But perihelion precession alone would give a period of 66,462 years instead 45,000. (See *Appendix 10.1* for a simple calculation yielding 66,462.) There are also cyclic eccentricity variations as well as the evaporative and precipitative properties of Titan's climate system that contribute to Titan's seasonal period.

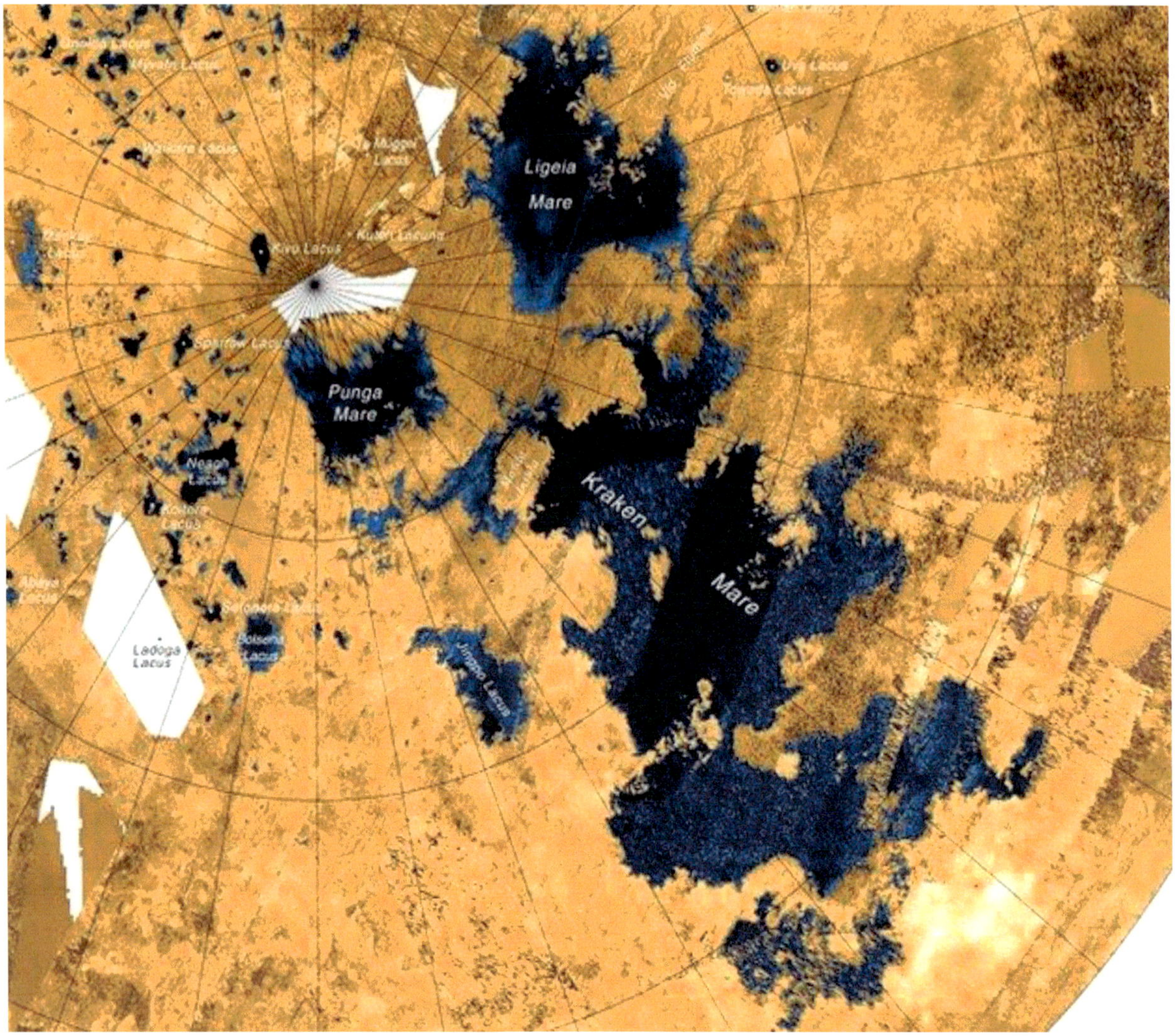

Figure 2. A radar map of Titan's north polar region, showing Kraken Mare, Ligeia Mare, and Punga Mare. What is labeled Kraken Mare had been referred to as Kraken I, with the Kraken Throat joining the large sea Kraken II (not labeled) below Kraken I. Credit: NASA/JPL-Caltech/USGS

Figure 3. The Kraken Throat joining Kraken I and II.

The *Kraken Throat* is officially named *Seldon Fretum* (Latin: fretum, strait). It is named for Hari Seldon, the fictional, intellectual hero who is First Minister of the Galactic Empire in Isaac Asimov's Foundation series. Think of it as Titan's equivalent of the Strait of Gibraltar.

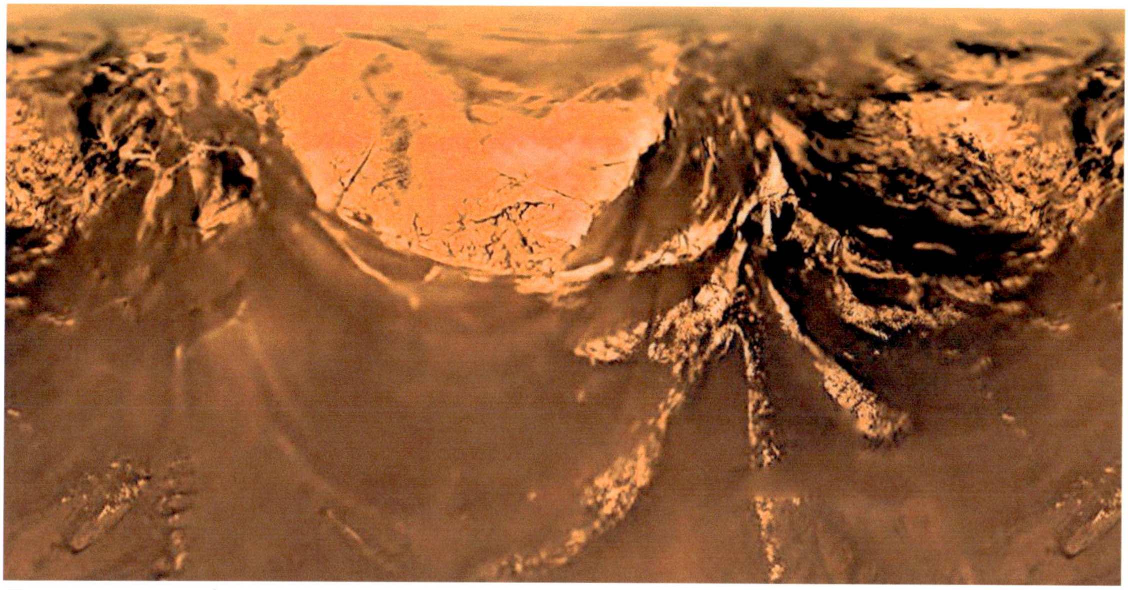

Figure 4. View of mountains as Huygens descends to the surface of Titan. Image taken at an approximate altitude of 15 kilometers (nine miles). Credit: ESA/NASA/JPL/University of Arizona

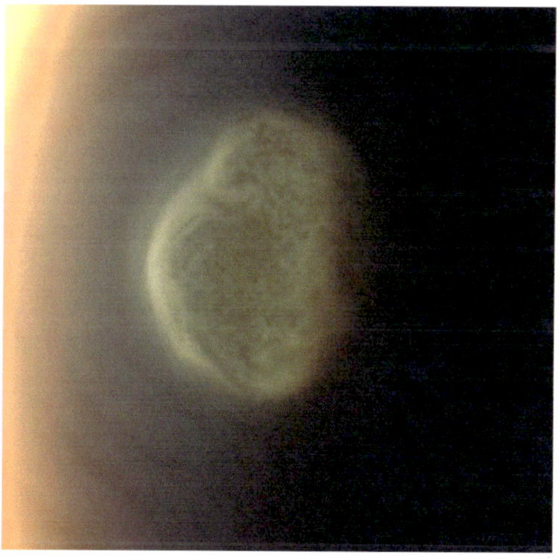

Figure 5 (left). This surface view of Titan has been colorized to look the way it would to our eyes if we were on Titan. The "rocks" are actually blocks of incredibly cold ice. Credit: ESA/NASA/JPL/University of Arizona
Figure 6 (right). Cassini captured this image of Titan's South Polar Vortex on June 27, 2012. Credit: NASA/JPL-Caltech/Space Science Institute

A swirling mass of gas around Titan's north pole was observed in 2006 during Titan's northern winter. In 2012, during the southern winter on Titan, a southern vortex formed.

Life on Titan?

Finally, while it is very cold on Titan, one cannot rule out the possibility there is life on this moon. If it exists it would be very different from life on Earth. Such a discovery would be proof of a second genesis in our solar system. While it is difficult to generalize from an instance of one, with two independent occurrences of life originating in the solar system one could say life most likely is present throughout the universe. Interestingly, the chemical vinyl cyanide (C_2H_3CN) has definitively been discovered in the atmosphere of Titan by the Atacama Large Millimeter/submillimeter Array (ALMA). And simulations have indicated that it is the best candidate molecule for the formation of cell membranes/vesicle structures in the hydrocarbon-rich lakes and seas of Titan.[12] While oxygen is a source of energy for living creatures on Earth, hydrogen could be the energy source for life on Titan.

But any life on Titan based on liquid methane or ethane would be very different than life based on liquid water. The H_2O molecule is what is called a polar molecule as it has a positive end and a negative end. As such, water only dissolves other polar molecules, such as amino acids, proteins, and DNA, allowing cells to use them.[13] Liquid methane and ethane, being non-polar molecules, cannot dissolve polar molecules. Thus any life on Titan would not be based on DNA as is all life on Earth. For further discussion see Topic 59.

References

1. Cable, Morgan L. (September 2020). What does Titan smell like? *Astronomy Magazine*, p 28-35.
2. Hayes, Alexander G. (October 2015). Secrets from Titan's seas. *Astronomy Magazine,* p 24-29. Also see [5].
3. McKay, Chris (March 2005). Talk on the Cassini mission to Titan, Bizaar Cafe, San Francisco.
4. Aharonson, Oded (November 2009). Titan's lakes. California Institute of Technology. https://web.archive.org/web/20180415114835/http:/web.gps.caltech.edu/~oa/titanlakes.shtml
5. Hayes, Alexander G. (April 27, 2016). The lakes and seas of Titan. *Annual Review of Earth and Planetary Sciences, 44,* p 57-83. https://geosci.uchicago.edu/~kite/doc/Hayes_2016.pdf
6. Barbieri, L, & Talamucci, F. (February 20, 2018). Calculations of apsidal precession via perturbation theory. arxiv.org. https://arxiv.org/pdf/1802.07115.pdf See Table 2 for the apsidal precessions of the planets.
7. Hörst, Sarah (February 2019). Titan's veil. *Sky & Telescope*, p 22-26.
8. Lopes, Rosaly (April 2012). Probing Titan's seas of sand. *Astronomy Magazine,* p 30-35.
9. Klesman, Alison (June 2018). The tides of Titan. Ask Astro, *Astronomy Magazine*, p 70.
10. K., L. (October 2012). Evidence that Titan harbors an ocean. *Astronomy Magazine*, p 17.
11. Turtle, Elizabeth (September 2017). Titan's surface. *The Planetary Report,* p 15-18.
12. Palmer, Maureen Y., et al. (July 28, 2017). ALMA detection and astrobiological potential of vinyl cyanide on Titan. AAAS, *Science Advances, 3*(7), e1700022. https://advances.sciencemag.org/content/3/7/e1700022
13. Cable, Morgan L. (January 2021). Looking for life in the universe. *Astronomy Magazine,* p 47-48.
14. Barnes, Jason (December 2008). Titan: Earth in deep freeze. *Sky & Telescope*, p 26-32.

If I tell you there's cheese on the moon, bring the crackers.

—Tyronn Lue, basketball coach

TOPIC 11. The Moon—a slice in it (Vallis Alpes) and a striking visual effect (the Werner X).

The Slice.
When I was a teenager in the 1950s I had a 60 mm Unitron refractor. And I was amazed when I looked at the features on the Moon and saw a slice cut right out of the Alps mountain range (Latin: Montes Alpes). It was as if the Moon were made of clay or cheese and someone had decided to create a straight, level walkway through the mountains. This is the Alpine Valley (Latin: Vallis Alpes) discovered in 1727 by Francesco Bianchini. My favorite crater on the Moon, Plato, is nearby. Plato is 109 kilometers (67.7 miles) in diameter, and its lava-filled floor is quite smooth. Charles Wood, a contributing editor of *Sky & Telescope* magazine, has called such craters "lunar skating rinks."[2] The Lunar Reconnaissance Orbiter has shown that the highest mountain in Montes Alpes is about 3.8 kilometers above the adjacent plains of Mare Imbrium and is 1.72 kilometers in elevation, where the lunar level of zero elevation is a sphere with radius 1,737.4 kilometers.[3] (From this it follows that the Imbrium plain is well below zero elevation.) The flat floor of the Vallis Alpes is about 160 kilometers long with a maximum width of about 10 kilometers.

Three images below show the area: Figure 1 is the huge Mare Imbrium with the Vallis Alpes in the upper right; Figure 2 is a closeup of Plato, the Alps, and the Vallis Alpes; and Figure 3 is a closeup of the valley itself, showing the thread-like Rille, a linear depression (not a protrusion) in the middle of the valley ranging from 0.5 to one kilometer wide.

Figure 1. Mare Imbrium with Plato, the Alps, and the Alpine Valley at the top. Credit: NASA Lunar Reconnaissance Orbiter

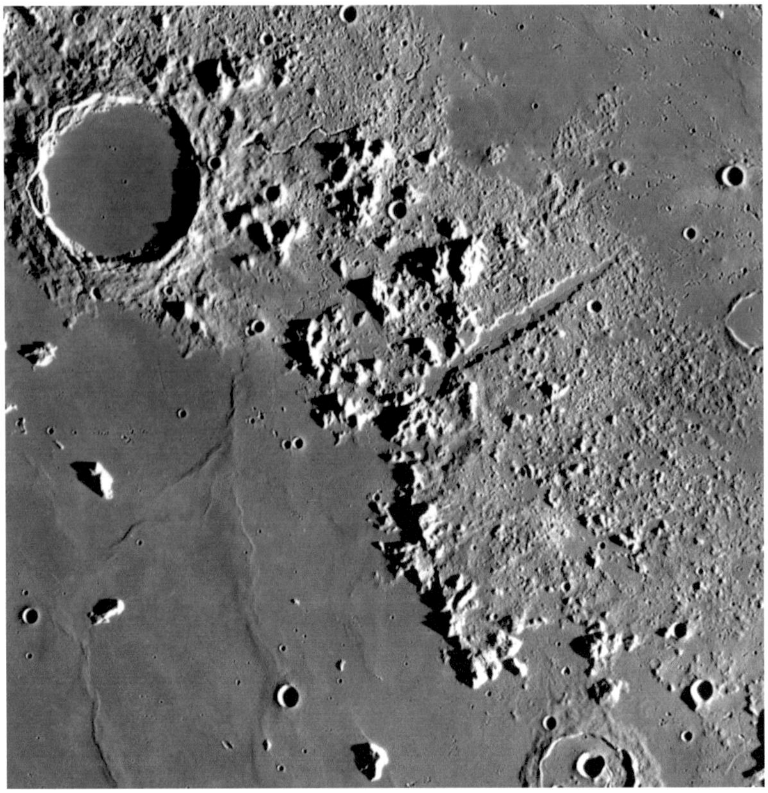

Figure 2. Plato and the Alpine Valley. Credit: NASA Lunar Reconnaissance Orbiter

Figure 3. Oblique view of Vallis Alpes with the Rille visible at the valley's center. Mare Imbrium is in the background. Credit: NASA Lunar Orbiter 5

The leading belief is that the Valley is a *graben* (German: ditch or trench). In geological terms a graben describes a valley with a distinct escarpment (a steep slope or long cliff) on each side caused by the downward displacement of a block of land between two parallel faults. Another possibility is that the Vallis Alpes was produced by stress fractures, where either the mantle expanded or the surface contracted after solidification of the regolith.

Will tourists visiting the Moon in the future be driven through Vallis Alpes?

A striking visual effect—the Werner X.
When the Moon's terminator (the dark/light "line") is on the crater Werner, which occurs around first quarter phase, a prominent "X" is visible just northwest of Werner. The "X" is created by sunlight reflecting off of crater ridges and can be seen even with a small telescope. As the terminator approaches Werner the observer must be patient, but the reward is a quite striking visual effect lasting about an hour. This phenomenon is also called Purbach's X, after another adjacent crater. But, as Chapman[1] points out, while both craters Werner and Purbach are next to the X, when the X is visible Purbach is hidden in darkness while Werner is still visible. Figures 4-6 below show the effect and exactly where to look to see it.

Figure 4. A close-up of Werner X. Credit: Taken with a Meade C11 and Philips ToUcam PRO II webcam by Randy Attwood, 2007-01-26, 0138 UT. Image obtained from Chapman[1]

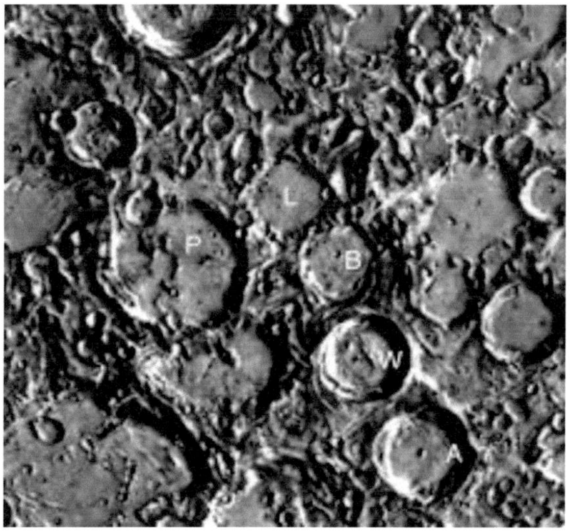

Figure 5. The lunar region centered on the topography that forms the Werner X: Purbach (P), La Caille (L), Blanchinus (B), Werner (W), and Aliancis (A) craters are indicated. The area shown is roughly 16 degrees square. The Werner X lies in between craters P, L, and B. In Figure 4, craters L and P are in darkness and are not visible. Credit: NASA Planetary Data System Imaging Node, U.S. Geological Survey, Flagstaff, AZ. Image obtained from Chapman[1]

Figure 6. Image shows the location of the Werner X on the Moon.

References

1. Chapman, David M. F. (February 2007). The lunar X files: a fleeting vision near the crater Werner. RASC Halifax Center. http://wasociety.us/Lunar-X.pdf
2. Wood, Charles (April 2021). Lunar skating rinks. *Sky & Telescope*, p 52-53.
3. DBpedia, About: Mont Blanc (Moon). dbpedia.org. https://dbpedia.org/page/Mont_Blanc_%28Moon%29

Let's call it a day
Let us say
It was just meant to be

—Peggy Lee

TOPIC 12. Strange things happen with the length of a day and of a year on Venus and Mercury. For example, three rotations on Mercury's axis = two Mercury years = one Mercury day. What is going on? The explanation is discussed along with other peculiarities.

On Venus one could ride a bike on the equator and easily keep up with the terminator (the longitude line separating day from night). You only would have to maintain a speed of 8.4 miles per (Earth) hour. So if you wanted to ride around the equator (a long trip), with the Sun remaining at sunrise the entire time, it's easily done. Here I am conveniently ignoring the fact that the temperature on the surface is hot enough to melt lead because of the Venusian greenhouse effect. See *Appendix 12.1* for the mathematical details.

While the apparent diameter of the Sun as seen from Earth or Venus doesn't vary much between aphelion (farthest approach) and perihelion (closest approach), that is not the case for Mercury. On Mercury the difference is nearly 53 percent. From Mercury, the apparent angular diameter of the Sun varies between 1.15 degrees and 1.76 degrees. The average is 1.4 degrees in comparison to a virtually constant 0.53 degrees for the Earth. So the diameter of the Sun as seen from Mercury can be over 3.3 times the size as seen from Earth—which would give a surface area for the Sun 11 times that as seen from Earth. Anyone want a quick sun tan?

Venus' surface pressure is 90 Earth atmospheres. The average temperature is 864° F (462° C) and lead melts at 621.5° F (327.5° C). And Venus has clouds of mainly sulfuric acid droplets.

On Venus one day is slightly longer than a half a Venusian year, and one rotation of Venus on its axis takes longer than one Venusian year.

One day (noon to noon) on Mercury equals two Mercury years which equals three rotations of Mercury on its axis.

On Venus the Sun rises in the west. This is because Venus' axis of rotation is tilted 177 degrees to its orbital plane (which means the angle between the axis and the *perpendicular* to the orbital plane is 177 degrees). Venus is flipped over on its axis and rotates in the opposite direction compared to all the other planets of the solar system except Uranus. The Earth's tilt is 23.5 degrees.

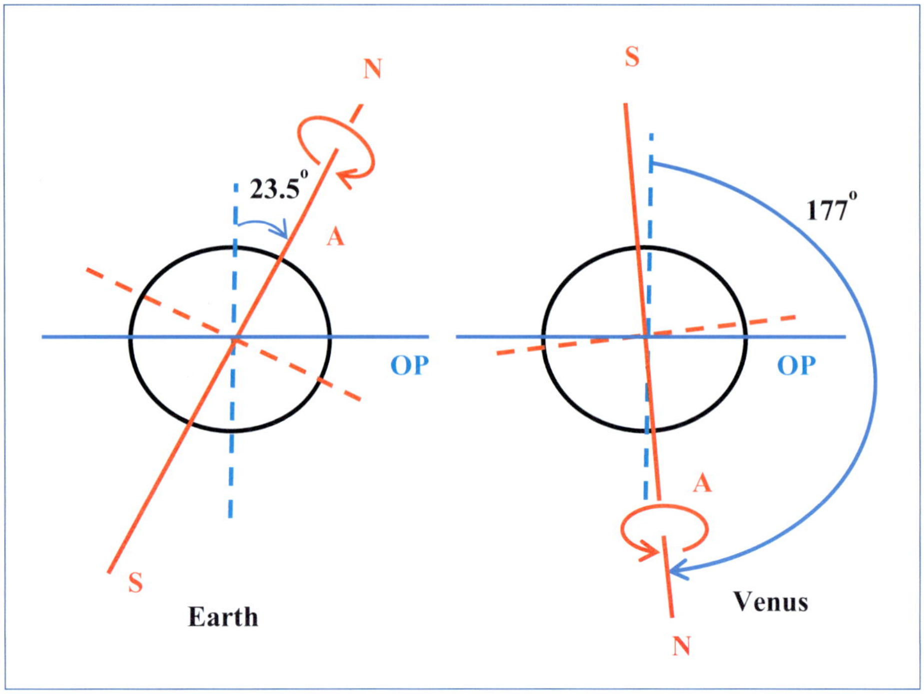

Figure 1. Relationship between the rotational axis A and the orbital plane OP for Earth and Venus.

Notation for the diagram:
A = rotational axis; OP = orbital plane.
N and S are the north and south poles respectively.
The perpendicular to the rotational axis, which indicates the equatorial plane, and the perpendicular to the orbital plane are given as red and blue dashed lines respectively.

The orbital plane of the Earth is, of course, the ecliptic. Venus' orbital plane is tilted about 3.4 degrees to the ecliptic. From the standpoint of someone on Earth or Venus, the Sun revolves around the planet in its orbital plane.

To understand rotational direction (as indicated by the circular red arrows) about an axis we use the right-hand rule: When you curl the fingers of your right hand around the axis of rotation in the direction of the rotation, your right thumb points in the direction of the north pole.

To explain the odd relations between the length of a day and a year on Venus and on Mercury, we first need some definitions.

The *sidereal day* or *axial rotation period* for a planet is the time period between successive passages of a star across the planet's meridian. This is the time it takes for the planet to make one rotation on its axis with reference to the background stars.

The *synodic day* or *solar day* for a planet is the time period between successive passages of the Sun across the planet's meridian (determining how long one day is for the planet). So it's the period from noon on one day to noon on the next day. (Note: The term *synodic period* has a different meaning than synodic day and will be discussed below.)

The *yearly* or *orbital period*, also called the *sidereal year*, of a planet is the time for the planet to go from one point in its orbit back to the same point as seen from the Sun with respect to the background stars.

Prograde motion in astronomy is, in general, orbital or rotational motion of an object in the *same* direction as the axial rotation of its primary. E.g., the Earth has a prograde orbit, indicating its orbital motion is in the same direction as the axial rotation of the Sun. Also, the Earth has prograde *rotational* motion, indicating its axial rotation is in the same direction as the axial rotation of the Sun.

Retrograde motion in astronomy is, in general, orbital or rotational motion of an object in the direction *opposite* the axial rotation of its primary.

All planets are in prograde orbits about the Sun. All planets except for Venus and Uranus have prograde rotational motion.

For the Earth, the sidereal and synodic days are almost the same, so one might mistakenly assume that one day on a planet (noon to noon) is the same as the sidereal day or axial period. But this would be a mistake. The synodic day (or solar day) is determined by using both the sidereal day and yearly period.

To give an extreme example to show that one day is not a sidereal day, suppose the Earth turned once on its axis every year (i.e., the sidereal day is one year and not almost 24 hours). Then the Earth would be tidally locked to the Sun and have one side always facing the Sun. If you are on the Earth where it is noon, it will always be noon (assuming a perfectly circular orbit). So you could say that you would have to wait forever for the next noon. One actual day would last indefinitely which is certainly not equal to a sidereal day of one year.

The following diagram reveals how the orbital period of a planet affects its day.

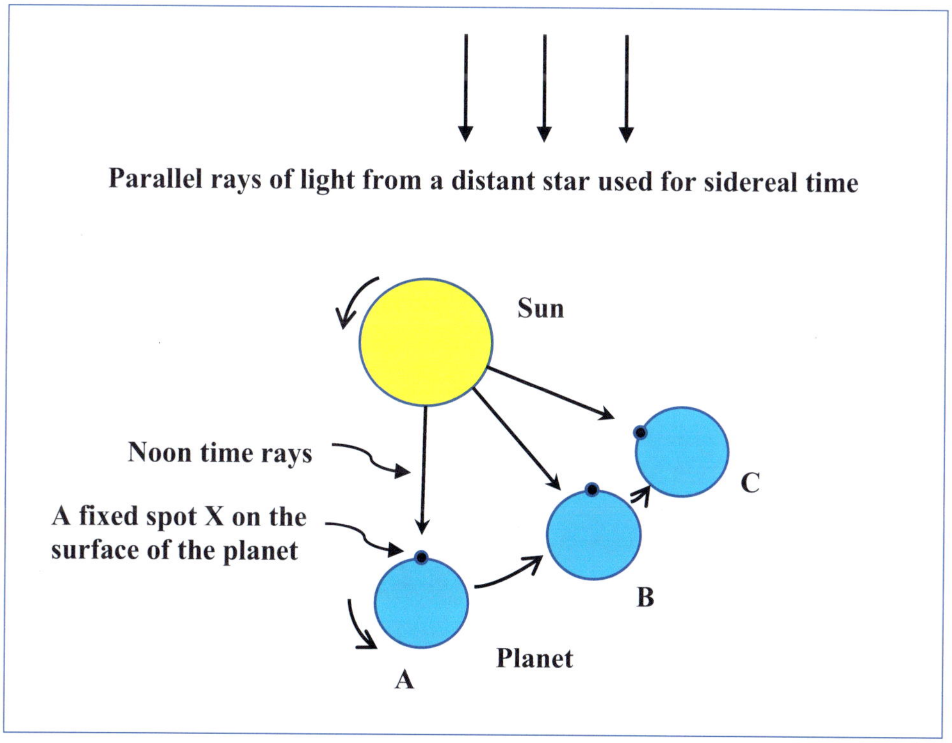

Figure 2. Synodic Day vs Sidereal Day.

The diagram depicts a planet with prograde rotational and orbital motion, and we assume the sidereal day is less than the yearly period. X is a spot on the surface of the planet, and when the planet is at position A it is noon at X. Position B for the planet is one sidereal day after A, so a distant star is in the same position in the sky in position B (from anywhere on the planet) as it was in position A. Position

C is where the planet is when it is again noon at X. So the time for the planet to go from A to B is one sidereal day and the time to go from A to C is one synodic day. The revolution of the planet around the star makes the synodic day longer than the sidereal day.

There is a formula relating the three periods, where s = sidereal day (in any unit of time), P = yearly period, and S = synodic day.

For prograde rotation of the planet,

$$1/s - 1/P = 1/S .$$

For retrograde rotation,

$$1/s + 1/P = 1/S .$$

Here is the argument justifying the equation for prograde rotation:
The number of synodic days in a year is the number of sidereal days in the year minus one. The subtraction by one is due to the one revolution of the Earth around the Sun in one year. Symbolically, this says $P/S = P/s - 1$, from which our equation follows. A similar argument gives the equation for retrograde rotation.

Below are the numerical values of the three periods for both Mercury and Venus:
(For Earth: One year = 365.2422 days; one sidereal day = 23.93447 hours; one average synodic day = 24 hours.)

Venus:
Sidereal day = *243.025* Earth days
One year = *224.701* Earth days
Synodic day (= 1 solar day on Venus) = *116.75* Earth days (*about half a Venusian year*)
Synodic period = *583.92* Earth days (This term will be defined below.)

Mercury:
Sidereal day = *58.646* Earth days
One year = *87.9691* Earth days
Synodic day (= 1 solar day on Mercury) = *175.94* Earth days
Synodic period = *115.88* Earth days

So three rotations on Mercury's axis = two Mercury years = one Mercury day.

Mercury rotates three times on its own axis for every two revolutions about the Sun. One says that Mercury has a *3:2 spin-orbit resonance.* Also, Mercury's synodic day (one solar day) is twice as long as one Mercury year (88 x 2 = 176).

We have introduced several definitions. At the risk of making things more complicated, we now present another: namely the *synodic period* referred to above. However there is a reward for doing this. A big piece of cheese? Well no. But with this definition we can explain why it was that for many years astronomers thought that Mercury was tidally locked to the Sun, presenting the same face toward the Sun throughout its orbit around the Sun. So a sidereal day for Mercury would have been the same as its orbital period: one year for Mercury or 88 Earth days. In 1965, by bouncing radar off Mercury, it was discovered that Mercury's rotation rate was actually about 59 Earth days, not 88. Below we explain how they had erred.

The synodic *period* is the time it takes for a body to return to the same angular position from the Sun *as seen from Earth*. Note that is very different than synodic day. For one thing, the definition of a planet's synodic day doesn't require observing anything from Earth.

The synodic period is usually used in referring to the Moon and the planets. The Moon's synodic period is the time between successive recurrences of the same phase. The synodic period of a planet would be the same as the time required for the Earth to overtake it as both go around the Sun—or, in the case of Mercury or Venus, for the planet in question to overtake the Earth. The synodic period for Venus or Mercury would also be the period of time between successive appearances of each as an evening star at a given angular distance from the Sun. The synodic period of an artificial satellite of the Earth is measured between its conjunctions (closest apparent approaches) with the Sun.

By managing to see some markings on the surface of Mercury when Mercury was at a particular angular distance from the Sun, astronomers noticed that these same markings were present the next time that Mercury was at the same position relative to the Sun. This meant that at that time, Mercury was presenting its same face toward the Earth as before, and thus was presenting its same face toward the Sun as before. And indeed, this would be the case if Mercury were tidally locked to the Sun.

But they were fooled by a coincidence: the synodic period for Mercury (115.88 Earth days) is approximately twice its actual sidereal day (58.64 Earth days). So Mercury had rotated almost exactly twice on its axis between the observations. And thus Mercury was presenting its same face toward the Sun as before, but for the second time. Halfway through the synodic period (when the Earth, Mercury, and the Sun were aligned differently and seeing markings on Mercury's surface was not possible) Mercury *also* had this face directed at the Sun.

There is a formula giving the synodic period. For superior planets like Jupiter,

$$1/\Sigma = 1/E - 1/P \text{ ;}$$

for inferior planets like Venus,

$$1/\Sigma = 1/P - 1/E \text{ ,}$$

where Σ = synodic period of the planet, P = orbital period of the planet, and E = orbital period for the Earth. E.g., for Venus: $1/\Sigma = 1/224.701 - 1/365.2422$ gives $\Sigma = 583.96$, which is close to the value of 583.92 given above.

While quite accurate, these formulae are not exact. For example, the synodic period of Venus actually varies between 576.6 and 588.1 days. For proofs of these formulae, see *Appendix 12.2*.

Another world, another day, another dawn.

—Douglas Adams, The Ultimate Hitchhiker's Guide to the Galaxy

TOPIC 13. Mercury and Venus have lengthy sunrises. It takes about 16 hours and 5.7 hours respectively for the Sun's disk to clear the horizon on Mercury and Venus once its upper limb first comes into view. In contrast, Earthly sunrises are only 2.13 minutes long.

Here are the times it takes the Sun's disk to clear the horizon on various planets once its upper limb first comes into view[1]

Mercury	16.1 hours	Jupiter	10.2 seconds
Venus	5.7 hours	Saturn	5.9 seconds
Earth	2.1 minutes	Uranus	4.8 seconds
Mars	1.4 minutes	Neptune	2.9 seconds

The values here are for an observer located on each planet's equator. See *Appendix 13.1* for mathematical details. It should be noted that the Sun always rises/sets vertically as seen from a planet's equator—see Topic 56.

References

1. Astro News. (May 2014). The sun also rises. *Astronomy Magazine*, p 12.

It is impossible to travel faster than the speed of light, and certainly not desirable, as one's hat keeps blowing off.
—*Woody Allen*

TOPIC 14. The discovery in 1676 that the speed of light is finite, the 1849 ingenious *terrestrial* method for measuring the speed of light using a rotating, toothed wheel, and how you can measure the speed of light in your kitchen using a microwave oven and marshmallows.

I. Ole Christensen Roemer showed in 1676 that the speed of light is finite by observing the orbital motion of Jupiter's moon Io.

The speed of light is so great that many people believed light travel time might be instantaneous. Aristotle and Rene Descartes believed this.

Descartes wrote to a friend in 1634:

> for me this was so certain that if it could be proved false, I should be ready to confess that I know absolutely nothing about philosophy.[1]

In his masterpiece on mechanics, *Two New Sciences,* Galileo discussed an experiment to determine if the speed of light was finite or instantaneous:

> I would have two men each take one light, inside a dark lantern or other covering, which he could conceal and reveal by interposing his hand, directing this toward the vision of the other. Facing each other at a distance of a few braccia, they could practice revealing and concealing the light from each other's view, so that when either man saw a light from the other, he would at once uncover his own.[1]

(A braccio (Italian: arm) was about 60 cm long.[1])

Galileo proposed, after the two men perfected their reaction times, that they try this out at various distances, including three miles (for a round trip of six miles). Galileo reported conducting the experiment himself at less than a mile. He concluded that he was not able "to make sure whether the facing light appeared instantaneously."[1]

The Italian-born French astronomer Domenico Cassini observed the moons of Jupiter between 1666 and 1668, publishing in 1668 reasonably accurate tables of their motions. In 1675 he found a difference in the motion of Jupiter's innermost moon Io that was strongly correlated with the distance between Earth and Jupiter. At first he attributed this finding to a finite speed for light, but then rejected the idea.

In 1671 the Danish astronomer Ole Roemer joined the astronomer Jean Picard on the island of Hven near Copenhagen, the former location of Tycho Brahe's observatory Uraniborg. They made many observations of Io's motion. In 1672 Roemer became Cassini's assistant in Paris. There he continued observing Io's motion and noted that the time of an eclipse of Io by Jupiter varied depending on where the Earth was in its orbit in relation to the position of Jupiter.

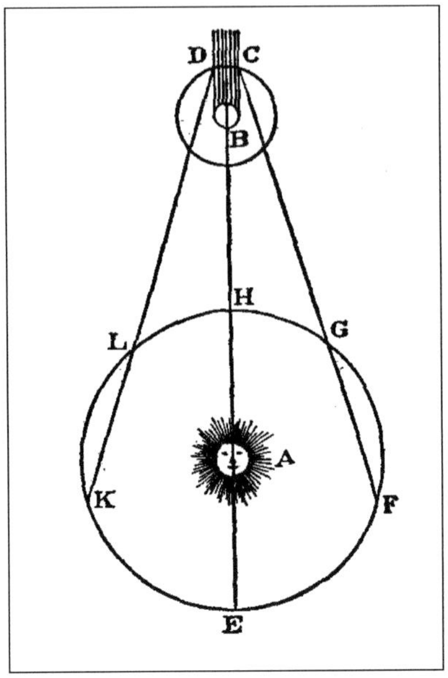

Figure 1. Illustration from the 1676 article on Roemer's measurement of the speed of light. Roemer compared the duration of Io's orbits as Earth moved toward Jupiter (F to G) and as Earth moved away from Jupiter (L to K). The times of immersions of Io into Jupiter's shadow (point C) and of emersions out of the shadow (point D) were recorded. Note that only ingresses into Jupiter's shadow can be seen from G and F; only egresses from the shadow can be seen from L and K.

Roemer correctly concluded that the time differences were due to differences in the distance between the Earth and Jupiter at the time of the eclipses. He estimated that light takes 22 minutes to traverse the diameter of the Earth's orbit.[2] (The correct value is 16.7 minutes.) In September 1676 he forecast that the eclipse of Io on November 9, 1676, would take place 10 minutes later than the time predicted that ignored any travel time for light.[2] He presented his results on the finiteness of the speed of light to the French Academy of Sciences in 1676, and the information was summarized in the *Journal des Scavans* in an article published on December 7 that same year. Roemer himself didn't publish his results and did not give a value for the speed of light.

To give an actual value to the speed of light he needed a more accurate estimate for the Earth's orbital diameter. If we use the current value of 93 million miles as the radius of the Earth's orbit, and use the above value of 22 minutes to traverse the diameter of the Earth's orbit, the speed of light comes out to be

$$93{,}000{,}000/(11 \times 60) = 140{,}909 \text{ miles/second.}$$

The actual value is 299,792.458 kilometers/second or 186,282 miles/second.

In 1690 the Dutch physicist, mathematician, astronomer, and inventor Christiaan Huygens calculated the velocity of light from Cassini's and Roemer's data, finding it to be

> ...more than 600,000 times larger than that of sound, which is not at all the same thing as being instantaneous, since there is the same difference as between something finite and something infinite....[4]

This results in Huygens' speed of light being 230,000 kilometers/second or about 143,000 miles/second. Huygens was the first scientist to give a numerical value to the speed of light. Neither Roemer nor Cassini attempted this.[3,5]

It is to be noted that in half a year there are a total of 102 eclipses of Io, giving a maximum delay of 16.5 minutes.

Roemer's great contribution to science was showing that the speed of light was actually finite.

An equation giving the speed of light illustrating Roemer's idea.
Let us give an equation giving the speed of light based on eclipses of Io by Jupiter. The essence of Figure 1 above is repeated below, but relabeled.

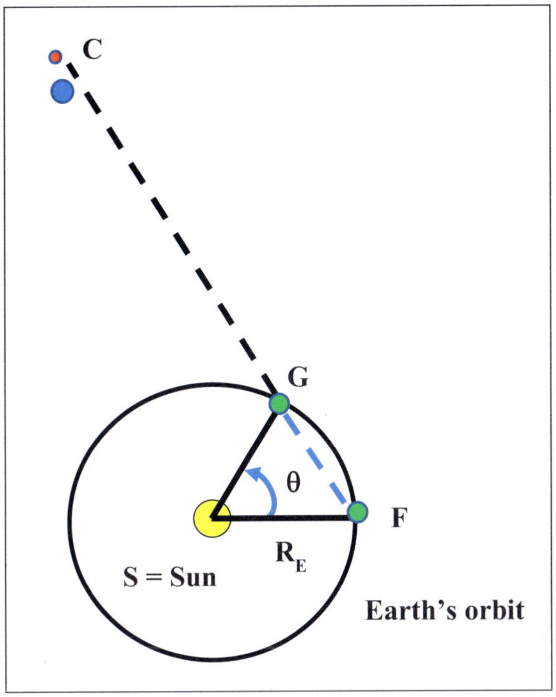

Figure 2. Light takes longer to go from Jupiter's moon Io (red dot) to F than to G. Here the two points F and G on Earth's orbit are assumed to be in a straight line. Jupiter is the blue dot and Earth is the green dot. As seen from Earth, Jupiter and Io are close to each other and along the same line of sight. So consider the dashed line as ending at the Jupiter "system."

Let $\Delta t = T_F - T_G$, the *light travel time difference*, with T_F equal to the time it takes light to travel from Io (point C) to F and T_G the time it takes light to travel from C to G.

R_E is the radius of the Earth's orbit, and P is the synodic period of Io around Jupiter. (Here Jupiter plays the role of Earth in our definition of synodic period.) Io's synodic and sidereal periods are 1.76925 and 1.76914 Earth days respectively. Io is eclipsed in every trip around Jupiter.

Let us also say that the second-observed eclipse of Io (at G) is the *Nth* such eclipse after the first-observed eclipse (at F). We will want to have the second-observed eclipse be some time after the first in order to give the Earth a chance to significantly change its distance from Jupiter. We will assume that the change in the distance is mainly due to the change in Earth's position around the Sun and not from a change in the position of Jupiter around the Sun. This is reasonable as the time difference between the two observations will be no more than half an Earth year, and Jupiter takes about 11.9 Earth years to orbit the Sun. Thus the Jupiter system would only have moved about 15 degrees if the observations are six months apart. At G the Earth and Jupiter are closer to each other.

One final symbol: Let ω_E be the angular orbital velocity of the Earth. Its value is

$$\omega_E = 360/(365 \text{ x } 24 \text{ x } 60^2) \text{ degrees/second.}$$

If T is the time for the Earth to move from F to G, then $\theta = T\omega_E$. And $T = N\text{x}P$. Thus we have

$$\theta = NP\omega_E .$$

In Figure 2, *where the points F, G, and C are in a straight line*, the equation for the speed c of light is

$$c = 2R_E\sin(NP\omega_E/2)/\Delta t\ . \qquad (1)$$

A proof of this formula is in *Appendix 14.1*. Going through the elementary steps in the proof facilitates understanding what is going on. Seeing the derivation, which only uses algebra and trigonometry, removes the magical pull-the-rabbit-out-of-the-hat nature of seeing the equation without justification.

At this point the reader may note that Figure 2 is a rather restricted situation, namely the two observations of Io were conveniently made along the same line of sight—i.e., the lines FC and GC are the same. Io does orbit Jupiter every 42 hours. So when Earth is close to G, within less than 42 hours Io would be at C. Hence requiring the lines FC and GC to be the same is not unreasonable. How can the speed of light be determined in the general case where FC and GC are not necessarily the same lines? The interested reader will find the answer in *Appendix 14.1.*

II. Fizeau's terrestrial method for measuring the speed of light (1849).

> An experiment is a question which science poses to Nature,
> and a measurement is the recording of Nature's answer.
> —Max Planck, Scientific Autobiography, 1949

The French scientist Armand Fizeau was born in 1819. His father, a physicist and professor of medicine, on his death left Armand a considerable fortune. Free to follow his interests, he decided to pursue scientific research, choosing to study astronomy under Francois Arago at the Paris Observatory.[6]

Fizeau was the first person to measure the speed of light in a terrestrial laboratory. The ingenious device he contrived included a toothed wheel, a lens, and a mirror 8.63 kilometers away from his main equipment).[7]

He also measured the speed of light in water. It was less than in air, but to everyone's surprise the speed was the same in both still water and running water at various speeds. It took Einstein's 1905 paper on special relativity to explain why—it has to do with the speed of light being a constant in a vacuum and the classic "addition of velocities" equation having to be modified.[6]

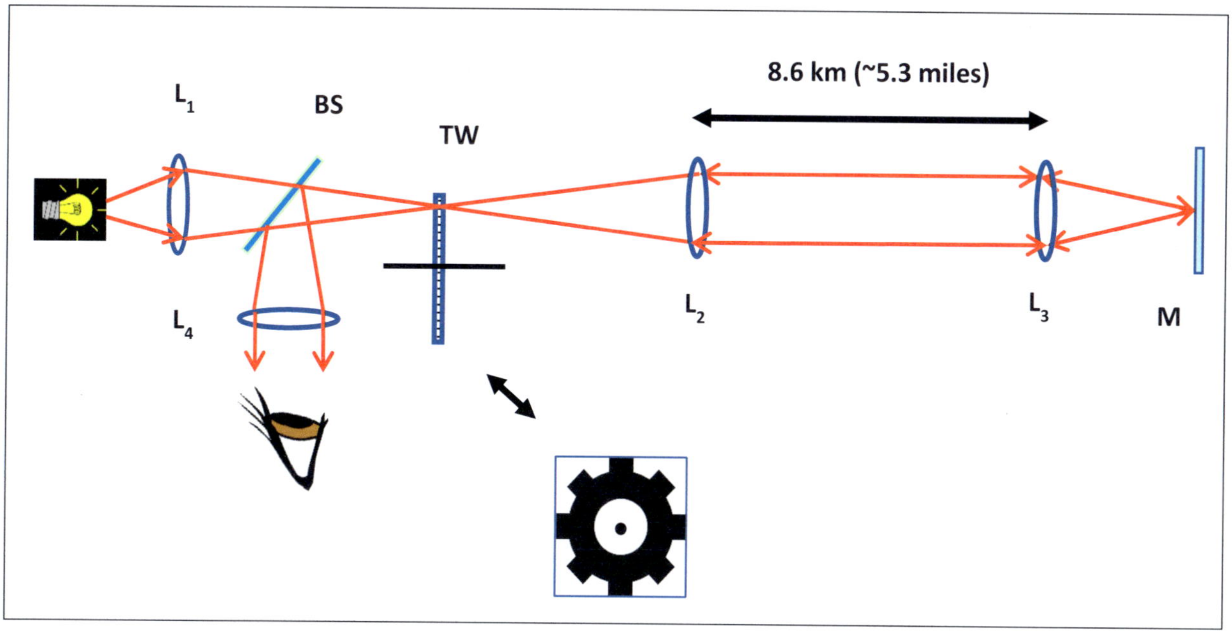

Figure 3. An illustration of Fizeau's 1849 experiment for determining the speed of light.

Figure 3 depicts a brilliant light source, four lenses labeled L_1 to L_4, a mirror *M*, a toothed wheel *TW*, and a beam splitter *BS* which is a partially reflective mirror that allows most of the light striking it to pass through the mirror and a small amount to be reflected.

Light from the light source is focused by lens L_1 onto the edge of the toothed wheel. The toothed wheel is spinning at a fast rate. If the light beam reaches a gap between the teeth, it can pass onward to lens L_2. From here it travels 8.63 kilometers to lens L_3 and then onto mirror M. The light then reflects off M and travels backward through L_3 and L_2, reaching the toothed wheel a second time. If it again can pass through a gap between the teeth, it then reaches the beam splitter BS where some of the light is reflected toward lens L_4 and into the eye of an observer.

The distances from the light source to L_2 and from L_3 to M are negligible, and, to be successful at producing the desired result, the actual toothed wheel would need many more teeth than in Figure 3. In fact, Fizeau's wheel had 720 teeth.[8]

Here is how the experiment is done. We assume the gaps have the same length as the teeth in the wheel. Let L be this length. Let N be the number of teeth in the wheel. The light from the source is broken up by the spinning, toothed wheel into regularly spaced bursts that travel 8.63 kilometers to the mirror M and are reflected back 8.63 kilometers to the toothed wheel. First the speed of the wheel is adjusted until all the light bursts are blocked with no light reaching the observer. This means all light passing through any gap will hit the next tooth on its return. The speed is slowly increased until all the light bursts are observed, i.e., the returning light is now passing through the gap just after the tooth that was blocking it. This means that the light has traveled 17.26 kilometers in the time it takes the toothed rim of the wheel to move length L. The wheel has a total of 2N teeth plus gaps, i.e., the rim of the wheel is broken up into 2N strips each of length L. Assume the wheel requires an *increase* of ω cycles/second in its rotation rate to go from no light seen to all bursts seen. It thus rotates $2N\omega$ strips faster in one second. So it takes $1/2N\omega$ seconds to rotate length L. Thus the speed of light is

$$c = (17.26)2N\omega \text{ kilometers per second.} \qquad (1)$$

Fizeau used N = 720, 17.26 kilometers as the light-travel distance,[8] and measured ω as 12.6 cycles/second. Using equation (1) gives his value as 3.13 x 10^5 kilometers/second, about 4.3 percent too high.

III. How you can measure the speed of light in your kitchen using a microwave oven and marshmallows.[9]

1. If your microwave oven has a revolving plate, remove it.
2. Place mini-marshmallows in a row across a plate.
3. Insert the plate in the microwave oven so that it cannot rotate.
4. Run the microwave for about 13 seconds (time will depend on the microwave)
5. Remove the plate. You should notice that there are evenly spaced places in the marshmallow row where the marshmallows are more melted or burned than in other places.
6. Measure in centimeters the distance *D* between two consecutive melt marks.
8. Look for a sticker on the microwave that tells you the frequency *f* in Hertz (Hz). Most microwaves are about 2,450 MHz. One MHz is one million Hz.
9. The speed of light c = 2Df in centimeters per second.

The speed of light, as noted above, is about 3.00 x 10^5 kilometers/second or 3.00 x 10^{10} centimeters/second.

I conducted this experiment. For me, f = 2,450 MHz and D = 7 centimeters, which gives a value of 3.43 x 10^{10} centimeters/second, about 14 percent too high. But then again, the only cost was for a few marshmallows—quite a low expense for a scientific experiment.

Why does this work?
The electromagnetic spectrum includes waves at the frequencies of the visible spectrum, which our brains interpret as colors. Blue photons are at the high end of the visible frequency range, and red photons are at the lower end. Frequencies higher than blue are the ultraviolet, x-rays, and gamma rays. Frequencies lower than red are infrared, microwaves, and radio waves. Some animals can see infrared and some ultraviolet. Goldfish are the only known members of the animal kingdom that can see both.

A microwave oven creates a *standing wave* in the oven. The peaks of the wave correspond to peaks in energy. These cause the burn or melt spots in our row of marshmallows. As Figure 4 below shows, the distance between consecutive peaks is one half a wavelength. So twice this distance is one wavelength. The speed of any wave (electromagnetic or otherwise) is its frequency times its wavelength. All electromagnetic waves travel at the same speed, which is the speed of light.

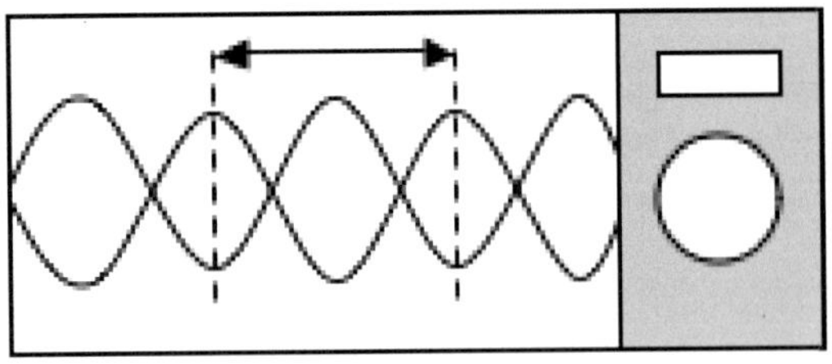

Figure 4. A standing wave in a microwave oven. The length of the arrow is one wavelength long. Credit: University of Wisconsin Department of Physics[8]

References

1. Foschi, Renato, & Leone, Matteo (2009). Galileo, measurement of the velocity of light, and the reaction times. *Perception, 38*(8), p 1251-1259. https://journals.sagepub.com/doi/pdf/10.1068/p6263

2. van Helden, Albert (1983). Roemer's speed of light. *Journal for the History of Astronomy, 14*(2) no. 40, p 137-140. http://adsabs.harvard.edu/full/1983JHA....14..137V

3. Bobis, Laurence, & Lequeux, James (2008). Cassini, Roemer and the velocity of light. *Journal of Astronomical History and Heritage, 11*(2), p 97-105. https://web.archive.org/web/20090717220337/http://www.bibli.obsp.m..fr/Bobis%20and%20Lequeux.pdf

4. Huygens, C. (1690). Traite de la lumiere. In *Euvres completes. t.19*, p 467. (This reference occurs within reference 3 above.)

5. Chester, Geoff (August 2009). Cosmic clocks. Ask Astro, *Astronomy Magazine,* p 49.

6. American Physical Society. (July 2010). This month in physics history. July 1849: Fizeau publishes results of speed of light experiment. *19*(7). https://www.aps.org/publications/apsnews/201007/physicshistory.cfm

7. Fizeau, H. (1849). Comptes rendus de l'Académie des Sciences (Paris). *29,* p 90-92.

8. White, Chris (April 2013). Measuring the speed of light: Investigating Fizeau's method. https://www.pcs.cnu.edu/~dgore/Capstone/files/WhiteC.pdf

9. The Wonders of Physics. Measure the speed of light using your microwave. Department of Physics, University of Wisconsin, Madison, WI. wonders.physics.wisc.edu. https://wonders.physics.wisc.edu/measure-the-speed-of-light/

TOPIC 15. On September 12, 1170 CE, Mars transited the face of Jupiter—a very rare event. A naked-eye merger of two planets in 2 BCE and a near merger in 2020.

Transit of Mars across Jupiter.
The transit of Mars across Jupiter was observed (with the naked-eye—it would be a few hundred years later before the invention of the telescope) by an English monk known as Gervase of Canterbury. He wrote (translated from Latin):

> On the Ides of September, at midnight, two planets were seen in conjunction to such a degree that it appeared as though they had been one and the same star; but immediately they were separated from each other.

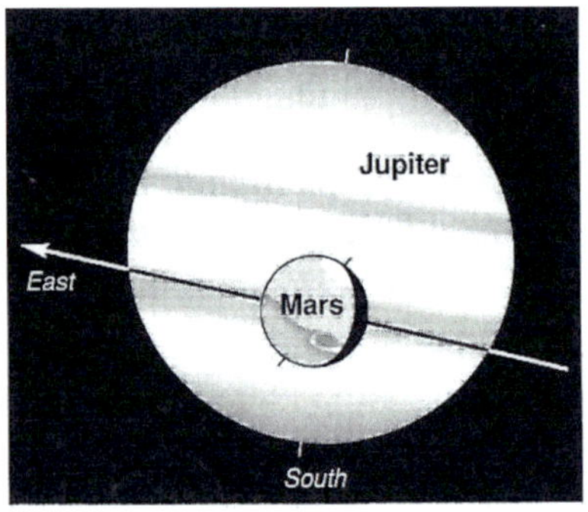

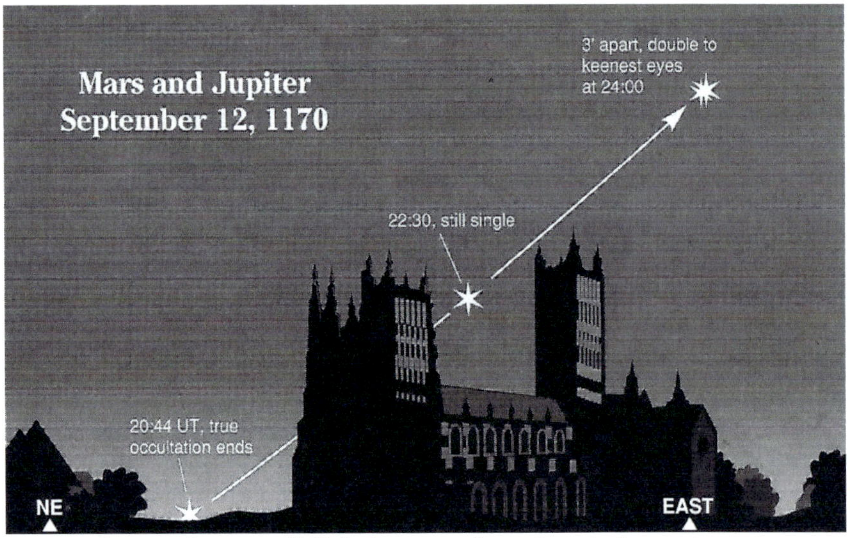

Figures 1 and 2. The above two images, from reference [11] below, are based on information generated by a computer calculation by Donald Olson, Regents Professor Emeritus, Texas State University; Russell Doescher, Professor of Physics, Texas State University; and Steven Albers, Senior Research Associate, NOAA Earth System Research Laboratory.

Less than four months after the transit, on December 29, 1170, Thomas Becket, archbishop of Canterbury, was murdered in Canterbury Cathedral (seen in the above illustration) by four knights of King Henry II.

The transit was also observed by astronomers at the Chinese Imperial Observatory in Beijing.

Venus and Jupiter merged as a single naked-eye star in 2 BCE.[1,2,3,4]
Imagine yourself in present-day Israel, on the beach, looking west, on June 17, 2 BCE, at 6:11 p.m. At about 45 degrees elongation from the Sun would be a brilliant "star," the combined light from Venus and Jupiter. They would appear as one bright object with their surfaces separated by only about five arc seconds. Venus would be at half phase, although it would take a telescope to observe this. But no telescopes would be available until about 600 years later. Venus was at magnitude about -4.3 and its diameter was 25.5 arc seconds, while Jupiter's magnitude was -1.8 and had a diameter of 32.2 arc seconds. This phenomenon was visible all over the Middle East, and much farther south Venus partially occulted Jupiter.

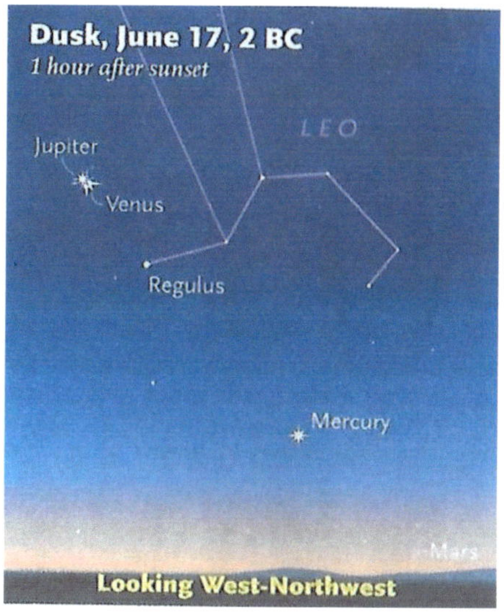

Figure 3. The Venus Jupiter Conjunction of 2 BCE as seen from Babylon. Credit: Aaron Adair, see [9]

Jupiter and Saturn came within only *seven arc minutes* of each other on December 21, 2020. This was the closest angular approach of these two planets, as seen from Earth, since July 16, 1623, almost 400 years ago. At that time they were less than 13 degrees from the Sun and most likely were unobserved.[5] The last closer approach of Jupiter and Saturn that was easily observable (49 degrees

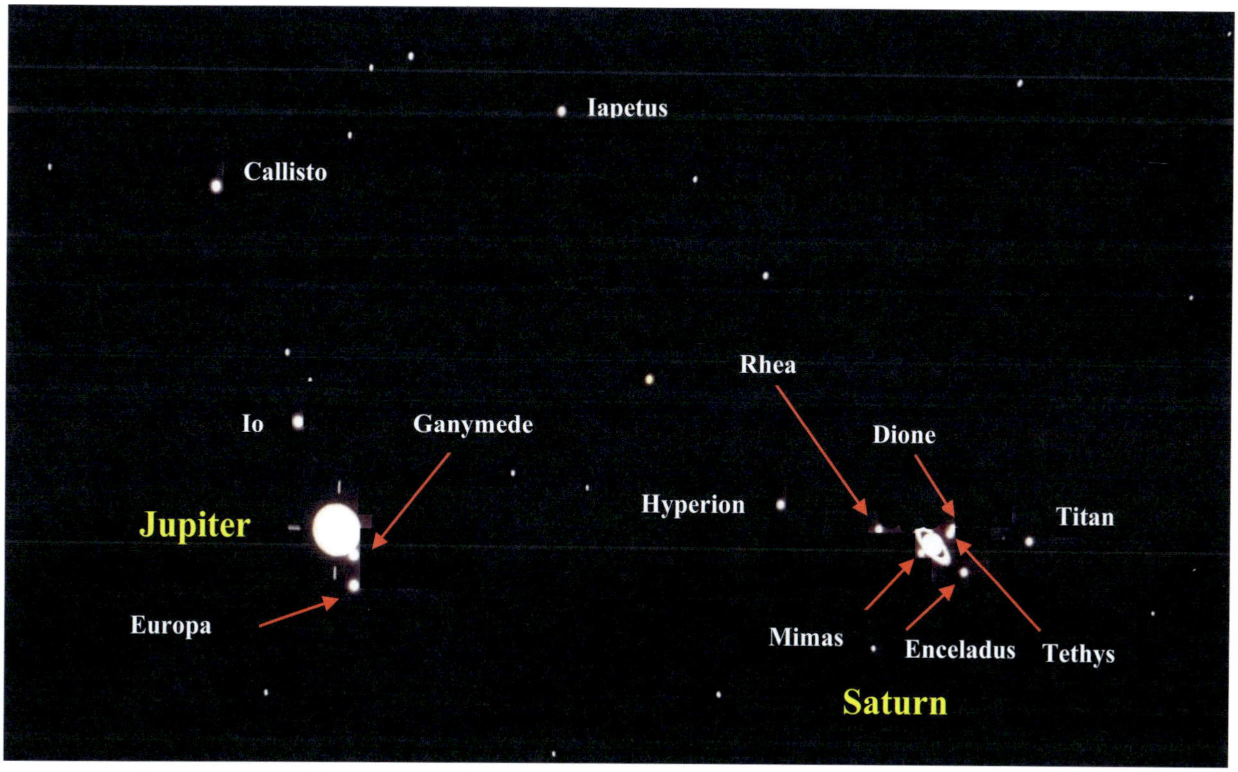

Figure 4. Conjunction of Jupiter and Saturn on December 21, 2020, as seen at 7:52 p.m. Pacific Standard Time showing the two planets with their major moons. Credit: astronomy app SkySafari (labels by author)

from the Sun) was March 5, 1226 CE.[2,6] The image in Figure 4 above shows the two planets accompanied by their major moons when near closest approach (as seen from the Earth) on December 21, 2020. It's a beehive, resembling M44, the Beehive cluster in Cancer.

Figure 5. Close approach of Jupiter and Saturn on December 21, 2020. Credit: Akarsh Simha

Two other references for conjunctions of Jupiter and Saturn: [7, 8]. How about the future? On November 22, 2065, Venus occults Ganymede and then transits Jupiter. The downside is that they will be only about eight degrees from the Sun.[10] So using binoculars or a telescope would be too dangerous.

References

1. http://www.bogan.ca/astro/occultations/occltlst.htm and https://www.bogan.ca/astro/occultations/2bcocclt.htm Here Larry Bogan discusses mutual planetary occultations past and future.
2. http://stevealbers.net/albers/ast/conjun/conjun.html Steven Albers gives a list of mutual planetary occultations from 44 BCE to 2251 CE.
3. Sinnott, Roger (December 1986). Computing the Star of Bethlehem. *Sky & Telescope*, *72*(6), p 632.
4. DeYoung, James, & Hilton, James (April 1987). Star of Bethlehem (Letters). *Sky & Telescope*, p 357. A discussion of the 2 BCE conjunction/occultation of Venus and Jupiter. The authors are affiliated with the Nautical Almanac Office, U.S. Naval Observatory, Washington, D.C. 20390.
5. Hunt, Jeffrey L. (February 20, 2020). *1623: The great conjunction of Jupiter and Saturn, when the curves line up.* (Blog) https://whenthecurveslineup.com/2020/02/20/1623-the-great-conjunction-of-jupiter-and-saturn/
6. Schaaf, Fred (December 2020). A very rare Jupiter and Saturn pairing. *Sky & Telescope,* p 46-47.
7. Church, John A. (March 1991). Great conjunctions of Jupiter and Saturn. *Sky & Telescope*, p 305-307. A list of remarkable Jupiter-Saturn conjunctions from 3780 BCE to 2656 CE.
8. Berman, Bob (December 2020). The best conjunction ever. *Astronomy Magazine*, p 14. Ever-witty writer Bob Berman discusses the December 20, 2020 conjunction and others.
9. Adair, Aaron M. (December 2007). Science, scholarship, and Bethlehem's starry night. *Sky & Telescope*, p 26-29. A depiction of the 2 BCE conjunction.
10. Pazourek, Joe (January 2007). Do the planets ever transit or occult each other? Ask Astro, *Astronomy Magazine.* https://astronomy.com/news-observing/ask%20astro/2007/01/do%20the%20planets%20ever%20transit%20or%20occult%20each%20other
11. Olson, Donald, Doescher, Russell, & Albers, Steven (August 1992). A medieval mutual planetary occultation. *Sky & Telescope*, p 207-208. A most interesting article about the Mars/Jupiter event with a brief discussion of other planetary occultations. The authors used a specialized Fortran-era computer program to verify the naked-eye event actually was a transit of Mars across the face of Jupiter. And the computation yielded specific information on the transit, allowing the creation of the two images which are in their article. (Albers is a Senior Research Associate, NOAA Earth System Research Laboratory; Olson is an astrophysicist at Texas State University; and Doescher is a physicist at Texas State University.)

...I discovered further movement, and only then did I conclude that Mercury had come in on his splendid wings.

—Pierre Gassendi commenting on the first observed transit of Mercury, 1631

TOPIC 16. Transits of Mercury and Venus across the Sun.

For a transit to occur (say with Venus), the Sun, Venus, and Earth must be in a line. This means that Venus must be at one of its two *nodes* where its orbit crosses the ecliptic. Transits of Mercury occur about 13 to 14 times a century. The last one was in 2016 and the next is in 2032. The timing depends on the ratio of the orbital periods and the relative tilt of the orbits.

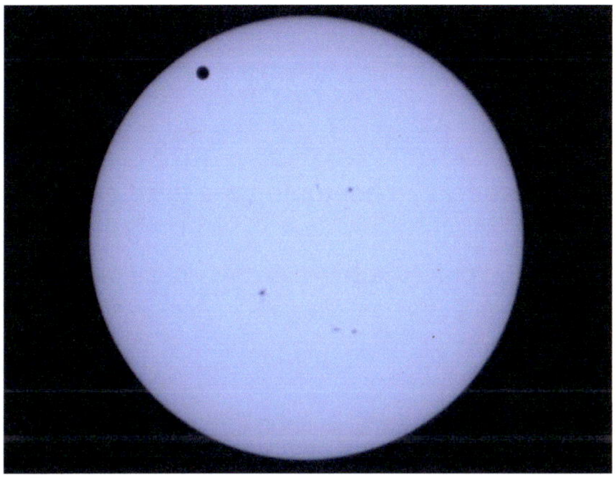

Figure 1. The June 5-6, 2012, Venus transit photographed by Don Pettit on board the International Space Station. Credit: NASA

Transits of Venus occur in a pattern that generally repeats every 243 years, with pairs of transits eight years apart separated by long gaps of 121.5 years and 105.5 years. The last two transits were in 2004 and 2012; the next two are in 2117 and 2125. The number 243 is from the addition

$$8 + 121.5 + 8 + 105.5 = 243 .$$

The eight-year gap is due to the fact that the orbital periods of Earth (365.256 days) and Venus (224.701 days) are close to a ratio of eight to 13, i.e., Venus completes 13 orbits around the Sun in almost the same time that the Earth completes eight orbits. (Note that both eight times the orbital period of Earth and 13 times the orbital period of Venus are about 2,922 days.) As a result, Venus and Earth line up in the same positions with respect to the Sun.

But why after two transits eight years apart doesn't a third transit occur after another eight years? It is because the ratio of the two orbital periods is not quite 8/13 since Venus arrives at the eight-year rendezvous (that is, at its appropriate node) about 2.45 days earlier than Earth (i.e., the Earth has not quite reached the point where it would be lined up with the Sun and the node of Venus). Is anything exact in celestial mechanics? Venus arrives too early for a transit to occur. The next transit occurs either 105.5 years or 121.5 years later at the opposite node of Venus' orbit. Once again, a pair of transits will occur separated by eight years. It is worth mentioning that 100 percent exact alignment is not required as from the Earth the Sun's disk is so much larger than Venus' disk.

An example of the 243-year cycles:[1]

A pair of transits occurred in

1631 and 1639,

(that's eight years apart) then after 122 years the pair

1761 and 1769

occurred (another eight years), then with the passage of 105 years the first season ended (giving a total of 243 years) and the next season began with the pair

1874 and 1882

followed by 122 years later with the pair

2004 and 2012

followed by 105 years ending the second cycle (with 2117 and 2125 beginning the third cycle).

To complicate matters further, occasionally one of the eight-year "double-transits" may disappear for several centuries because one of them is a near miss (so instead of a pair we only get one).

References

1. Nasa Eclipse Website—Six Millennium Catalog of Venus Transits: 2000 BCE to 4000 CE. https://eclipse.gsfc.nasa.gov/transit/catalog/VenusCatalog.html

The crashes people remember,
but drivers remember the near misses.

—Mario Andretti

TOPIC 17. Have any stars come close to our solar system in the past? What about the future? Currently, what are the closest stars to our solar system?

About 70,000 years ago Scholz's star, a red dwarf similar to Proxima Centauri, came within about 52,000 AU or about 0.8 light years from the Sun. That compares with Proxima Centauri, currently the closest star to our Sun, at about 4.2 light years. No other star is known to have come as close to the Sun as Scholz's star. Fifty two thousand AU would place it within the outer region of the Oort cloud, a sphere of billions of comets circling the Sun. This cloud is thought to extend from between 2,000 and 5,000 AU out to perhaps 100,000 AU from the Sun.[9] A star passing within the inner region of the Oort cloud would perturb some comets to head toward the inner solar system, possibly causing a mass extinction on Earth. If Scholz's star did perturb any comets 70,000 years ago, they would take about two million years to reach the inner solar system—so they have not yet arrived.

Would any homo sapiens on Earth have seen Scholz's star? After all, homo sapiens have been around for about 250,000 years, and earlier homo species extend back more than three million years. The answer is no. The star would have been about 10th magnitude. With the unaided eye we can only see stars that are about sixth magnitude or brighter. However, being a magnetically active red dwarf, Scholz's star could have emitted flares, and it is possible such a flare would have been bright enough and lasted for minutes or hours to have been noticed by some of our ancestors.

Interestingly, Scholz's star (about .08 solar masses) has a brown dwarf companion (of about .06 solar masses) and they orbit each other with a period of about four years. Where is Scholz's star now? It is some 20 light years distant in the constellation Monoceros; its apparent magnitude is about 18.3—quite faint.

Figure 1. Could Scholz's star have flared bright enough for our ancestors to have noticed it? Credit: José A. Peñas/SINC

What about future close encounters?

An orange star named Gliese 710, approximately 60 percent of the Sun's mass, is calculated to come as close as 0.22 ± 0.05 light years or about 14,000 AU to the Earth in roughly 1.35 million years.[3,4,5] Scientists have said this will be "the strongest disrupting encounter in the future and history of the Solar System."[3,5] Comets in the inner Oort cloud will be perturbed and sent heading for the inner solar system posing a threat to the Earth. At its closest, the star would have apparent magnitude of about -2.7, which is brighter than Mars at opposition (when Mars is closest to Earth). And its proper motion would reach one arc minute per year, which in 60 years would be one full degree—very noticeable to the unaided eye against the background stars.[3,4,5] Currently Gliese 710 is in the constellation Serpens Cauda, approximately 62 light years distant, with apparent magnitude 9.7.

Let us now consider the closest stars to the Sun at the *present* time.
None of these stars are currently as close to the Sun as Scholz's star came or Gliese 710 will come. Of course Alpha Centauri (a binary, consisting of two stars, A and B) together with the red dwarf Proxima Centauri currently constitute the closest star system to our solar system, with Proxima being only 4.2 light years away. If we were in the Alpha Centauri system and looked for our Sun, we would find it halfway between Cassiopeia and Perseus, about five degrees NW of the Double Cluster.[6] Viewed from this system the constellation Centaurus would lose a first-magnitude star and the Perseus/Cassiopeia region would gain one. Alpha Centauri A has about 10 percent more mass than the Sun and, like our Sun, is of spectral type G. Alpha Centauri B (not to be confused with Beta Centauri) has about 10 percent less mass than the Sun and is of spectral type K. In 2016 a planet slightly larger than Earth was discovered orbiting Proxima—see Topic 30 on exoplanets for more details.

Barnard's star is the second-closest to us. Barnard's "Runaway Star" is a red dwarf and has the largest proper motion of any star in the sky, moving 10.31 arc seconds per year in position angle 356 degrees—almost due north. Since its discovery in 1916 the star has moved over 12 arc minutes, almost half the apparent diameter of the Moon. One of the reasons for the large proper motion is that it is only 6.0 light years away. Its mass is about 14 percent of the Sun's and it is located in the constellation Ophiuchus.

The third-closest is the binary system Luhman 16 (also known as WISE 1049-5319), a pair of brown dwarfs, which is 6.5 light years away in the southern constellation Vela. And the fourth-closest system, WISE 0855-0714, is a single sub-brown dwarf slightly over seven light years away in Hydra. Astronomer Kevin Luhman announced the discovery of the former in 2013 and the latter in 2014.[7,8] Both discoveries used the Wide-field Infrared Survey Explorer (WISE) Earth-orbiting satellite. One can only wonder how many more brown dwarfs are close to us but remain undiscovered.

And how about the future? Well, the red dwarf Ross 128, currently the 12th-closest system to us, in about 80,000 years will be the closest star to the Sun, about 6.2 light years away.[7] It is now 10.9 light years away in Virgo, has magnitude 11.13, and a mass about 15 percent that of the Sun. Ross 128 has high proper motion and, like Proxima, is known to have a planet orbiting it. For more on Ross 128's planet, see Topic 30.

Figure 2. Alpha Centauri (A/B) is on the left, Beta Centauri is on the right, and a small red circle encloses Proxima Centauri, being at about 5:30 o'clock from Alpha Centauri.
Figure 3. Enlargement of red circle centered on Proxima Centauri. Credit: CC BY-SA 3.0

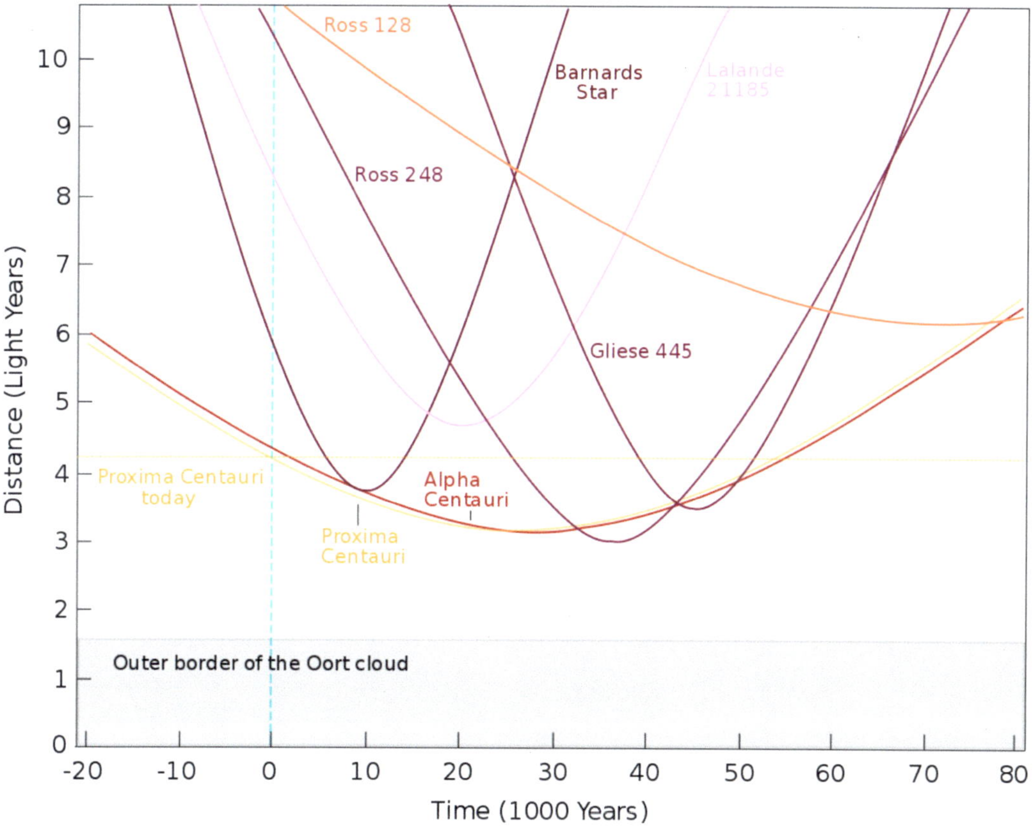

Figure 4. The vertical dashed blue line indicates the current time (time 0). The horizontal dotted yellow line indicates the current distance of Proxima Centauri from the Sun, about 4.2 light years. Credit: CC BY-SA 3.0 (Creative Commons)

References

1. Mamajek, Eric, et al. (February 12, 2015). The closest known flyby of a star to the solar system. *The Astrophysical Journal Letters.* https://iopscience.iop.org/article/10.1088/2041-8205/800/1/L17

2. University of Rochester Newscenter. (February 16, 2015). A close call of 0.8 light years. https://www.rochester.edu/newscenter/scholz-star/

3. The predictions for Gliese 710 are from calculations based on Gaia DR2 data. Gaia is a global space astrometry mission. The Gaia space satellite orbits the Sun at the L2 Lagrangian point and will make the largest, most precise three-dimensional map of our galaxy by surveying more than a thousand million stars, monitoring each of the stars about 70 times over a five-year period. It will precisely chart their positions, distances, movements, and changes in brightness. Launched in 2013, it is expected to function until at least 2022.

4. Bailer-Jones, C.A.L., Rybizki, J., Andrae, R., & Fouesnea, M. (2018). New stellar encounters discovered in the second Gaia data release. *Astronomy & Astrophysics, 616* (A37). arXiv:1805.07581. Bibcode:2018A&A...616A..37B. doi:10.1051/0004-6361/201833456. https://www.aanda.org/articles/aa/abs/2018/08/aa33456-18/aa33456-18.html

5. Berski, Filip, & Dybczyński, Piotr A. (November 1, 2016). Gliese 710 will pass the Sun even closer. *Astronomy & Astrophysics, 595* (L10). Bibcode:2016A&A...595L..10B. doi:10.1051/0004-6361/201629835. ISSN 0004-6361. https://www.aanda.org/articles/aa/abs/2016/11/aa29835-16/aa29835-16.html

6. Kaler, James, A. (September 1998). Discussion of the Alpha Centauri system. Ask Astro, *Astronomy Magazine*, p 97.

7. Dorminey, Bruce (July 2020). Stellar neighbors close-up. *Astronomy Magazine,* p 30-35.

8. Balouci, Sharazade (July 2013). Closest brown dwarf system discovered. *Sky & Telescope,* p 12.

9. NASA, Oort cloud. https://solarsystem.nasa.gov/solar-system/oort-cloud/in-depth/

Mars tugs at the human imagination like no other planet.

—John Noble Wilford, Mars Beckons

TOPIC 18. Mars and its moons.

I. Interesting facts about Mars itself including the recent *very* close approach of a comet to Mars.

Figure 1. The Red Planet Mars. Hubble Space Telescope 1999. Credit: Steve Lee (University of Colorado), Jim Bell (Cornell University), Mike Wolff (Space Science Institute), and NASA, Public domain, via Wikimedia Commons

In Figure 1, four major surface features are prominent on images of different parts of the Martian sphere. Clockwise from upper left: Acidalia Planitia is the large dark area in the north, Tharsis is a massive volcanic region straddling the equator, Elysium is the central, bright area just above the equator, and Syrtis Major Planum is the dark area pointing north. (Latin: planitia, low plain; planum, high plain or plateau.) The Tharsis region is a major bulge on the Martian surface. Elysium Mons is the second-largest volcanic region on Mars and is just north of Elysium Planitia.

Stars are visible during the day from the Martian surface.
The Martian atmosphere is so thin that if you were to look straight up in the daytime you would see stars and the blackness of space. You may have seen images of a pink Martian sky taken by rovers on the surface. But these photographs depict the sky near the horizon and so are taken looking through more of the atmosphere.[4]

Meteorites from Mars found on Earth.
Dozens of Martian meteorites have been discovered in Antarctica. How can we be sure they are from Mars? Gas trapped inside the meteorites has the same isotopic and chemical compositions as in the

Martian atmosphere as determined by Viking.[6]

The Opportunity lander.
The Opportunity Lander set down on Mars in 2004 and, though its mission was planned for only 90 days, it lasted more than 15 years. The lander traveled over 45.16 kilometers (about 28 miles).[7] As successful as Opportunity was, humans on the surface could have traveled that far, all the while doing research, in a few weeks. We also need boots on the (Martian) ground.

The polar caps on Mars.
Mars' southern polar cap is larger than its northern cap, and the southern Martian winter is longer and colder, occurring at aphelion when the planet is farthest from the Sun. Winter in the north occurs at perihelion, Mars' closest point to the Sun. At aphelion Mars is about 21 percent farther from the Sun than when it is at perihelion. Compare that to Earth where the difference is only 3.4 percent.

An additional factor producing a longer southern winter is that Mars (and indeed any planet) travels more slowly when farther from the Sun than when it is closer. At both Martian poles carbon dioxide condenses in winter to form a cap about a meter thick, overlaying a water ice cap that is about three kilometers thick. In winter the northern polar cap extends to about the 80 degrees latitude circle. In winter the southern cap extends to about the -45 degrees latitude circle and contains almost a third of the atmospheric CO_2. With the arrival of summer at either cap, CO_2 sublimates at that cap, enters the atmosphere, and, in a seasonal promenade, migrates to the other pole where it condenses. So we have a seasonal migration of CO_2. The northern cap loses virtually all of its CO_2 in its summer, but the southern cap, even at midsummer, retains a residual amount of CO_2, with some patches of water ice visible.[6]

Trying to grasp the size of Olympus Mons.
Olympus Mons (Latin: Mt. Olympus) is one of the largest volcanos in the solar system. If someone could name only one feature on Mars it most likely would be Olympus Mons. In the late 1800s it was seen as an albedo or reflective feature and named Nix Olympica (Latin: olympic snow).

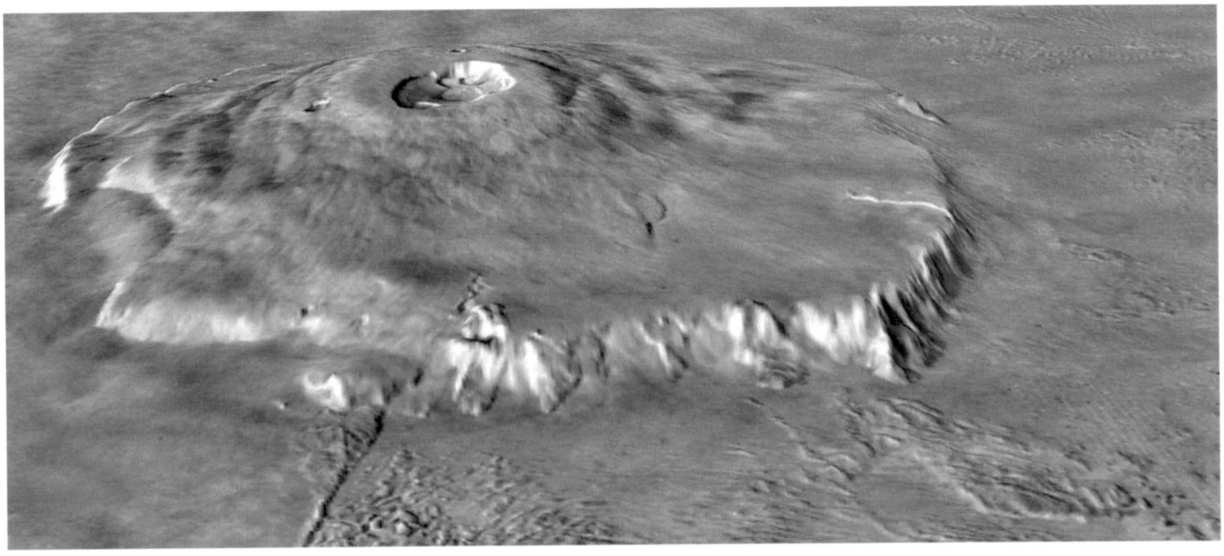

Figure 2. Olympus Mons from a Viking image mosaic overlaid on Mars Orbiter Laser Altimeter (MOLA) data. Note: The vertical exaggeration is 10 to 1. So it is much flatter than it appears here. Credit: NASA/MOLA Science Team (original uploader was Nephelin at de.wikipedia), Public domain, via Wikimedia Commons

To try to grasp its enormous size, take a look at a map of Washington State, about 240 miles wide and 360 miles long. Olympus Mons, about 370 miles in diameter, would cover considerably more than the entire state. Washington's Mt. Rainer is 14,400 feet high and is less than 100 feet from being taller than the highest mountain in the lower 48 states, Mt. Whitney in California (although Rainer is much more massive than Whitney). Mt. Rainer's footprint is of course only a fraction of the area of the state, in sharp contrast to that of Olympus Mons. Olympus Mons is about 13 miles (over 68,000 feet) above the surrounding plains of Tharsis. Like Mauna Kea in Hawaii, it is a shield volcano with very gently sloping sides. It is rather flat on top. In fact, standing near the summit you would not realize you were on the top of a mountain—the terrain would appear to be relatively flat from horizon to horizon. And even at a considerable distance from the volcano you could not grasp the volcano as a whole due to the planet's curvature and that of the volcano itself. It is shaped like a giant cow patty.[8] Someone on the Martian surface could not comprehend the mountain in its entirety.

Korolev Crater—Mars' year-round skating rink.

Korolev Crater, 82 kilometers (51 miles) in diameter, contains a permanent sheet of water ice in its interior. The ice persists because of what is called a "cold trap." The ice lies two kilometers (1.2 miles) below the crater's rim, with the rim being two kilometers above the surrounding plain. Air passing over the crater is cooled, sinks, and becomes trapped in the crater, maintaining a thin, chilly

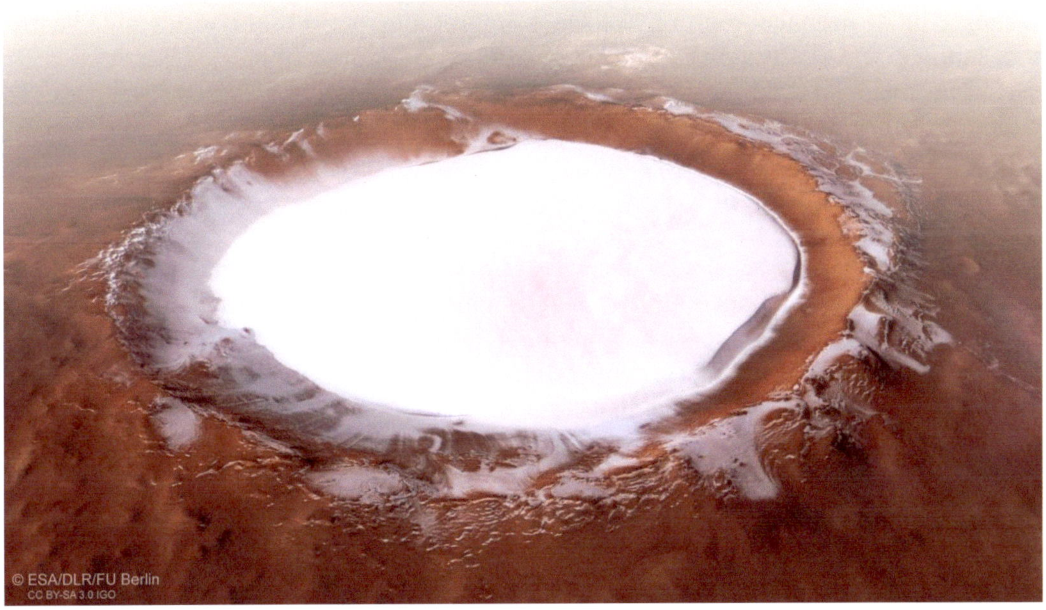

Figure 3. This image from ESA's Mars Express shows an oblique perspective view of Korolev crater generated using a digital terrain model and Mars Express data gathered during more than five orbits. The image has a resolution of roughly 21 meters per pixel and was created using data from the nadir and color channels of the High Resolution Stereo Camera (HRSC). The nadir channel is aligned perpendicular to the surface of Mars, as if looking straight down at the surface. Credit: ESA/DLR/FU Berlin -CC BY-SA 3.0, http://m.esa.int/spaceinimages/Images/2018/12/Perspective_view_of_Korolev_crater

barrier over the ice, preventing it from melting and evaporating.[15,16,17] The ice disk is 1.8 kilometers (1.1 miles) thick and is about 60 kilometers (37 miles) in diameter. The crater is located 73 degrees north latitude (so quite high) and 165 degrees east longitude and is named after Sergei Korolev (1907-1966), the head Soviet rocket engineer and designer during the 1950s and 1960s space race.

A wide-ranging tilt to Mars' axis of rotation, high orbital eccentricity compared to Earth, and variability in its eccentricity imply wide changes in sunlight over long time periods at any given point on Mars.

Today, the Earth's axis is tilted 23 degrees 26 minutes from the plane of its orbit around the Sun. But this tilt changes. During a cycle averaging about 41,000 years, the tilt of the axis varies between 22.1 and 24.5 degrees.[19,20] The Earth's tilt is stabilized by its large moon.

But on Mars the situation is different. Mars' axis is currently tilted 25 degrees 12 minutes, almost the same as Earth's. But over timescales of about 10 to 20 million years[18] it can range from 13 to 40 degrees, leading to dramatic changes in sunlight for any given location on the surface. (The range for Mars' axial tilt varies with other sources—some giving a tilt higher than 40 degrees. But the conclusion is that the range is much greater than for Earth's axial tilt.)

Mars' current orbit is more eccentric than the Earth's. Thus the difference between its perihelion and aphelion is considerably greater than the Earth's: .29 AU for Mars vs only .034 AU for Earth. Furthermore, the eccentricity of Mars' orbit can shift so that Mars is as much as 50 percent farther from the Sun at aphelion than at perihelion.[9] So the amount of sunlight reaching Mars at perihelion is much greater than at aphelion.

Dramatic changes in the orbital eccentricity of Mars suggest long periods of considerable yearly variability in sunlight. And extreme axial tilt suggests that, for periods of at least hundreds of thousands of years, a large part of the planet would have the Sun beating down on it for about half the year and experience almost continuous darkness for the other half of the year. Clothes designers for future inhabitants of a terra-formed Mars should pay attention.

A recent close call with a comet.

On October 14, 2014, a comet passed very close to Mars. At its closest approach to Mars (2:28 p.m. EDT) the comet's nucleus was about 87,000 miles from the Red Planet,[1,2] only about one-third the distance between the Earth and its Moon and one-tenth the distance that any known comet has come to Earth. Mars traveled through the comet's coma of gas and dust for hours. It must have been a stunning site from the surface of Mars. If the comet had come this close to Earth the meteor display possibly would have surpassed the 1833 meteor storm on Earth—see Topic 9.

Discovered on January 3, 2013, by astronomer Robert McNaught at the Siding Spring observatory in Coonabarabran, Australia, Comet Siding Spring (officially named C/2013 A1) came from the Oort cloud, a spherical shell of what is possibly more than a trillion comets enveloping the entire solar system. The cloud extends outward perhaps about 100,000 AU,[21] a third of the way to the nearest star system, Alpha Centauri. When will Comet Siding Spring revisit Mars? In about 740,000 years.[1]

Thinking about Comet Siding Spring's close approach to Mars brings to mind the close approach of Halley's Comet to the Earth in 1910. The Earth passed through Halley's tail, and some people were worried that the Earth's atmosphere might be poisoned. It didn't happen. But it was an awesome display (see Figure 6 below). Consider, however, while Halley's Comet approached within 13.9 million miles of Earth, Comet Siding Spring came within only 87,000 miles of Mars. I would love to have been there on October 14, 2014.

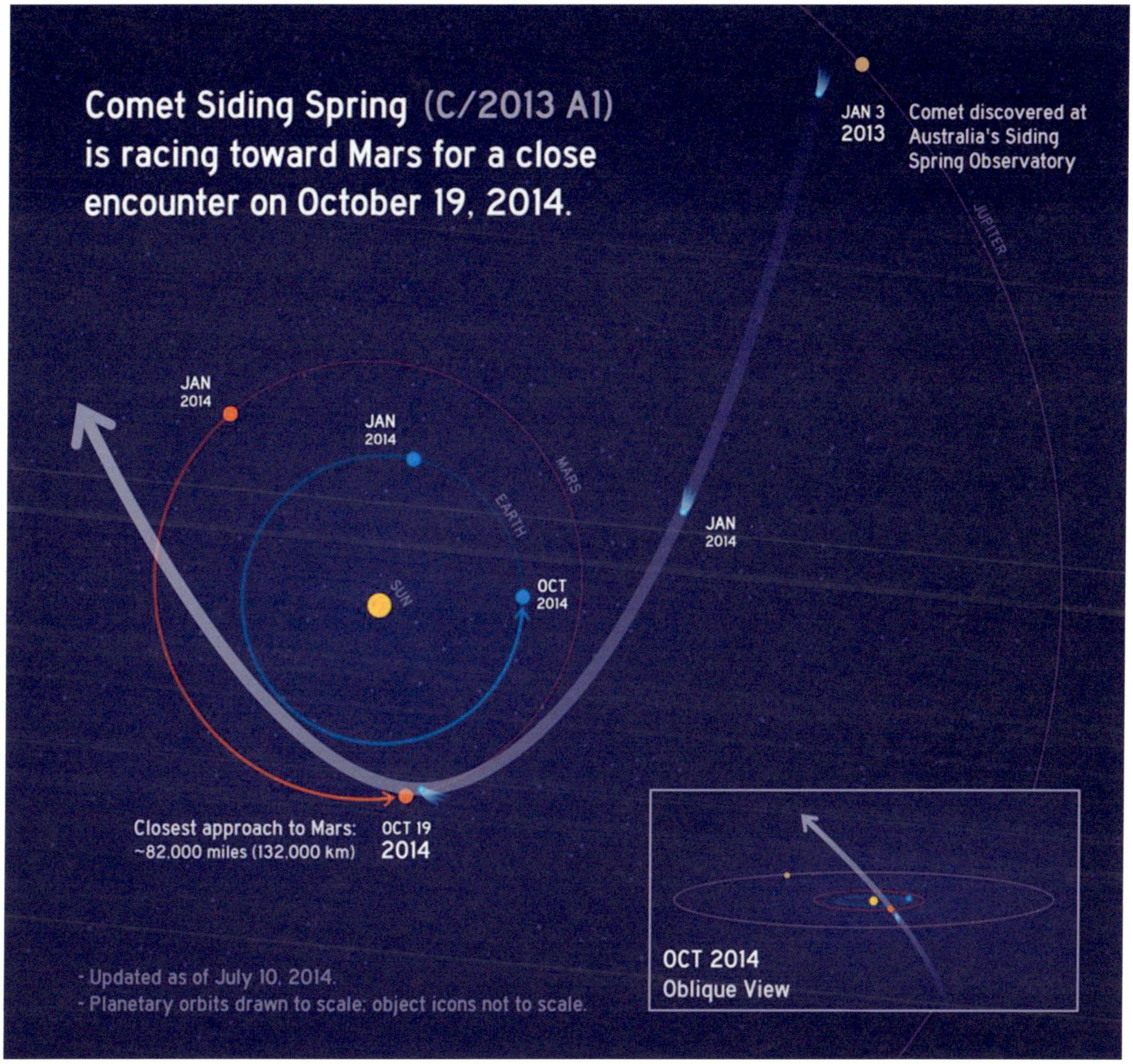

Figure 4. Very close approach of comet C/2013 A1 to Mars. Credit: NASA

Figure 5. Composite Hubble Space Telescope image of comet C/2013 A1 and Mars. Credit: NASA/ESA/J.-Y. Li(PSI)/C.M. Lisse (JHUAPL) /The Hubble Heritage Team (STSci/AURA)

Felicia Chou of NASA describes this image[2] in the four paragraphs below:

The comet image is a composite of Hubble exposures taken between October 18, 8:06 a.m. to October 19, 11:17 p.m. Hubble took a separate image of Mars at 10:37 p.m. on October 18. The Mars and comet images have been added together to create a single picture to illustrate the angular separation, or distance, between the comet and Mars at closest approach. The separation is approximately 1.5 arc minutes, or one 20th of the angular diameter of the full moon. The background star field in this composite image is synthesized from ground-based telescope data provided by the Palomar Digital Sky Survey, which has been reprocessed to approximate Hubble's resolution.

The solid, icy comet nucleus is too small to be resolved in the Hubble picture. The comet's bright coma, a diffuse cloud of dust enshrouding the nucleus, and a dusty tail are clearly visible.

This is a composite image because a single exposure of the stellar background, Comet Siding Spring, and Mars would be problematic. Mars actually is 10,000 times brighter than the comet, so it could not be properly exposed to show detail in the Red Planet. The comet and Mars also were moving with respect to each other and could not be imaged simultaneously in one exposure without one of the objects being motion-blurred. Hubble had to be programmed to track on the comet and Mars separately in two different observations.

NASA used its extensive fleet of science assets, particularly those orbiting and roving Mars, to image and study this once-in-a-lifetime comet flyby. In preparation for the comet flyby, NASA maneuvered its Mars Odyssey orbiter, Mars Reconnaissance Orbiter (MRO), and the newest member of the Mars fleet, Mars Atmosphere and Volatile Evolution (MAVEN), in order to reduce the risk of impact with high-velocity dust particles coming off the comet. Other NASA space observatories also joined Hubble in observing the encounter, along with ground-based telescopes on Earth. The new composite image was taken with Hubble's Wide Field Camera 3.

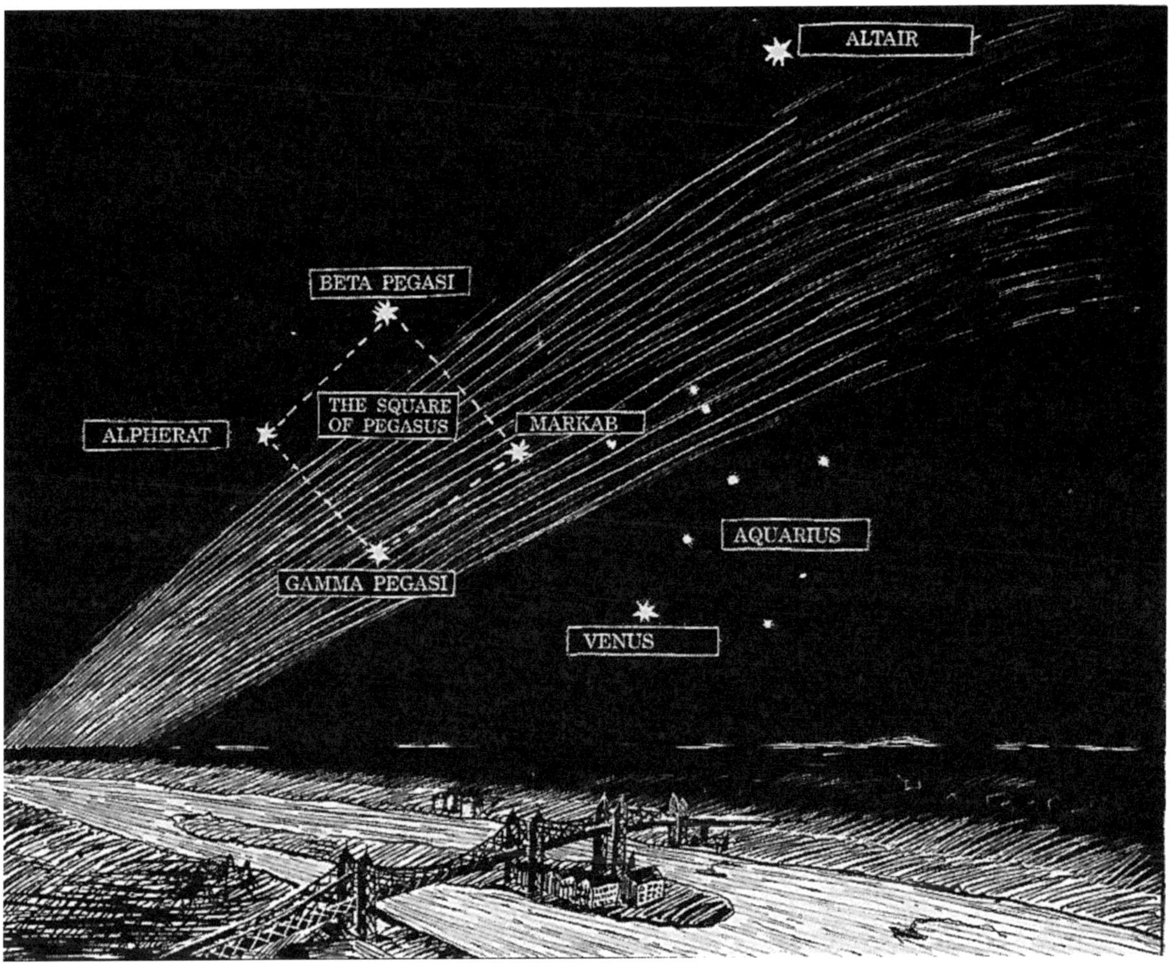

Figure 6. View of Comet Halley's tail as sketched between 2:30 a.m. and 3:15 a.m. on May 19, 1910, from the Times Tower in New York City. Earth passed through the tail on May 18. Credit: This illustration in Sky and Telescope magazine, May 1997, p 51[3] is a faithful rendition of the illustration that appeared in the N.Y. Times as drawn by Mary Proctor on the morning after closest approach.

II. The Martian moons Phobos (Greek: fear) and Deimos (Greek: terror or panic).

A little history.

American astronomer Asaph Hall III (1829-1907) was about to give up his frustrating search for a Martian moon one August night in 1877, but his wife, Angeline Stickney, (1830-1892) urged him on. He discovered Deimos the next night and Phobos six nights after that.[12,13]

Hall was born in Goshen, Connecticut. His father, a clockmaker, died when he was 13 leaving the family in financial difficulty. Hall left school at 16 to become a carpenter's apprentice. Later he studied mathematics at New York Central College in McGrawville, New York, a progressive school enrolling students of modest means, free African slaves, and women. It was there he met Angeline Stickney. She was able to attend with financial help from her sister Ruth and by teaching at the college. She studied science and mathematics, including mathematical astronomy, and was an outstanding student. While at college she became a passionate supporter of women's suffrage and an abolitionist. Hall met Angeline when he was a student in a class that she taught (she was two years ahead of Hall).

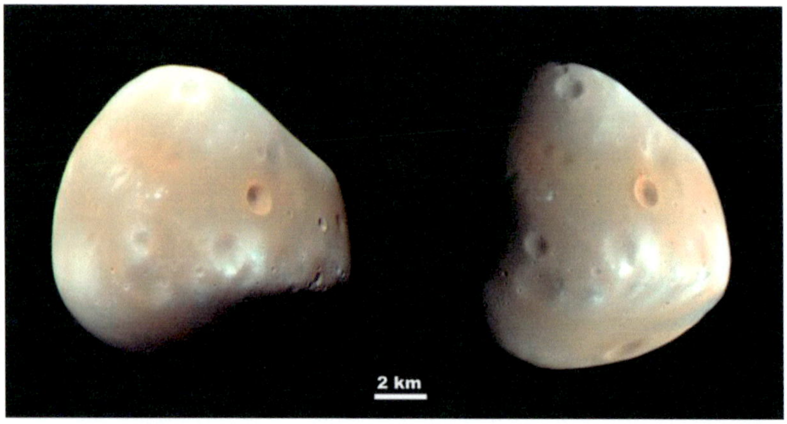

Figure 7. These color-enhanced views of Deimos, taken on February 21, 2009, are the result of images taken by the High Resolution Imaging Science Experiment (HiRISE) camera on NASA's Mars Reconnaissance Orbiter. Credit: NASA/JPL-Caltech/University of Arizona

Figure 8. View of Phobos taken by the High Resolution Imaging Science Experiment (HiRISE) camera on NASA's Mars Reconnaissance Orbiter on March 23, 2008. Credit: NASA/JPL/University of Arizona

They married in 1856. Unfortunately, as was common in those days, she had to give up having a career. Hall became an expert at computing orbits and was appointed an assistant professor at the U.S. Naval Observatory in 1862, attaining the rank of professor within a year. Using the observatory's 26-inch refractor, the largest in the world at the time, he discovered Phobos and Deimos. In 1973 Phobos' largest crater, nearly half of the diameter of Phobos itself, was named Stickney in his wife's honor.

Figure 9. Angeline Stickney Hall (left) and Professor Asaph Hall at the U.S. Naval Observatory in 1899.

Figure 10. Hall used this 26-inch refractor at the U.S. Naval Observatory to discover Phobos and Deimos.

Properties of the two moons.

Phobos is the only moon in the solar system that revolves around its primary faster than the primary rotates on its axis. Because it is so close to Mars, Phobos zips around its planet in seven hours 39 minutes, about three times in one Martian day of about 24 hours 37 minutes. It would be seen racing across the sky in about four hours, rising in the west and setting twice in the east. Here "time to cross the sky" means the time from moonrise to moonset as seen from the Martian equator. Observed from the Earth's equator our Moon takes about 12.5 hours to cross the sky.

In contrast, Deimos orbits Mars in about 30.3 hours. Rising in the east and setting in the west, it takes about 2.5 Martian *days* to move cross the sky.[14]

Phobos' mean diameter is about 14 miles; Deimos' is about 7.7 miles. Neither are spherical. From the surface of Mars, when Phobos is overhead it subtends an angle of 13 arc minutes; on the horizon the angle is about 8.9 arc minutes. Deimos subtends an angle of 1.8 arc minutes. (See *Appendix 18.1* for calculations.) The average angle that the Sun subtends at Mars is 21 arc minutes.[14] The Sun's size varies more as seen from Mars than from Earth as Mars' orbital eccentricity is greater than that of Earth. Again, in comparison, from the Earth the Sun and our Moon each subtend an angle of about 30 arc minutes—which allows for total solar eclipses. From Mars there can be no total solar eclipses.

If you were on Mars and poleward of 70 degrees (i.e., north of 70 degrees N or south of -70 degrees S) you would never see Phobos because it orbits so close to Mars. Likewise, you would be unable to see Deimos poleward of 80 degrees.[5]

A layer of regolith, the dusty residue from asteroid impacts over millions of years, covers Phobos to a depth of at least five meters (16 feet).[5]

What about transits of the Martian moons across the face of the Sun and eclipses of the moons themselves in Mars' shadow? Can Phobos be seen to occult Deimos?
Transits of the Sun occur for both moons, but neither comes close to covering the face of the Sun for a total solar eclipse. Transits by Deimos last only a few tens of seconds, but typically several would occur during a period of two weeks.[14] Just how many depends on the observer's latitude and the Martian season. For Phobos the solar intensity can drop by about a third.

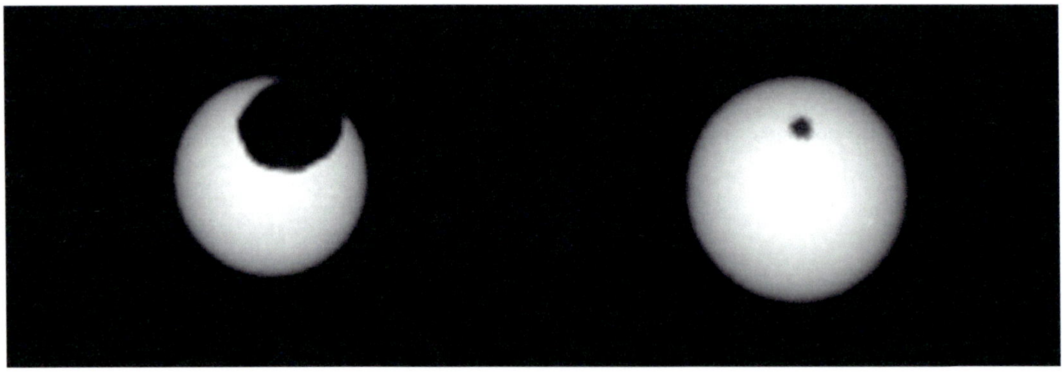

Figure 11. As viewed from the Martian surface, Phobos (left) and Deimos (right) transit the Sun. Credit: NASA/JPL/Malin Space Science Systems

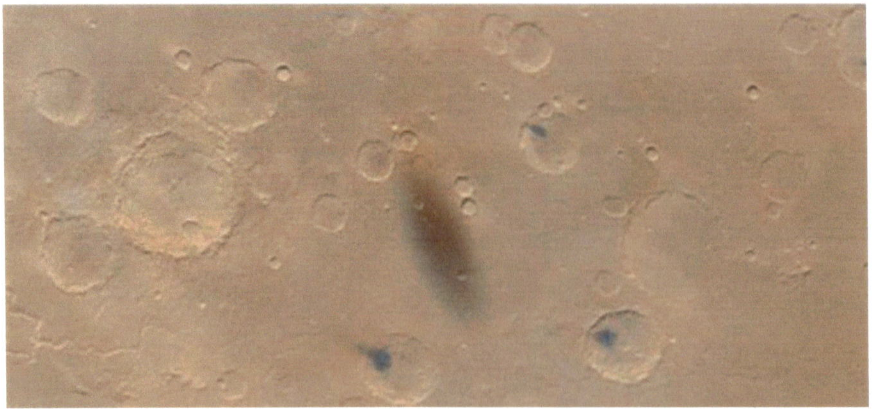

Figure 12. Phobos casts a dark spot, its penumbral shadow, on the Red Planet's surface. Observers within the shadow could see a partial or annular eclipse. Credit: NASA/JPL/Malin Space Science Systems

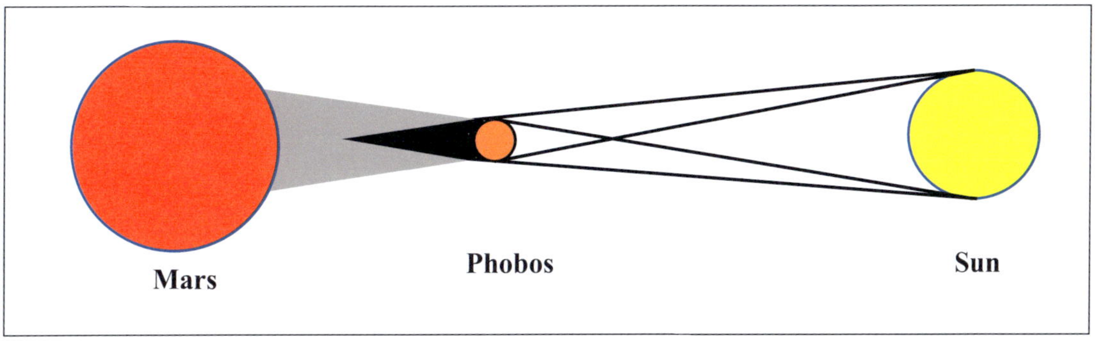

Figure 13. Phobos' black umbral shadow cone never reaches the surface of Mars—there are no total eclipses of the Sun. Phobos' grey penumbral shadow cone can reach the surface of Mars, resulting in a partial or annular eclipse of the Sun—or what, viewed from Mars, could be called a transit of Phobos across the face of the Sun.

From Mars, both moons experience the equivalent of lunar eclipses. For Phobos a lunar eclipse lasts about an hour; for Deimos up to two hours. Furthermore, at times near the Martian equinoxes, lunar eclipses occur every night that the moons are visible and on every revolution they make.[14]

Phobos can occult Deimos, but only from a narrow latitude band near the Martian equator.[14]

Grooves on Phobos.
Phobos' many parallel striations or grooves on its surface were first seen by the Viking orbiters in 1976. As Phobos gets closer to Mars, tidal effects are stretching Phobos into a football shape. It has been suggested[5,10] that these parallel grooves may be stress lines from tidal effects and that in perhaps 50 million years Phobos could break up. Evidence has been given to support another theory: the grooves were produced by bouncing ejecta boulders from the impact that created the large Stickney crater (visible in Figure 14 below).[5,11] Some grooves, however, don't appear to emanate from Stickney.

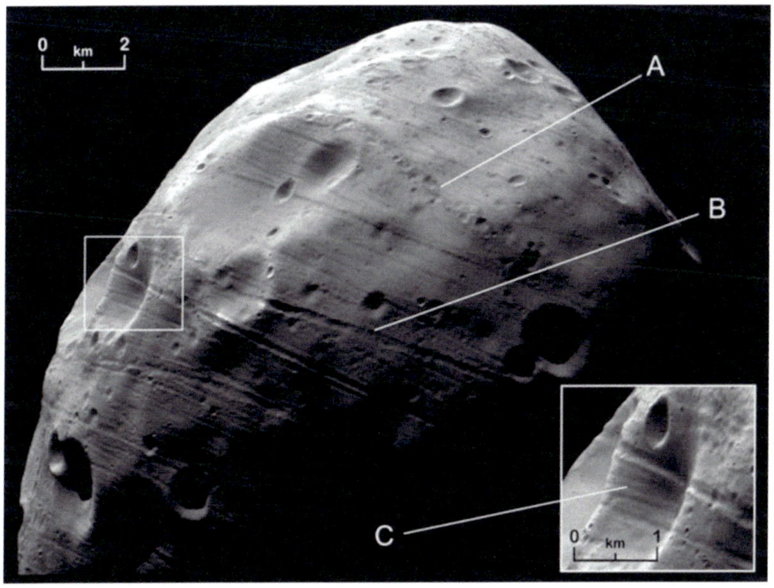

Figure 14. Phobos with its parallel grooves. A indicates a string of pits; B, a groove; C, a groove in a crater. Credit: See reference [11]

References

1. NASA. (December 19, 2019). *C/2013 A1 Siding Spring*. solarsystem.nasa.gov https://solarsystem.nasa.gov/asteroids-comets-and-meteors/comets/c-2013-a1-siding-spring/in-depth/

2. NASA. (October 24, 2014). Close encounters: Comet Siding Spring seen next to Mars. mars.nasa.gov https://mars.nasa.gov/news/1740/close-encounters-comet-siding-spring-seen-next-to-mars/

3. Schaefer, Bradley E. (May 1997). Comets that changed the world. *Sky & Telescope,* p 46-51.

4. The Planetary Society (February 26, 2021). https://www.planetary.org/the-downlink/a-perseverance-extravaganza

5. Barbuzano, Javier (April 2021). The mystery of the Martian moons. *Sky & Telescope,* p 34-40.

6. Carr, Michael H. (2011). The surface of Mars. *Sky & Telescope,* p 3-16. An announcement of the third edition of a globe of Mars produced in cooperation with NASA and the U.S. Geological Survey, where Carr is a scientist.

7. Grossman, Lisa (March 16, 2019). Farewell Opportunity. *Science News,* p 7.

8. Beish, Jeff (April 30, 2009). Images of some Martian volcanos. How far could I see standing on Olympus Mons? https://web.archive.org/web/20090827141153/http:/www.dustymars.net-a.googlepages.com/VOLCANO.htm

9. Lakdawalla, Emily (September 2013). The history of water on Mars. *Sky & Telescope,* p 21.

10. Crockett, Christopher (December 12, 2015). Martian moon headed for breakdown. *Science News,* p 11.

11. Ramsley, Kenneth R., & Head, James W. (January 2019). Origin of Phobos grooves: Testing the Stickney Crater ejecta model. *Planetary and Space Science, 165*, p 137-147. The labeled image is Figure 19 in a 2012 research paper "The origin of Phobos grooves from ejecta launched from impact craters on Mars: Tests of the hypothesis" by the same authors. https://www.semanticscholar.org/paper/The-origin-of-Phobos-grooves-from-ejecta-launched-Ramsley-Head/cc7f803fedd655613d0f7a4c1ecbee0b6a9a803d/figure/18

12. NASA. Mars moons. solar system.nasa.gov https://solarsystem.nasa.gov/moons/mars-moons/in-depth/

13. Hall, Asaph (1878). *Observations and orbits of the satellites of Mars*. Government Printing Office. https://en.wikipedia.org/wiki/Asaph_Hall (quoted in Hall, Angelo (1908). *An astronomer's wife: The biography of Angeline Hall.* Nunn and Company, p 112. https://archive.org/details/cu31924029854738

14. Christou, Apostolos (August 2004). Are either of the moons of Mars large enough to eclipse the Sun as seen from the surface of Mars? Ask Astro, *Astronomy Magazine,* p 78-79.

15. German Aerospace Center (DLR) (December 20, 2018). A winter wonderland in red and white—Korolev Crater on Mars. https://www.dlr.de/content/en/articles/news/2018/4/20181220_korolev-crater-on-mars.html

16. Sample, Ian (December 21, 2018). Mars express beams back images of ice-filled Korolev crater. *The Guardian.* https://www.theguardian.com/science/2018/dec/21/mars-express-beams-back-images-of-ice-filled-korolev-crater

17. Mastroianni, Ernie, & Klesman, Alison (May 2019). Fresh ice. *Astronomy Magazine,* p 9.

18. NASA. Phoenix Mars mission. http://phoenix.lpl.arizona.edu/mars102.php

19. Talcott, Richard (October 2009). Seeking ground truth on Mars. *Astronomy Magazine,* p 37.

20. Guinn, Jim (August 2001). What do we know about the onset of ice ages and their relation to Earth's orbit and tilt with respect to the Sun? Ask Astro, *Astronomy Magazine,* p 73.

21. NASA Science Solar System Exploration. (December 19, 2019). Oort cloud. https://solarsystem.nasa.gov/solar-system/oort-cloud/in-depth/

I reckon I seen my fair share. Happens on rare occasion.
The last glimpse of sunset, a green flash shoots up into the sky.
Some go their whole lives without ever seeing it.
Some claim to have seen it who ain't....

—Joshamee Gibbs, Pirates of the Caribbean: At World's End

TOPIC 19. The green flash, zodiacal light, and the gegenschein.

I. The green flash.

This illusive effect can sometimes be seen for a few seconds at sunrise or sunset. Just as the top of the Sun reaches the horizon, its first/last rays glow green.

Figure 1 (left). Green Flash. Credit: John Lindsey, Contributed to the Santa Ynez Valley News, July 9, 2019.
Figure 2 (right). Green Flash. Credit: Brocken Inaglory—Own work, CC BY-SA 4.0

Figure 2 was photographed in San Francisco, September 17, 2006. The photographer described this sequence of images as "a very complex mock mirage. It seemed that the Sun could not get enough of playing with [thermal air layers over the Pacific Ocean] and produced a different shape every moment."

In the above, the term "mock mirage" was used. A *mock mirage* is an inverted image produced by a thermal inversion below eye level. A thermal inversion occurs when warmer air lies above colder air. So in this situation, off in the distance near the horizon, the surface air had a warmer layer of air above it, but a line from the observer's eye to the horizon is above this distant, warm layer of air.

The most common type of green flash occurs with no inversion, i.e., warmer air is close to the ground or water and cooler air is above the warm air (which is frequently the case over water).

A thorough discussion of the various types of green flashes and much more is found in reference [1]. Here we will restrict the discussion to the basic process responsible for a green flash.

The cause of the green flash.
When an observer sees the bottom of the setting Sun's image touch the western horizon, the Sun itself is actually just below the horizon with its *top* touching the horizon. This phenomenon occurs because

light from the Sun passing through the Earth's atmosphere is refracted (or bent) slightly down. See Figure 3. The Sun's image appears to the observer to be about half a degree higher than its actual position. And since the Sun's angular diameter is also about a half a degree, we have the result illustrated below.

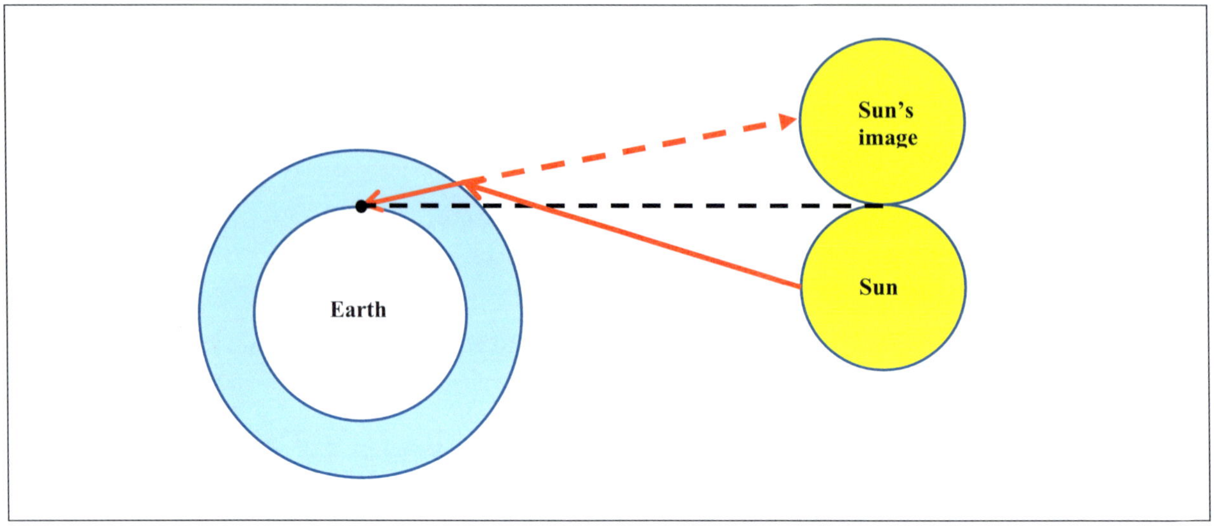

Figure 3. The location of an observer is indicated by the black dot. The image has the Earth rotated so that the observer is on the edge of the globe, and the Sun is moving down. The dashed black line is the observer's horizon line. A light ray (in red) from a spot on the Sun reaches the Earth's atmosphere and is refracted down into the observer's eye. The red dashed line indicates where the observer perceives this light ray to be originating.

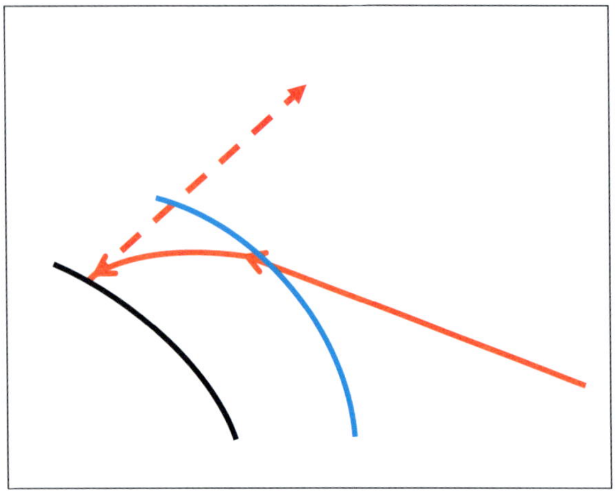

Figure 4. Figure 3 is a simplification in that it shows the light ray (solid red) refracting when it reaches the upper atmosphere and traveling in a straight line to the observer's eye. If the atmosphere were a solid piece of glass, this would happen. But atmospheric density is not uniform and becomes denser as the ray moves downward. So the path of the ray through the atmosphere is an ever-steepening curve. Figure 4 gives a more accurate picture: the Sun is seen in the direction of the red dashed line, a different direction from that in Figure 3.

Even considering the expanded description in Figure 4, an important element has been left out. The amount of the bending of light due to refraction depends on the frequency of the light—its color. Bending increases as we move from the red end of the spectrum to the blue end (Figure 5). The result is overlapping red, orange, yellow, green, and blue images of the Sun (Figures 6 and 7). (Here I have only mentioned five colors, but of course the frequency range is continuous.) Where these images overlap we get white light or, in the case of the Sun, yellow light. But a slight sliver of red (Figures 5, 6, and 7) is at the bottom of the overlap; at the top of the overlap is a slight sliver of blue, just above a sliver of green (Figures 5, 6, and 7). The blue ray has been left out of Figures 5 and 6 because Rayleigh scattering of blue light (which gives us a blue sky) weakens the blue sliver (produced by refraction) and so a blue flash is very rarely seen. (In this context, Rayleigh scattering is the instantaneous absorption and emission of predominantly blue wavelength photons by atmospheric nitrogen molecules. The emissions occur in all directions.)

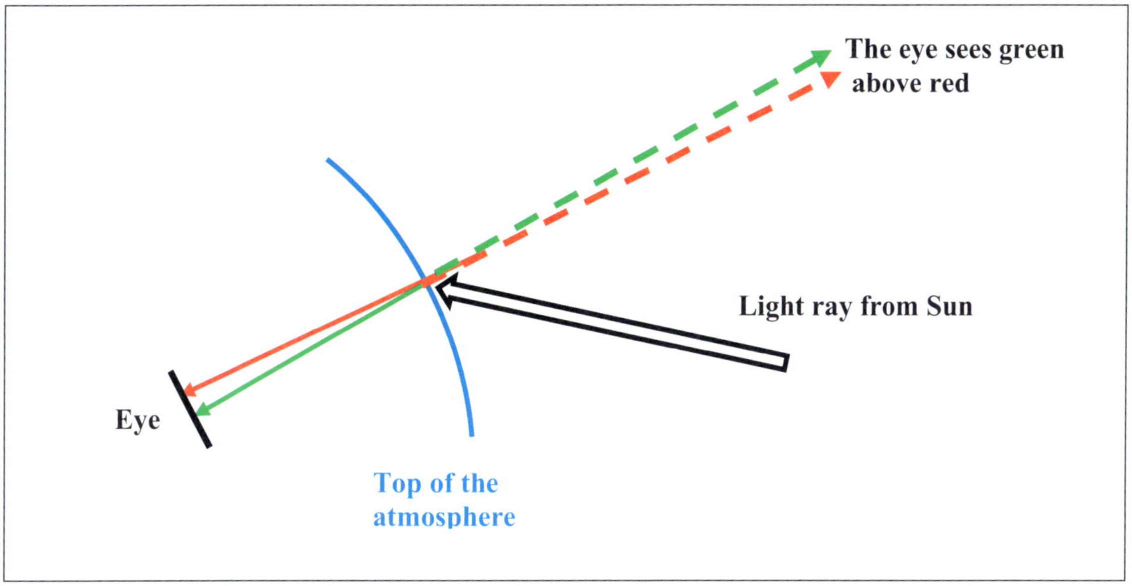

Figure 5. Light from the Sun is refracted in the atmosphere, with the green part of every ray refracted more than the red part. So an observer on Earth will see green at the top of the Sun and red at the bottom. Refraction is greatest when the Sun is on the horizon.

Figures 6 (left) and 7 (right). An illustration of the refracted image of the Sun (left) and an image of the Sun setting behind the Golden Gate Bridge in San Francisco. In the atmosphere different-colored disks of light from the Sun overlap each other imperfectly. Our eyes see the rim of the green disk at the top and the edge of the red disk at the bottom. Credit (right): Brocken Inaglory - Own work, CC BY-SA 3.0, https://commons.wikimedia.org/w/index.php?curid=4008651

The best opportunity to see a green flash at sunset is either from sea level or from a hill/mountain with a clear view of the western horizon. Viewing the sunset over an ocean or a large body of water is one possibility. The air's transparency is critical. While water vapor absorbs much more light at the red end of the visible spectrum, with high humidity green will be absorbed enough to eliminate a green flash. Green flashes are also correlated with a heavily distorted Sun (because of air turbulence between the Sun and the observer).

Of course the same phenomenon occurs at sunrise. But in this case one has to know exactly where and when to look for the first appearance of the Sun's disk. There is a physiological component to sunset flashes which does not occur at sunrise ones. The effect[1] is primarily bleaching of the red-sensitive retinal pigment, though there is also some neural adaptation. Looking at the low, orange-red Sun for a few seconds is enough to make most flashes appear green at the stage where they photograph as yellow on color film. That's why it's difficult to take good color photographs: most people trip the shutter too soon at sunset.

An intriguing thought: If an observer were at the north or south pole and the Sun were at either equinox (i.e., directly over the equator), the Sun would circle the sky at the horizon. Could this result in intermittent or extended green flashes throughout the day? Variations in the topography at the horizon (only possible at the south pole) might help.

If you observe the green flash sitting down (when near sea level) you can actually prolong the effect by standing up.[2,3]
A surprising fact is that a very slight change in elevation (a few feet) can change the time of the green flash by several seconds, especially when the observer is at sea level. So an observer could prolong the time the green flash is visible just by standing up from a seated position! A change from two feet to six feet above sea level results in a green flash four seconds later.[2] But if the observer is 100 feet above sea level, a rise of four feet produces the flash less than a second later.[2] See *Appendix 19.2* for the basic reason for this surprising result.

In July 2002, former astronaut Owen K. Garriott was on board the Russian oceanographic vessel Akademik Mstislav Keldysh, which was conducting research on hydrothermal vents, when a green flash was observed by shipmates on decks seven and eight.[3] As quoted in [3], Garriott described what occurred:

> Some observers were on an open portion of Deck 7 and others on Deck 8, nearby but 2.85 meters higher. Five of us on Deck 7 saw the usual red disk of the Sun change to solid green just before disappearance. We broke into spontaneous applause. The people on Deck 8 were still a seeing red disk, and one of them looked down at us, wondering why we were so excited. He missed the green flash, but his two partners did see it—several seconds after we had.

On July 22, 2002, observers were stationed on Decks 5-9 and timings were taken in order to quantify the delays relative to the time of the observation of a green flash from deck 5. The results (which are given in [3]) confirmed this delay phenomenon. Garriott and his shipmates made a calculation similar to the one given in *Appendix 19.2* and compared the predicted delays with the observed delays.

See [4, 5] for two interesting articles on the green flash.

II. The zodiacal light.

The zodiacal light is a wedge-shaped area of light emanating from the horizon just after sunset or before sunrise. It is created by sunlight reflecting off dust orbiting the Sun in the band of the zodiac. The zodiacal light in Figure 7 is an example of the forward-scattering of sunlight. While the dust band extends completely around the ecliptic (the center line of the zodiac), the glow fades with distance from the Sun.

Figure 7. Zodiacal light. Image taken in Tres Piedras, New Mexico, January 27, 2019, 6:57 p.m. Credit: Mike Lewinski

Seeing the zodiacal light requires a moonless night with a dark sky; the effect is best viewed when the ecliptic is as close as possible to being vertical to the horizon. In mid-latitudes, a good time to view the zodiacal light after sunset is in the spring; before sunrise best viewing is in autumn. The gegenschein is the exception (see below).

It is believed more than 85 percent of the dust comes from fragmentations of nearly dormant Jupiter-family comets.[6] Dormant means that the comet is not now actively outgassing, but could in the future. The Jupiter family of comets has periods of fewer than 20 years. They are so named because their current orbits are primarily determined by the gravitational influence of Jupiter.

For an equation giving the angle between the ecliptic and the horizon in terms of an observer's latitude, sidereal time, and the mean angle of inclination (23 degrees 26 minutes) of the Earth's axis to the ecliptic, see *Appendix 19.1*.

III. The gegenschein.

The gegenschein (German: counterglow) is part of the zodiacal light and occurs 180 degrees from the Sun. It is a round area of faint zodiacal light brighter than the zodiacal light on either side of it. This faint glow is back-scattered light from the Sun (as is light from our full moon).

Figure 8. The gegenschein above the VLT (Very Large Telescope) in Chile. Credit: ESO/Y. Beletsky-ESO; CC BY 4.0, image taken by Yuri Beletsky in October 2007

The gegenschein is best seen in months when the point opposite the Sun is in a constellation of the zodiac that is high in the night sky, e.g., Aries for those in the northern hemisphere and Capricorn for our friends in Australia and other locations in the southern hemisphere. (I note that Sagittarius, while farther south than Capricorn, has a bright part of the Milky Way running through it which tends to obscure the glow.) According to Don Olson, emeritus professor of physics at Texas State University, Danish astronomer Theodor Johann Christian Ambders Borsen was the first to give a clear and accurate description of the gegenschein in publications from 1854 and 1855.[8]

References

1. Young, Andrew T. (2020). https://aty.sdsu.edu/ The author is an adjunct professor of astronomy at San Diego State University. This site is an outstanding, comprehensive, and carefully researched source of information on green flashes. Professor Young also has two brief green flash discussions in *Sky & Telescope*: November 2001, page 104 and December 2001, page 96.

2. Smith, Bradford A. (May 1997). Prolonging the green flash. *Sky & Telescope,* p 111-112.

3. Flanders, Tony (December 2004). The green light observed. *Sky & Telescope,* p 97-99.

4. Schaefer, Bradley E. (February 1992). The green flash. *Sky & Telescope,* p 200-203. Schaefer is Distinguished Professor and Alumni Professor in the Department of Physics and Astronomy at Louisiana State University.

5. Schaaf, Fred (July 18, 2006). Catch the green flash! *Sky & Telescope,* https://skyandtelescope.org/observing/celestial-objects-to-watch/catch-the-green-flash/

6. Nesvorný, David, Jenniskens, Peter, Levison, Harold F., Bottke, William F., Vokrouhlický, David, & Gounelle, Matthieu (April 20, 2010). Cometary origin of the zodiacal cloud and carbonaceous micrometeorites. Implications for hot debris disks. *Astrophysical Journal, 713*(2), p 816–836. https://iopscience.iop.org/article/10.1088/0004-637X/713/2/816

7. Meeus, Jean (1999). *Astronomical Algorithms* (2nd ed.). Willmann-Bell. p 99. (Referenced in the Appendix)

8. Olson, Don (May 1997). Who discovered the gegenschein? *Sky & Telescope,* p 111-112.

Time is the wisest counselor of all.

—Pericles

TOPIC 20. What does the GPS (Global Positioning System) have in common with black holes?

I. GPS.

The GPS system consists of a network of satellites in circular orbits around the Earth. Each satellite is at an altitude of about 20,000 kilometers above the ground and orbits in about 12 hours. To provide adequate coverage, at least 24 satellites are used.

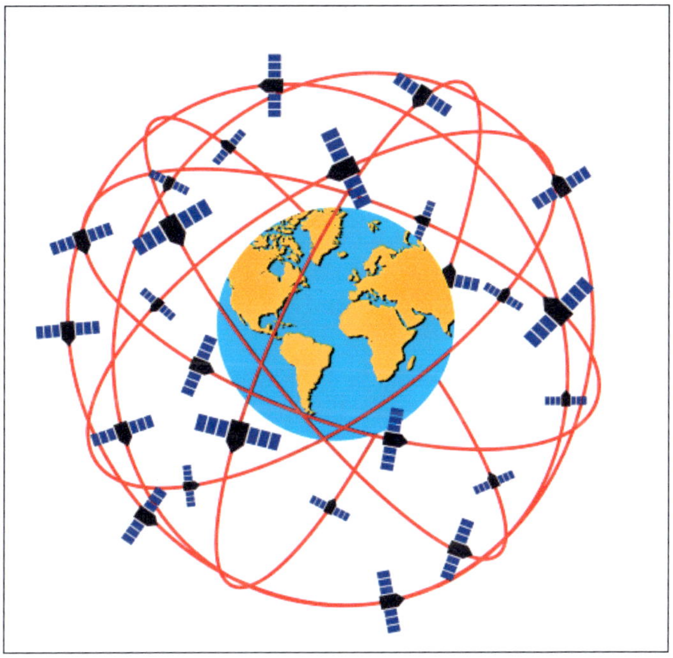

Figure 1. GPS system using 24 satellites. Credit: CC BY-NC-SA 2.0

For a person, call her Alice, to know her position on the Earth's surface, signals are received by Alice's GPS receiver from at least four satellites that are above her horizon. The network is designed so that at all times at least six satellites are above the horizon from any point on the ground. Key to the system's accuracy is knowing precisely the time of an atomic clock on board each of the satellites. No problem, one might say. Just synchronize all the satellite atomic clocks with one on the ground. Then the time on any satellite clock would always be the same as the ground-based clock. But special and general relativity cannot be ignored. If any such system were built, it would prove useless. And people would be puzzled as to the reason. But of course Einstein gave us relativity well before the GPS system was created.

Speed's effect on time.
Special relativity tells us that the orbital velocity of the satellite will affect the rate at which a clock ticks on the satellite *as seen by Alice on the ground.* To Alice, the satellite clock would be ticking slower than her clock, and the greater the satellite's orbital velocity, the slower the satellite clock would be ticking. In a more general context, special relativity tells us that if two observers, call them A and B, are traveling at a constant velocity with respect to each other, then A will see B's clock ticking slower

than A's clock, and B will see A's clock ticking slower (by the same amount) than B's clock. And both are correct! Special relativity says that time is dependent on the observer, and it does not flow continuously throughout the universe independent of observers. GPS satellites are traveling at about 14,000 kilometers per hour. This speed will cause Alice to perceive that a satellite clock is ticking slower by about seven microseconds per day.[1,2]

A reader familiar with special relativity might see a discrepancy in the above discussion. Special relativity is generally thought to apply only to two objects when the relative velocity between them is a constant. But here the relative velocity between an observer on the Earth's surface and a satellite in orbit around the Earth is not a constant, i.e., each accelerates with respect to the other. However special relativity can, in fact, handle a situation where acceleration is present. See *Appendix 20.1* for the details.

General relativity, however, is needed to fully handle the situation where gravity is present—read on.

Gravity's effect on time.
Consider a satellite orbiting Earth. Our observer, Alice, is on the Earth's surface. We temporarily ignore the effect on time due to the motion of the satellite and only consider the effect of gravity on time at the two locations. General relativity tells us that the farther the satellite orbits from the center of the Earth, and thus the weaker Earth's gravitational field on the satellite, the faster is the ticking of the satellite's clock as seen by Alice in comparison to her Earth clock. And anyone on the satellite would view time passing slower on the surface of the Earth compared to a clock in the satellite.

The point here is that since the gravitational field at the satellite is weaker than the field at the surface of the Earth, time passes slower on Earth's surface than at the satellite (as seen by *both* Alice and an imaginary observer on the satellite). This effect causes a GPS satellite clock to tick faster by about 45 microseconds per day than a clock on the surface of the Earth.[1,2] See *Appendix 20.2* for the relevant equation and a calculation.

Combining the two relativistic effects.
Combining the two effects we see that the satellite clock gains about 45 - 7 = 38 microseconds per day compared to a clock on the ground. To have a functional GPS system, the times of the clocks must be known to an accuracy of a few nanoseconds. Thirty-eight microseconds are 38,000 nanoseconds!

Without compensating for the two relativistic effects on time, the GPS would become inaccurate in only two minutes. And errors in the navigational system would accumulate at a rate of about 10 kilometers each day.[1,2] That would mean goodbye GPS. Here we see relativity manifest in everyday life.

II. Black Holes.

A black hole is the remnant remaining after a very massive star gravitationally collapses. Its mass is thought to reside at a point called the *singularity*, which is at the center of the black hole. The black hole's *event horizon* is the spherical shell surrounding the singularity inside of which nothing can escape, including light itself. Scientists don't really know yet what the inside of a black hole is like. And while the singularity is frequently discussed as being a point, scientists believe once a single theory combines general relativity and quantum mechanics yielding a quantum theory of gravity, that the singularity will not be a point. To define it as a point at present merely says that relativity theory breaks down in this situation. The radius of the black hole, called the *Schwarzschild radius*, is defined as the distance from the singularity to the event horizon.

What would happen to an astronaut who is moving into the black hole at the event horizon? *At the event horizon* the surface gravity of a black hole is inversely proportional to the mass of the black hole. (See reference [4] and also *Appendix 20.3*.) So the *less* massive the black hole, the stronger the gravitational force on the object *at the event horizon*. This might seem counter-intuitive, but it is because the event horizon is closer to the singularity for a less massive black hole (the radius is directly proportional to the black hole's mass).

Thus if the black hole is super massive, tidal forces on an astronaut's body at the event horizon might not stretch her into a long, thin noodle, a process called spaghettification (no joke, that is the term used). But inside the black hole, as she approached the singularity, spaghettification would occur. For a less massive black hole, spaghettification would occur at or before arriving at the event horizon. At any rate, it would be a bad day for the astronaut.

The issue we really wish to discuss here, however, is the black hole's gravitational field and its effect on time. This phenomenon was discussed above, but in the case of the GPS. On her ship, the astronaut would not notice any change in the passage of time. But because she is in a very strong gravitational field, someone observing her clock from the safety of Earth would perceive it to be ticking very slowly. So slowly, in fact, it would almost seem as if time on the ship had come to a halt—had frozen. To the astronaut, it might only be one hour before she passed through the event horizon. In contrast, the great grandchild of a person on Earth might not notice any change in the astronaut's distance from the event horizon from that noticed many years earlier by the great grandparent. At this point it should be clear that the answer to the question raised in Topic 20's title "What does the GPS have in common with black holes?" is "the effect that gravity has on time."

In reality, another issue would also affect what a human actually could see from the Earth. A photon moving upward in a gravitational field will have its frequency decreased, which means its wavelength is increased. Hence visible light coming from the astronaut's ship would be shifted toward—and beyond—the red end of the spectrum in an effect called *gravitational redshift*. Why this occurs is closely related to why gravity affects time. Intuitively, one could say the photon losses energy as it moves upward in a gravitational field. This is because quantum mechanics states that the energy of a photon equals a constant (the Planck constant) times the photon's frequency. As an equation this is $E = h\nu$, where h is Planck's constant and ν is the frequency of the photon. Hence if the energy decreases, then frequency decreases, and vice-versa. So, because light coming from the ship is redshifted, the light arriving at Earth would not be in the range of the visible spectrum. The ship would not be seen at all by an Earthbound human.

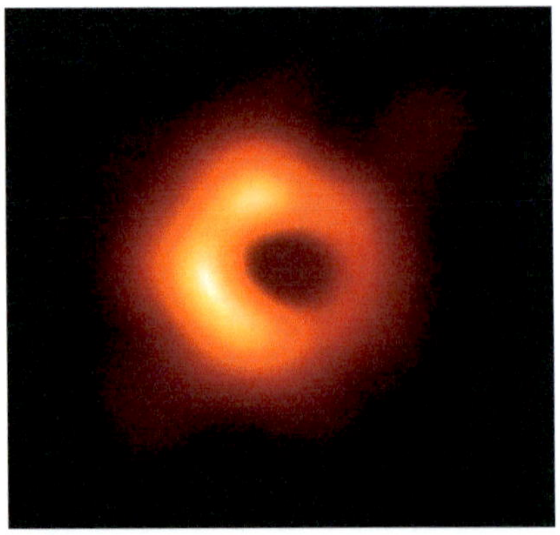

Figure 2. In 2019, using the Event Horizon Telescope, astronomers obtained this image of the black hole at the center of galaxy M87. The hole is outlined by emission from hot gas swirling around it under the influence of strong gravity near its event horizon. Credit: Event Horizon Telescope collaboration et al.

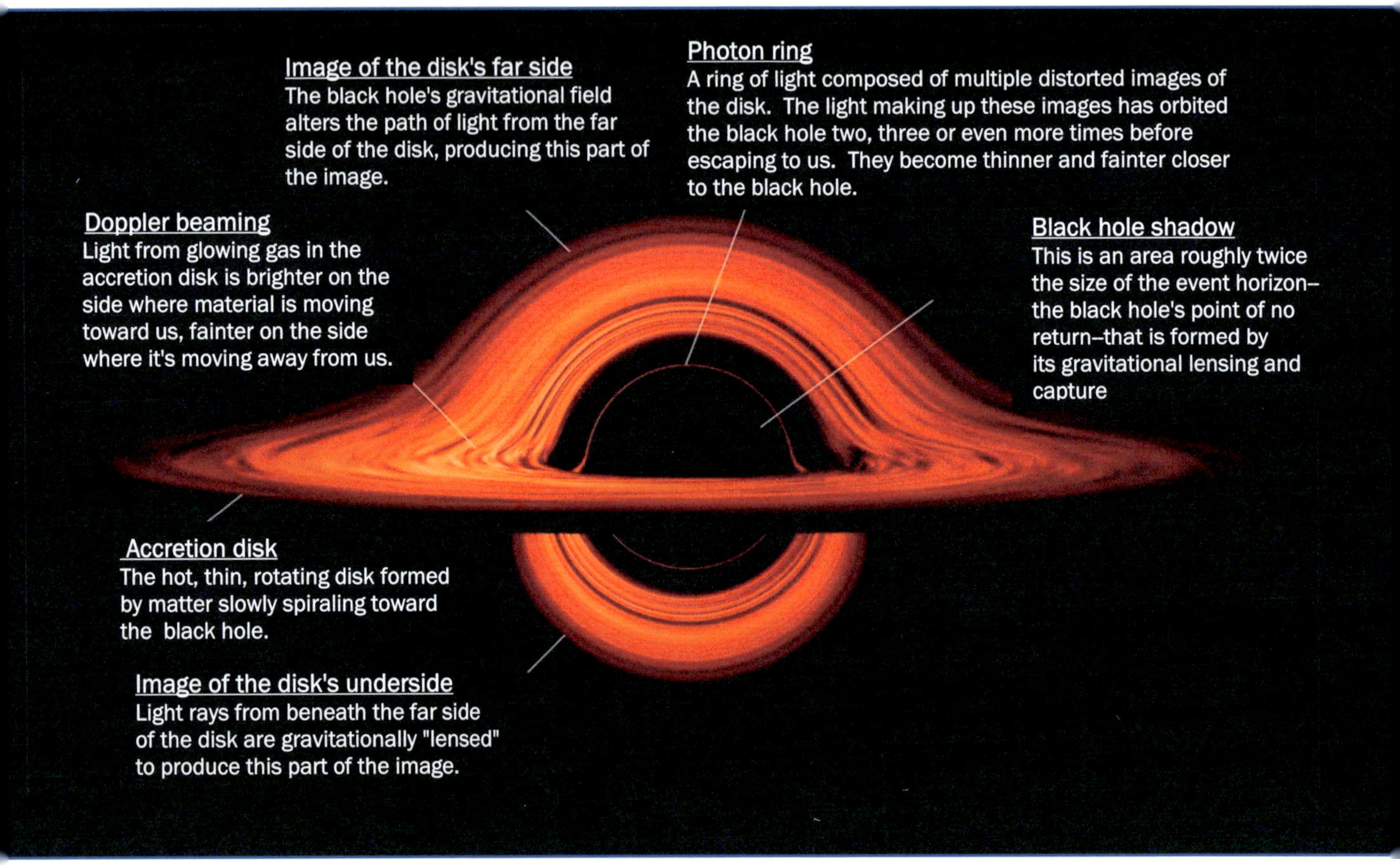

Figure 3. This graphic, resembling a WWI army helmet with a strap, is an overview of the main observable features near a black hole. Credit: NASA Goddard Space Flight Center/Jeremy Schnittman (Text was altered to be larger and bolder by the author)

A black hole has an accretion disk of hot gas swirling around its center just outside the event horizon. This gas emits radiation ranging from radio to x-rays. The detection of x-rays helps astronomers determine if an unseen object might be a black hole. All this light is coming from the accretion disk. The accretion disk in Figure 3 doesn't resemble a disk because the black hole's intense gravity bends (or gravitationally lenses) the light it emits. The top of the "helmet" shape is actually the disk's far side where the light that we see is bent upward. The "strap" originates from lensed light emitted from underneath the disk's far side. So we observe a warped disc: it's not seen as flat. Strictly speaking, gravity doesn't bend light rays. What actually occurs is the black hole's mass warps the space-time continuum around it, and light rays of all wavelengths are forced to travel the curved paths of this curved space-time region. Here, as in Einstein's General Theory of Relativity, "space-time" instead of "space" is used, placing us in a four-dimensional universe—three dimensions for space and one for time.

In Figure 3, looking down from above the "helmet," the gas in the accretion disk is moving counter-clock-wise. Thus gas on the left side of the accretion disk is moving toward us and gas on the right side is moving away from us. This movement causes the frequency of light emitted by gas coming in our direction to be Doppler shifted. Hence the light we see from the disk's left side is brighter than on the right. In Figure 2, the actual image (i.e., not a depiction), we see the same thing: the left, bright region is from light emitted by gas moving toward us, while the right, dimmer region is light from gas moving away from us.

The dark, circular region in Figure 2 is about 2.5 times the diameter of the actual black hole, which lies in the center. The bright light of the "helmet" silhouettes the black hole. Indeed, many sources refer to the circular black area as the black hole's "shadow." The light imaged in Figure 2 is at radio frequencies and is emitted by electrons corkscrewing along intense magnetic field lines. Such light is called synchrotronic radiation.

But there is more. See Figure 4. It is important to realize not all that exists in the vicinity of a black hole is visible to us. And, as we have seen, what is visible can appear warped. Consider both the *photon ring* and the *photon sphere*. The *photon sphere* is a vanishingly thin sphere just barely outside the event horizon, surrounding it, and closer to the event horizon than is the inner edge of the accretion disk. Only photons, not matter (specifically fermions), are on this sphere. They originate mainly from the accretion disk and move on this sphere in great-circle orbits. Such photons may curve down and pass through the event horizon, never to be seen again. Or they may escape, moving tangentially off and away from the sphere. The *photon ring* forms the boundary of the photon sphere as seen from Earth. So it is perpendicular to our line of sight. The only photons moving on the sphere that we can see from Earth are from points on this ring, as only from these points on the sphere can photons be directed (tangentially off the sphere) at us. And note that these photons directed at us from this ring are not orbiting in this *ring*, but pass perpendicularly through it and head toward Earth. In the Figure 3 depiction the photon ring is the thin, inner circle; in the Figure 4 depiction it is a circle of light on the photon sphere. It is predicted that all black holes should have a photon ring.[11]

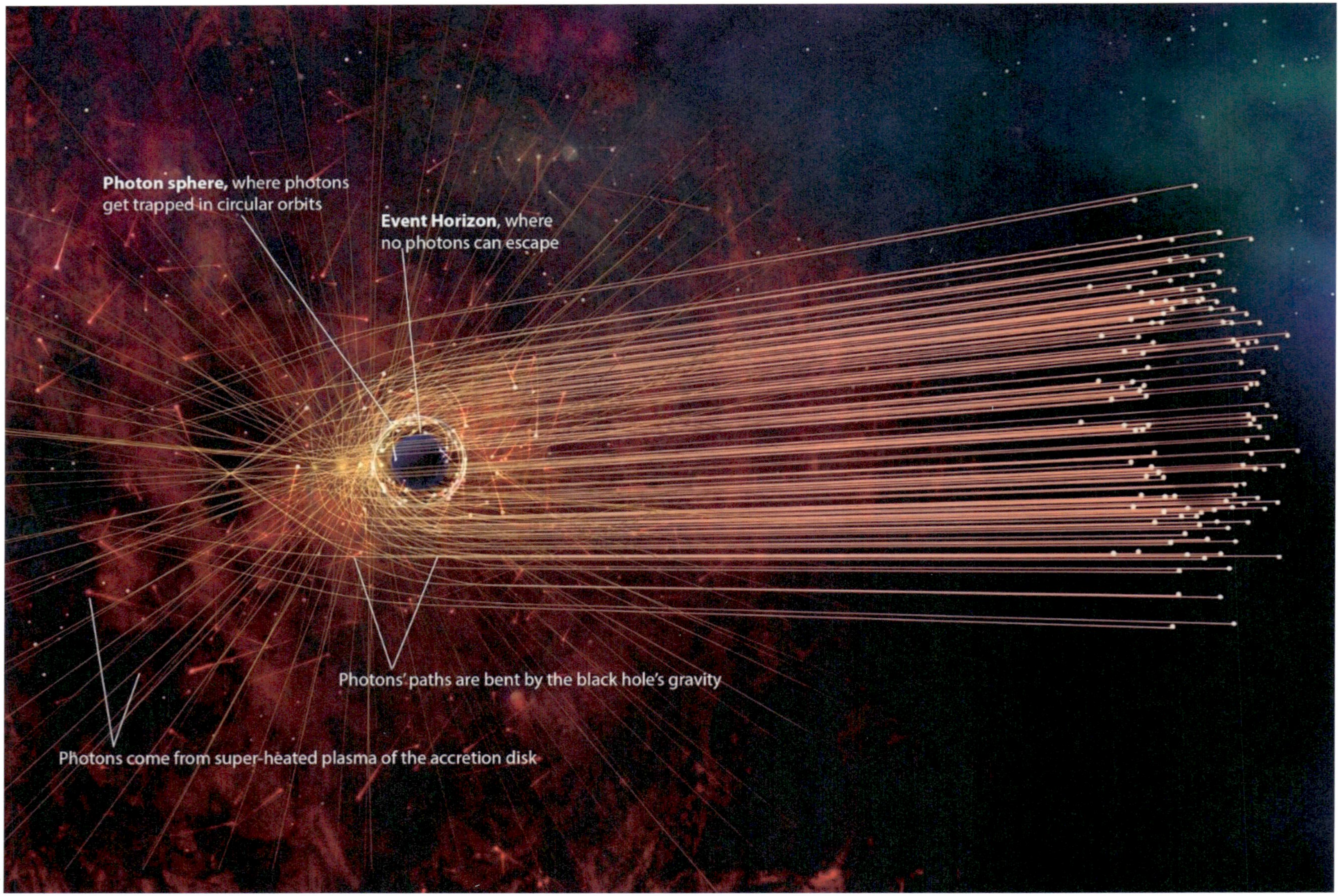

Figure 4. Depiction showing the photon ring. Credit: Nicolle R. Fuller/NSF

Let us now discuss the actual black hole imaged in Figure 2. It is not, as you might assume, the massive black hole at the center of our own Milky Way galaxy, called Sagittarius A* (abbreviated Sgr A* and pronounced Sagittarius A-star). Rather it is the super massive black hole at the center of the elliptical galaxy Messier 87 (M87), about 55 million light years from Earth and almost 2,000 times farther away than is Sgr A*. M87's black hole is about 1,000 times as massive as Sgr A*. Its radius is thought to be about 62 AU, one and one-half times the average distance of Pluto from the Sun.[6]

Figure 5. The galactic core of Messier 87 as seen by the Hubble Space Telescope with the blue plasma jet clearly visible. Credit: Nasa/PD Hubble

Why did astronomers choose to image a black hole so distant? A critical reason for favoring M87's black hole over our very own is that we can image it over four days of observations.[3] Gas in Sgr A*'s accretion disk moves around its circumference a thousand times faster than does gas around the circumference of M87's accretion disk.[3] So, compared with Sgr A*, changes in brightness in M87's black hole are less often and less dramatic. Sgr A* can alter its brightness in a matter of minutes,[7,8] while similar changes in M87's black hole would take about a week.[8] You might compare imaging the black hole in M87 to that in Sgr A* with painting a picture of an experienced model who will sit still instead of trying to paint a fidgeting child. (On May 12, 2022, the first image of Sgr A* was released. The image was based on radio interferometer data taken in 2017.)

The image in Figure 2 was created using the Event Horizon Telescope (EHT), a network of eight radio telescopes around the world. Signals received from each scope were merged to form one image with a quality approximately equivalent to an image from a radio telescope the size of the Earth. Several spacecraft and satellites were also used to observe M87 at various wavelengths. If the EHT observed changes in the structure of the black hole's environment, data from these supplemental sources could help determine what was happening. One doesn't just click a button and instantly produce an image. As noted above, it took about four days' observations to form an image, so any changes around the environment of M87's black hole during this time would cause concern.

Finally let us say something about the obvious blue jet steaking out of M87 seen in Figure 5. The jet, powered by matter streaming onto the black hole's spinning accretion disk, is several thousand light years long. It is thought the black hole's powerful magnetic field lines become twisted as they extend outward from its poles. Electrons and subatomic particles corkscrew up these field lines and form the jet of material traveling at nearly light speed. We will discuss two types of electromagnetic radiation emitted in this jet: *synchrotron radiation* and *Cherenkov radiation*.

Synchrotron radiation is electromagnetic radiation given off by any charged particle (here an electron) moving in a curved path or accelerating in a straight-line path. It is thought that most of the radio emissions from active galactic nuclei (all of which have a black hole at their center) are synchrotron radiation.

Cherenkov radiation in the current discussion is produced when electrons actually travel faster than the speed of light through the material *in the jet*. Cherenkov radiation does *not* violate the law that no object with mass can travel faster than the speed of light *in a vacuum*. More generally, Cherenkov radiation is emitted whenever a charged particle passes through a dielectric medium at a speed greater than the phase velocity of light *in that medium*. Cherenkov radiation consists of a continuous spectrum of wavelengths extending from the ultraviolet region into the visible part of the spectrum peaking at about 420 nanometers. Only a negligible amount of photon emissions is found in the infrared or microwave regions.[10] A more familiar example of such radiation is when electrons from an underwater nuclear reactor travel through the water faster than the speed of light *in water*, giving off an eerie, blue light. Cherenkov radiation is analogous to the shock wave that results from a supersonic object traveling through air. The electromagnetic "disturbance" caused by the electrically charged particles' interaction with atoms in the medium generates a "shock cone" of light.[9] This radiation is named after the Soviet physicist Pavel Cherenkov who shared the 1958 Nobel Prize in physics for its discovery in 1934. For more on M87's jet, see Topic 22, The illusion of superluminal speed.

References

1. Pogge, Richard W. (January 12, 2016). Real-world relativity: The GPS navigation system. Astronomy Dept., Ohio State University. http://www.astronomy.ohio-state.edu/~pogge/Ast162/Unit5/gps.html
2. Will, Clifford M. Einstein's relativity and everyday life. Physics Central. American Physical Society. https://physicscentral.com/explore/writers/will.cfm
3. Carlisle, Camille M. (September 2019). The face of a black hole. *Sky & Telescope*, p 18-23.
4. Carroll, Sean M. (September 2019). *Spacetime and geometry. An introduction to general relativity.* Cambridge University Press, p 248.
5. Schutz, Bernard (2009). *A first course in general relativity.* (2nd ed.). Cambridge University Press. p 282. (Referenced in the Appendix.)
6. Grossman, Lisa, & Conover, Emily (April 27, 2019). First picture of a black hole wows. Image of M87's behemoth matches general relativity. *Science News*, p 6.
7. Grossman, Lisa (December 21, 2019 & January 4, 2020). 2019 year in review. First black hole image made its debut. *Science News, 196*(11), p 22-23.
8. Klesman, Alison (January 2020). Top 10 space stories of 2019. Astronomers shine light on a black hole. *Astronomy Magazine,* p 30-31.
9. Chen, M.C. (2014). Cherenkov light. volume 15, section 15.24.3.2. In Heinrich Holland & Karl Turekian (Eds.), *Treatise on Geochemistry* (2nd ed.) Elsevier Science. https://www.sciencedirect.com/topics/chemistry/cherenkov-radiation
10. L'Annunziata, Michael F. (editor), *Handbook of Radioactivity Analysis* (second ed.), 2004. Academic Press. ISBN 978-0-12-436603-9. The reference is from the introduction to Chapter 9, Cherenkov Counting, L'Annunziata. https://www.sciencedirect.com/book/9780124366039/handbook-of-radioactivity-analysis
11. Yiu, Yuen (June 24, 2020). *All black holes should sport light rings.* Astronomy.com. https://astronomy.com/news/2020/06/all-black-holes-should-sport-light-rings. An interesting article.

Do not Bodies act upon Light at a distance,
and by their action bend its Rays;
and is not this action strongest at the least distance?

—*Isaac Newton, Opticks 1704 p 132*

TOPIC 21. Gravitational lensing with five examples. Three definitions of cosmological distance needed to understand our relationship to very distant objects.

Consider an object, e.g., a quasar, that is very far from Earth.

Three relevant distances.

1. The quasar's distance from Earth at the time the light left the quasar.
2. The distance the light travels before reaching Earth.
3. The quasar's distance from Earth at the time the light reaches Earth.

Because the universe is expanding, the first distance is less than the second which is less than the third. It is common to read that an astronomical object is "D light years distant," for some number D. Readers should translate this to mean "D is the distance light travels before reaching Earth." In other words, it is distance in the sense of the second type of distance listed above. If the light source isn't just a point, unlike a quasar, the first distance determines the angular size of the light-emitting object as seen from Earth. Even though the space between the object and the Earth has been expanding as the light travels to Earth, it is the first and not the second distance that determines angular size.

See *Appendix 21.1* for formulae for the three types of distance. And those interested in cosmology should see reference [3] for an outstanding Great Courses lecture series on cosmology. In the appendix of the transcript books (which one can buy along with the course) Professor Mark Whittle of the University of Virginia gives mathematical derivations of many of the equations in his lectures. By carefully following his proofs one can see where these "magical" cosmology equations come from and gain a deeper understanding of cosmology. Without following the mathematics, the equations are like pulling rabbits out of a hat. Follow the mathematics and the magician's secrets are revealed.

Gravitational lensing.

Gravitational lensing was mentioned in Topic 20 in discussing the first image taken of a black hole. There the mass of a black hole bends space-time near itself so much that light emitted by gas in its accretion disk is bent, creating a warped, fun-house image. Here we discuss lensing for a different situation.

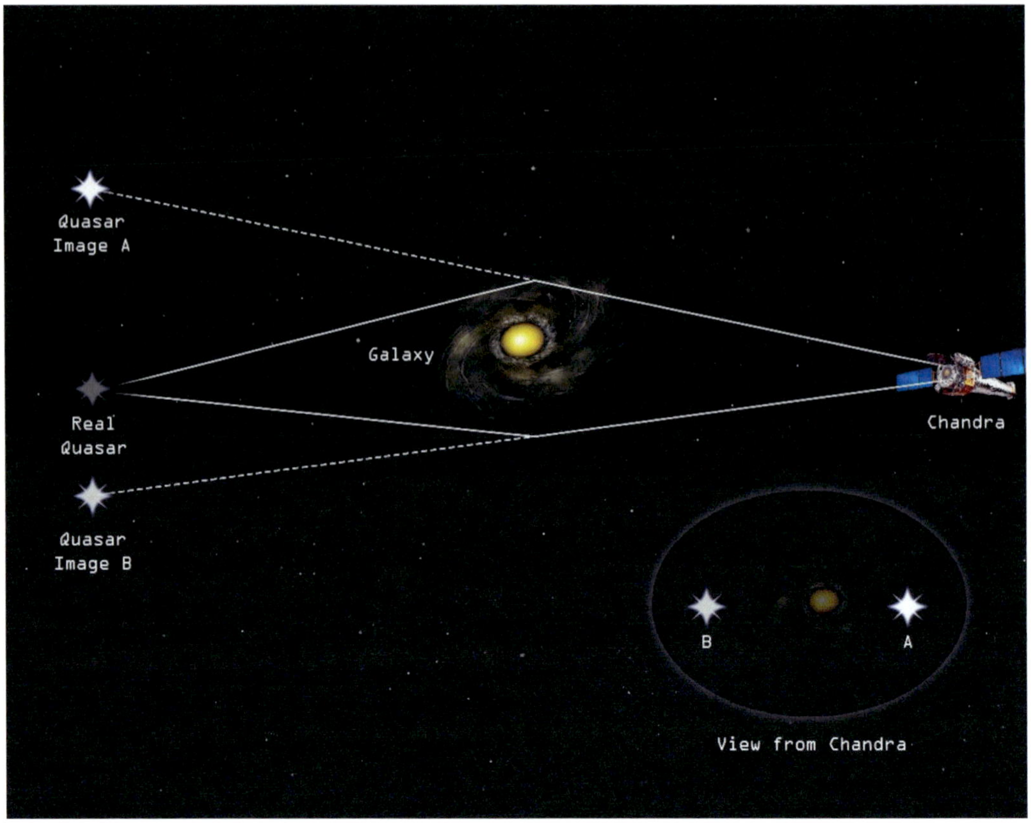

Figure 1. Gravitational lensing. Credit: Chandra X-ray Observatory

Suppose light from a very bright source, for example a quasar billions of light years away, is headed straight at us, and between the quasar and Earth is a massive galaxy (or galaxy cluster). Light rays from the quasar will pass around the galaxy and be bent because of warped space-time near the galaxy. This curved space-time is of course caused by the mass of the galaxy, and light rays must follow its curvature. Think of the galaxy as like a very large Coke bottle in the path of the light. Parallel rays coming toward one spot on the edge of the intervening galaxy are all bent the same amount and stream toward Earth remaining parallel to each other. Parallel rays coming toward a different spot on the edge of the galaxy are likewise bent and stream toward Earth. An earthbound telescope can take the parallel rays from each of these spots and create two images of the quasar. In the same way more than two images can be created.

Also, instead of different star-like images, circular streaks, called *Einstein rings*, can be created. Put a black dot on a piece of white paper. Take a wine glass with a wide base and look through the glass with the base facing down and well above the dot. You should see black arcs. This illustrates how Einstein rings are created. See Example 3 below.

Microlensing is the term used when a small object, e.g., a planet, lenses an object, e.g. a star, that is behind it. This is one method for trying to discover new planets.

Five examples of gravitational lensing.

1. Einstein's Cross.

Discovered in 1985, the quasar Q2237+0305 (or QSO J2240+0321) is in the constellation Pegasus. The central object in Figure 2 is the core of a massive 15.2 magnitude galaxy (Huchra's Lens = CGCG 378-15 = PGC 69457) whose distance from Earth is about 400 million light years.[1] The galaxy's gravity is bending the light of the more distant quasar, forming four surrounding images. The quasar's redshift is $z = 1.69$[4] and thus its light has been traveling toward Earth for about 9.8 billion years (which means the quasar was much closer at the time its light was emitted). While the light is traveling to us, the distance between the quasar and Earth is increasing because of the expansion of space. And since the light left the quasar, the expansion of space has moved the quasar even farther away from Earth. The symbol z is a measure of the Doppler shift in the wavelength of light. $z > 0$ for light emitted by an object receding from us; $z < 0$ for light from an object that is approaching us. Due to the expansion of the universe, very distant objects are all receding from us and have their light shifted toward the red end of the spectrum. z can be used to give a measure of the time that light from a distant object has been traveling toward us—the light travel time.

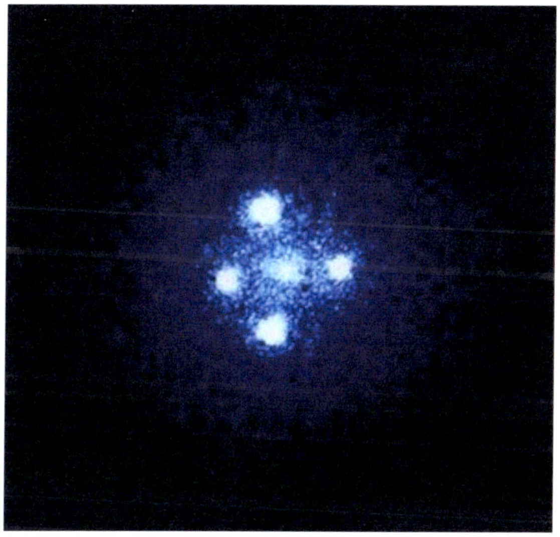

Figure 2. Einstein's Cross (QSO 2237+0305). Credit: NASA, ESA, and STScI

Each of the quasar's four images is visually only about one arc second from the center of the galaxy's nucleus, with the galaxy itself measuring 52 arc seconds x 20 arc seconds. Only the core of this large galaxy (about two arc seconds in diameter in Figure 2) can be seen through an amateur telescope. The magnitudes of the four quasar images are 17.4, 17.4, 18.4, and 18.7.[1] Viewing the four quasar images, which are faint and close to the galaxy, requires a large telescope (one with a mirror at least 25 inches in diameter) and a dark sky with exceptional seeing conditions.

2. The Twin Quasar (Q0957+561A and Q0957+561B) in the Constellation Ursa Major.

About 14 arc minutes north of the large, bright galaxy NGC 3079 is the first-discovered gravitationally lensed quasar, found in 1979. Known as the Twin Quasar, it appears in large amateur telescopes as a double star. The magnitudes of the two images are 16.5 and 16.7, and they appear six arc seconds apart. The quasar's redshift is $z = 1.41$.[5] Light reaching Earth left the quasar 9.1 billion years ago,[5] and it is currently just under 14 billion light years[6] distant. As shown in Figure 3, the lensing galaxy is visually almost directly in front of Q0957+561B, the quasar's lower image.

Figure 3. The Twin Quasar. The quasar images are the two bright, vertically aligned spots, the upper spot being Q0957+561A *and the lower being* Q0957+561B. *Visually touching the edge of the lower spot at about 11:30 o'clock is the lensing galaxy. Credit: ESA/Hubble & NASA, Robert Gendler (http://www.robgendlerastropics.com)*

3. An example of Einstein rings.

Here we have the luminous, red galaxy (LRG), named LRG 3-757, lensing the light from a much more distant blue galaxy. The lens alignment is so precise that the background galaxy is distorted into overlapping rings instead of distinct stellar-like images. It was discovered in 2007 in data from the Sloan Digital Sky Survey; the Figure 4 image was subsequently taken by Hubble. The objects are in the constellation Leo, with the foreground galaxy having a red shift of $z = .4457$ and the lensed galaxy having a red shift of $z = 2.379$.[7] These redshifts translate to the foreground galaxy having a light travel time of 5.2 billion years and the lensed galaxy 10.3 billion years. Such lensing can allow astronomers to determine the mass and dark matter content of the foreground galaxy.

Figure 4. An Einstein Ring. A horseshoe-shaped image behind the red galaxy LRG 3-757.

4. Andromeda's Parachute.

The official designation of this quasar in the constellation Andromeda is J014709+463037. The word "parachute" is certainly evocative. The object was first observed in 2017 (quite recently) by C. T. Berghea, et al. and confirmed using the 10-meter Keck II telescope at Mauna Kea Observatory in Hawaii.[2] The four images of the quasar range in magnitude between 15.4 and 17.7, and the separation of the four varies from 1.3 arc seconds to 3.4 arc seconds, making the quasar visible in large amateur telescopes under good seeing conditions and a very dark sky. The red shift is z = 2.377 ± 0.007,[2] so the light travel time is about 10.3 billion years.

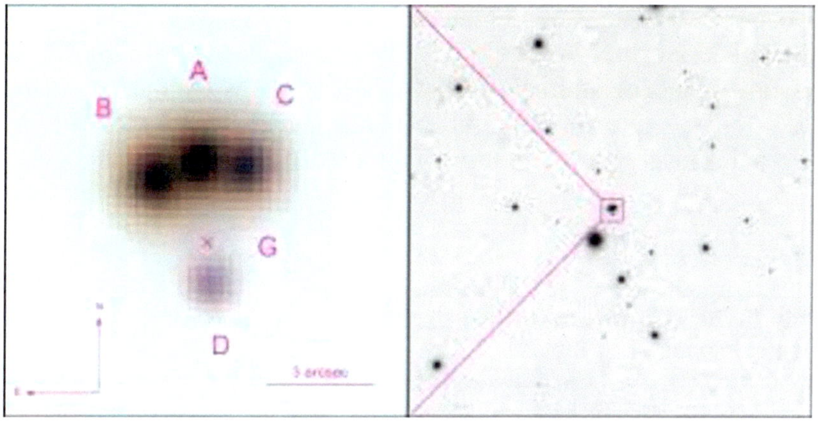

Figure 5. Andromeda's Parachute. Here a quasar is seen as four images labeled A, B, C, D and the lensing galaxy is marked with a (faint) x and labeled G. The left square is a blowup of the small box inside the right square. Alvin Huey's colored image is based on a black-and-white image from reference [8].
Credit: www.FaintFuzzies.com, Alvin Huey

5. A galaxy cluster lensing several galaxies far behind it allows the location of dark matter within the cluster to be determined.

The galaxy cluster Abell 1689 in the constellation Virgo is one of the largest and most massive of such clusters and contains more than 1,000 galaxies and more than 160,000 globular clusters. It is about 2.2 billion light years away. It lenses many galaxies, and this lensing can be used to determine the location and amount of dark matter in the galaxy cluster itself.

Figure 6. Dark matter revealed in the center of the galaxy cluster Abell 1689. Credit: image and description below: NASA,ESA, D. Coe (NASA Jet Propulsion Laboratory/California Institute of Technology and Space Telescope Science Institute), N. Benítez (Institute of Astrophysics of Andalucía, Spain), T. Broadhurst (University of the Basque Country, Spain), and H. Ford (Johns Hopkins University) https://spacetelescope.org/images/opo1037a/

This NASA/ESA Hubble Space Telescope image from June 2002 shows the distribution of dark matter in the center of the giant galaxy cluster Abell 1689, containing some 1,000 galaxies and trillions of stars.

Dark matter is an invisible form of matter that accounts for most of the universe's mass. Hubble cannot see dark matter directly; astronomers inferred its location by analyzing the effect of gravitational lensing, where light from galaxies behind Abell 1689 is distorted by intervening matter within the cluster.

Researchers used the observed positions of 135 lensed images of 42 background galaxies to calculate the location and amount of dark matter in the cluster. They superimposed a map of these inferred dark matter concentrations, tinted blue, on an image of the cluster taken by Hubble's Advanced Camera for Surveys. If the cluster's gravity came only from visible galaxies, the lensing distortions would be much weaker. The map reveals the densest concentration of dark matter is in the cluster's core.

References

1. Crinklaw, Greg (n.d.) *Focus on Einstein's Cross.* observing.skyhound.com https://observing.skyhound.com/archives/sep/Q2237+0305A.html

2. Rubin, Kate, et al. (July 18, 2017). *Andromeda's Parachute: A bright, quadruply lensed quasar at z = 2.377.* Cornell University. https://arxiv.org/abs/1707.05873

3. Whittle, Mark (2008). *Cosmology: The history and nature of our universe.* thegreatcourses.com https://www.thegreatcourses.com/courses/cosmology-the-history-and-nature-of-our-universe. See p 109, 110, and 187 in the transcript book Part 1 of 3 for this course. The entire lecture series is highly recommended for anyone interested in cosmology. Those who are mathematically inclined should buy the transcript book in addition to the course, and be sure to look in the math-physics appendix. (This reference is referred to above and in *Appendix 21.1.*)

4. The Set of Identifications, Measurements, and Bibliography for Astronomical Data (SIMBAD). http://simbad.u-strasbg.fr/simbad/sim-id?Ident=Einstein+Cross

5. Gendler, Robert (2006). *NGC 3079 and Lensed Quasar Q0957+561.* http://www.robgendlerastropics.com/NGC3079text.html

6. *Seeing double.* (January 20, 2014). https://esahubble.org/images/potw1403a/

7. Belokurov, V., et al. (December 10, 2007). The cosmic horseshoe: Discovery of an Einstein ring around a giant luminous red galaxy. *Astrophysical Journal, 671,* L9-L12, https://iopscience.iop.org/article/10.1086/524948/fulltext/

8. Berghea, C. T., Nelson, George J., Rusu, C. E., Keeton, C. R., & Dudik, R. P. (July 11, 2017). Discovery of the first quadruple gravitationally lensed quasar candidate with Pan-STARRS. https://arxiv.org/pdf/1705.08359.pdf

9. Bergstroem, Lars, & Goobar, Ariel (2004). *Cosmology and Particle Astrophysics,* (2nd ed.). Springer, p 76. https://www.springer.com/gp/book/9783540329244 (Referred to in the appendix of this topic.)

Take nothing on its looks:
take everything on evidence.
There's no better rule.

—*Charles Dickens*

TOPIC 22. The illusion of superluminal speed.

Some objects in the universe appear to be traveling faster than the speed of light. One such example of superluminal speed is the transverse speed of a blob of particles in the jet streaming away from the black hole in galaxy M87. By "transverse" is meant the direction perpendicular to our line of sight to the blob. But this is an illusion. It had better be, as Einstein's special theory of relativity says this is impossible. The blob emits visible light as well as radio waves. The jet is discussed in Topic 20. See also Figure 5 in that topic.

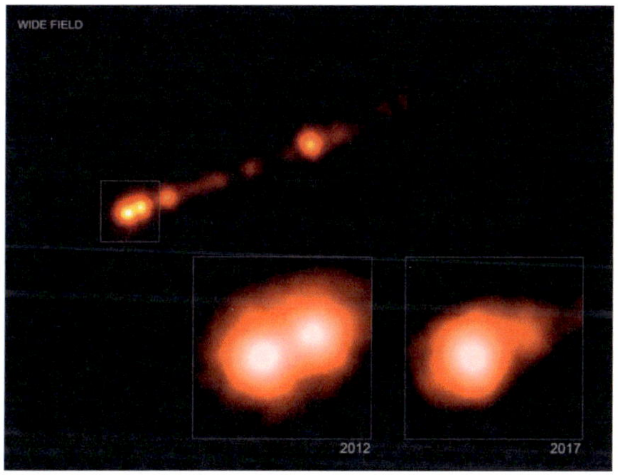

Figure 1. A blob in M87's jet. See the lower left end of the jet in the upper image. The bottom image gives two magnified views of this end of the jet. In the bottom on the left is the blob in 2012, and on the right is the blob in 2017. Credit: NASA/CXC/SAO/B. Snios et al.

The following example enables us to see why this is an illusion.[1]

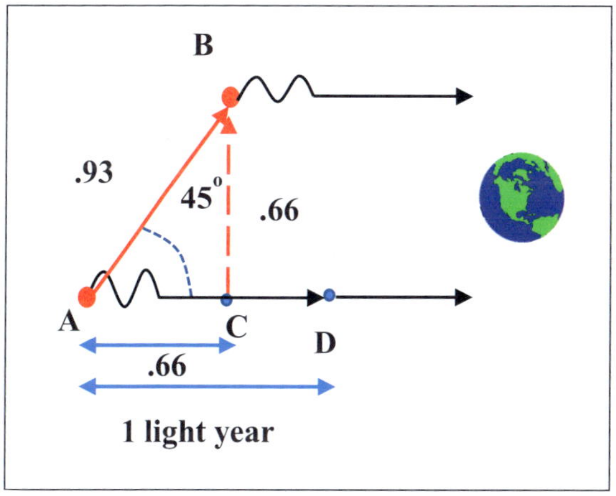

Figure 2. Superluminal speed for a 45 degree angle.

All light rays emitted by the blob (at A and B) in Earth's direction are parallel. Distances are not to scale. The distance from the blob to Earth is millions of light years. To a person on the Earth, points A and B are one and the same to the naked-eye; it would take a telescope to separate them.

Consider a blob traveling from point A to point B at 93 percent the speed of light and at an angle of 45 degrees to the direction to Earth. One year later a photon emitted at A (call it photon A) will be at point D which is one light year from A in the direction of Earth. In that year the blob will have traveled from A to B (at a 45-degree angle) a distance of .93 light years. That means that the blob has moved .66 light years to the right and .66 light years toward Earth. The distances are related by the Pythagorean theorem: $.93^2 = .66^2 + .66^2$ (amounts are rounded off).

Now consider a photon emitted by the blob at B (call it photon B) traveling toward the Earth. When photon A is at D, photon B is at B, a distance of 1 - .66 = .34 light years behind A. At Earth we will first see photon A, and then, .34 years later, photon B will arrive. As dictated by Einstein's special theory of relativity each photon is traveling at the same speed, denoted by the letter c. In particular the speed of photon B is not (contrary to Newtonian physics) the radial component of the blob's speed plus c. So to someone on Earth, the blob has moved horizontally to the right a distance of .66 light years in .34 years. But .66/.34 is almost 2. So from Earth the blob would be seen to be moving to the right at almost twice the speed of light. Mystery solved.

A general equation gives the percent γ of the speed of light that the blob would appear to have when the blob's velocity is β, expressed as a percent of the speed of light, and where its direction of travel makes an angle θ with a direct line from the blob to Earth:

$$\gamma = \beta\sin(\theta) / [1 - \beta\cos(\theta)] .$$

Note that with $\beta = .93$ and $\theta = 45$ degrees this equation does yield the result of $\gamma = 1.92 \sim 2$ obtained above. For the mathematically inclined, a derivation of this equation is found in *Appendix 22.1.*

What is the angle of M87's jet with respect to a line from it to Earth? The authors in reference [3] below state, "There is no one, simple way to determine that angle, so the reasoning leading to the choice of the angle used in this paper is explained here at some length." They adopt the value of $\theta = 17$ degrees and say that "all the recent model-dependent viewing angles lie within three degrees of this choice."

Using NASA's Chandra X-ray Observatory, astronomers have seen that the famous giant black hole in Messier 87 is propelling particles at speeds greater than 99 percent of the speed of light.[4] If we use $\beta = .99$ and $\theta = 17$ degrees, then using the formula we obtain $\gamma = 5.43$. Amazing. Some blobs in the jet appear to be moving at between five and six times the speed of light. But remember appearances can be—and in this case are—deceiving. For an interesting article on M87's jet including its "superluminal" motion see Banich's article in reference [2].

References

1. Naeye, Robert (March 1998). *Astronomy Magazine,* p 94.

2. Banich, Howard (April 2020). A curious straight ray. *Sky & Telescope,* p 30-35.

3. Walker, R. Craig, Hardee, Philip E., Davies, Frederick B., Ly, Chun, & Junor, William (March 15, 2018). The structure and dynamics of the subparsec jet in M87 based on 50 VLBA observations over 17 years at 43 GHz. *Astrophysical Journal, 855*(2), p 128. https://iopscience.iop.org/article/10.3847/1538-4357/aaafcc

4. NASA. (January 6, 2020). *Chandra x-ray: Famous black hole has jet pushing cosmic speed limit.* https://www.nasa.gov/mission_pages/chandra/images/famous-black-hole-has-jet-pushing-cosmic-speed-limit.html

For I dipped into the future,
far as human eye could see,
Saw the Vision of the world,
and all the wonder that would be.

—Alfred Lord Tennyson, Locksley Hall

TOPIC 23. For most people the Andromeda galaxy, M31, is the farthest object that can be seen from Earth with the naked eye. When was it first documented as a fuzzy, non-stellar spot in the sky? M33, the second-closest major galaxy to Earth, is much closer to M31 than we are. How would M31 be seen by inhabitants of a planet around a star in M33?

Some people claim to be able to see with the naked eye a few other deep-sky objects that are farther away (for example M33), but for most people the answer is M31. Anyone with normal good vision can see it.

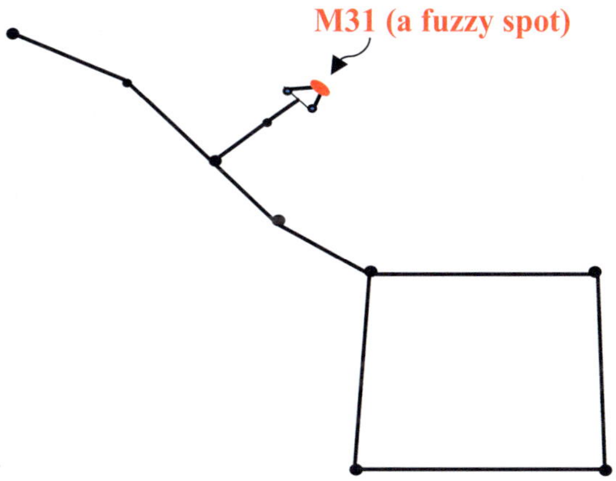

Figure 1. Finding the location of M31. First locate the Great Square of Pegasus. Look NE up the chain of stars of Andromeda. Look for the small triangle consisting of two stars at the base and a fuzzy "third star" at the top. M31 is the fuzzy third star.

M31 was observed, very probably before 905 CE, by Persian astronomers at Isfahan and was documented in 964 CE by the Persian astronomer Al Sufi (903-986 CE) in his *Book of Fixed Stars*. He referred to it as "A Little Cloud," mentioning it as lying before the mouth of the Big Fish, an Arabic constellation positioned across the body of Ptolemy's Andromeda.[1,2]

Figure 2. M31 and the two stars forming a triangle with it. Credit: A (rotated) small section of an image by Ted Van

Figure 3. M31 (north is up). Credit: Geoffrey Collins, MD, September 2013

M31 is 2.5 million light years from Earth and M33 is 2.6 million light years from Earth, but M31 and M33 are only 700,000 light years from each other. So M31 would be larger and brighter to the naked eye from a planet around a star in M33 than from Earth. It would be about 3.6 times larger (in length) and about 12.76 times brighter. This means that the apparent magnitude of M31 as seen from M33 would be 0.68, which is first magnitude. Of course the light is not a point source but is spread over a small area in the sky. This compares with a magnitude 3.44 for M31 as seen from Earth. (See *Appendix 23.1* for a mathematical derivation of these results.)

References

1. Kunitzch, P. (September, 1987). A medieval reference to the Andromeda Nebula. *The Messenger, 49*, p 42-43. http://adsabs.harvard.edu/full/1987Msngr..49...42K

2. Students for the Exploration and Development of Space (SEDS). (n.d.). *Al Sufi, Abd-al-Rahman (December 7, 903 - May 25, 986 CE.).* https://www.messier.seds.org/xtra/Bios/alsufi.html

There are more things in heaven and earth, Horatio,
Than are dreamt of in your philosophy.

—*Shakespeare, Hamlet*

TOPIC 24. Cepheid Variables—how we discovered that a known object (M31) in the night sky was not really of our "universe," the Milky Way galaxy.

The absolute magnitude of a star is defined as how bright it would appear (its apparent magnitude) at a distance of 10 parsecs (about 32.6 light years) from Earth. (Parsec is defined in *Appendix 24.2.*) Absolute magnitude is a measure of the star's luminosity (the total energy it radiates per second). The absolute magnitude of the Sun is -4.8.

A variable star, by definition, is one whose absolute magnitude varies with time. For the vast majority of variable stars there is no relationship between the star's period of variability (from being at its brightest to fading and returning again to being at its brightest) and the star's average absolute magnitude. In a breakthrough discovery in 1908, Henrietta Leavitt determined a relationship between the average absolute magnitude of a special type of star, now called a Cepheid variable, and the period over which it pulsates. (A little of her background is given below.) Her discovery is so important because by knowing the average absolute magnitude of a star and its apparent magnitude the distance of the star from the Earth is easily determined.

See *Appendix 24.1* for the defining mathematical relationship between magnitude and brightness (and luminosity and brightness). Magnitude is defined so that there is a magnitude difference of five if and only if one star is 100 times as bright as the other. Negative magnitudes correspond to brighter stars.

Many stars vary periodically in brightness, but Cepheids are a special class of such stars that gave astronomers a new method to determine distances to stars and other celestial objects (galaxies, globular clusters, etc.) that contain Cepheids. Cepheids are bright, yellow, giant and supergiant stars much more luminous and massive than the Sun. As these stars pulsate, their radii vary by as much as 10 or 20 percent. The period of a Cepheid generally ranges from one to 70 days[2,4] and produces a light curve with a distinctive shape. The key discovery was that the pulsation period of such a star is related to its average absolute magnitude (or equivalently its average luminosity). The longer the pulsation period the brighter the star.

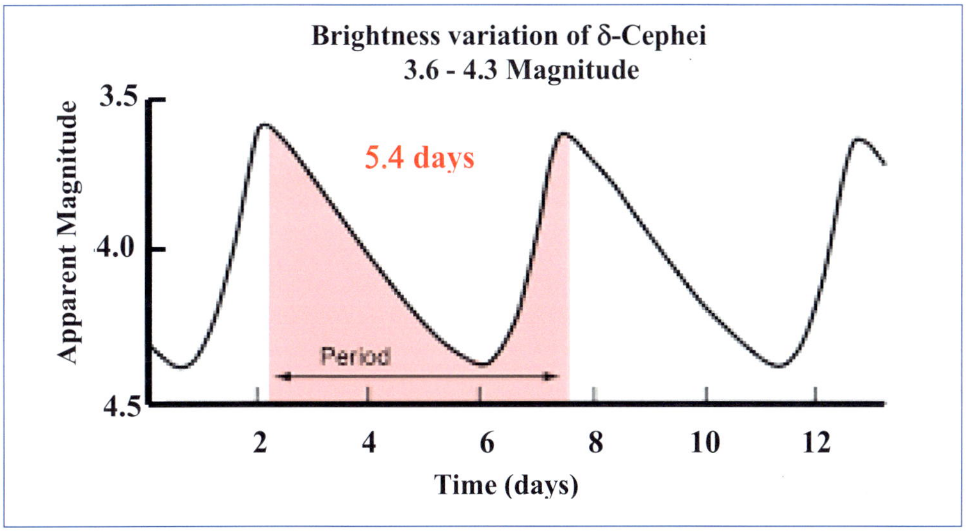

Figure 1. Delta Cepheid Light Curve. Credit: http://hyperphysics.phyastr.gsu.edu/hbase/Astro/cepheid.html

The first two Cepheids discovered.
On September 10, 1784, Edward Pigott, a wealthy, young British amateur astronomer, determined the star Eta Aquilae was a variable. At that time only five variable stars were known: Mira, Algol, P Cygni, Chi Cygni, and R Hydrae.[3]

That very same night Pigott's good friend, Richard Goodricke, discovered that Beta Lyrae varies in magnitude. (We now know that Beta Lyrae varies because it is an eclipsing binary, not a Cepheid.) In October Goodricke discovered that Delta Cephei also varies. Eta Aquilae and Delta Cephei turned out to be the first Cepheids discovered. Delta Cephei became the eponymous star for this class, giving Cepheids their name. Goodricke, who was deaf, died in 1786 at the young age of 21. Of course, at the time neither Pigott nor Goodricke knew that these two stars were of a class that would later take on such great significance.

The breakthrough discovery.
Henrietta Swan Leavitt (1868-1921) graduated from Radcliffe College and was enthusiastic about astronomy. In 1895 Edward Charles Pickering, director of the Harvard College Observatory, hired her to detect variable stars in the Large and Small Magellanic Clouds (LMC and SMC). This was to be done by examining glass photographic plates of the star clouds taken from Peru. By inspecting plates taken at different times she would try to detect variability in some of the stars.

She discovered Cepheids in the LMC and SMC. She plotted their period vs *apparent* magnitude. She assumed all the Cepheids that were in the same cloud were roughly the same distance from Earth, at least relative to the distance of the cloud from the Earth. Thus differences in *apparent* magnitudes would reflect differences in *absolute* magnitudes. To be more precise, there would be a constant k such that the *absolute* magnitude of any Cepheid in the cloud is k plus its *apparent* magnitude. The k would depend on the distance of the cloud from Earth. (From equation (1) below, $k = - 5[\log_{10} d - 1]$, where d is the common distance to all the stars in the cloud.)

In 1908 and 1912 she published her results, suggesting a way to determine the distance to any Cepheid variable, including those in the LMC and SMC. One would first use *parallax* (see *Appendix 24.2*) to determine the distances to several nearby Cepheids. (Cepheids in the LMC are too far away to be able to use parallax.) Then, measuring the apparent magnitude of those Cepheids, one can determine their absolute magnitudes from equation (1) below. We thus obtain the desired plot of average *absolute* magnitude vs period. Using this plot and knowing the periods of the Cepheids that were studied in the LMC and SMC, we can determine the distance to each of those Cepheids using equation (2) below. And we can do this without needing to know they are all in a cloud whose distance from Earth is large with respect to the distance of each of them from the Earth.

Leavitt later became head of Harvard's department of stellar photographic photometry.[1] When Magnus Gösta Mittag-Leffler of the Swedish Academy of Sciences tried to nominate her for the 1926 Nobel Prize in physics, he discovered she was no longer alive. Nobels are not awarded posthumously.

Example.
Let us say that the period of a Cepheid variable is 30 days and that its average apparent magnitude as seen from Earth is 12.4. Note that $\log_{10}(30) = 1.477$. Using the *average absolute magnitude-period plot*, Figure 2, its absolute magnitude M is about -5.2.

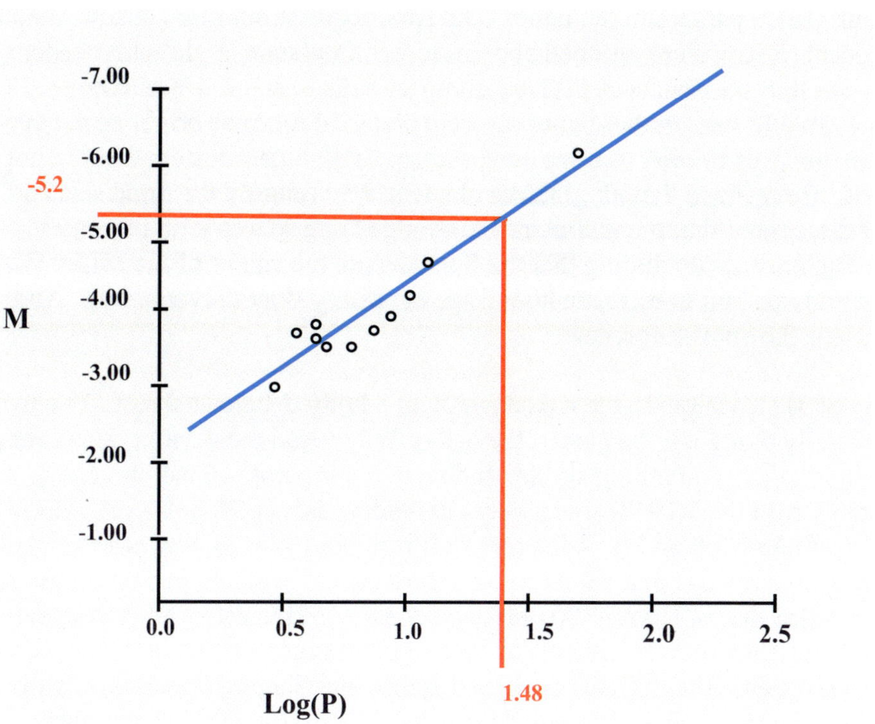

Figure 2. Period-luminosity plot for Cepheids. The average absolute magnitude M is plotted against Log(P) where P is the period in days. The line is a least squares regression fit to the data points.

The equation relating the absolute magnitude M of any star to its apparent magnitude m at a distance d is

$$M = m - 5[\log_{10}d - 1] . \qquad (1)$$

Solving this equation for d we get

$$d = 10^{[(m-M)/5 + 1]} \text{ with d given in parsecs.} \qquad (2)$$

Plugging in the values M = -5.2 and m = 12.4 gives d as approximately 33,100 parsecs.

Using parallax, Danish astronomer Ejnar Hertzsprung (1873-1967) was the first to determine the distances to nearby Cepheids. From this he concluded the distance to the SMC was about 10,000 parsecs. The true distance is about 62,000 parsecs. What happened? Determining the apparent magnitude of a distant star is not so straightforward as simply using a photometer to measure the star's brightness. Two difficulties must be taken into consideration: *reddening* and *dimming*. Dust in the intervening space between the Earth and a star will make the star appear dimmer and redder. (This reddening is not related to redshift from an expanding universe.) It was Henry Norris Russell, with his soon-to-become-prominent student Harlow Shapley, who improved the accuracy of these distance estimates in 1914. By the way, it is the aforementioned Hertzsprung and Russell who gave us the Hertzsprung-Russell Diagram discussed in Topic 49.

The Holy Grail: The discovery that M31 was not in our island universe.
Parallax can be used to determine distances to about 500 parsecs. Cepheids, being very bright, can be seen from far away. With the period-luminosity relationship, they can be used to determine distances to about 40 million parsecs.[2] This knowledge paved the way to determining how big our Milky Way

"universe" is since it is about 31,000 parsecs in diameter. The American astronomer Harlow Shapley (1885-1972), the former student of Russell mentioned above, studied Cepheids in globular clusters, determining their distances and thus the distances of these compact balls of stars. He also noticed that the few dozen brightest non-variable stars in these clusters were about 10 times as bright as the average Cepheid in the cluster.[5] He used this finding to infer the distances to globular clusters that did not contain any visible Cepheids. He realized that all globular clusters were roughly the same size and luminosity. And he further discovered that the globulars are arranged in a sphere with its center located in the constellation Sagittarius, concluding that the Sun was not the center of the Milky Way. His distance determinations allowed him to estimate how large the Milky Way universe was. And remember, at that time, this was the known universe.

The nature of the nebulosity of M31, visible to the naked eye, was a hotly debated subject. The two Magellanic clouds were generally thought to be part of the Milky Way. And most astronomers thought that M31 and other fuzzy objects were something similar and were not separate island universes. The American astronomer Heber Curtis (1872-1942) was a vocal dissenter. Edwin Hubble (1889-1953), using the new 100-inch Hooker telescope on Mt. Wilson in California in 1923-24, was able to resolve stars in M31 and in fact discovered several that were Cepheid variables. Using the period-luminosity relationship, in 1924 he concluded that M31 was 300,000 parsecs away. This was well beyond any object in the Milky Way. Hubble then found Cepheids in several other objects proving they, too, were island universes. So Curtis was right. The 300,000 parsecs distance and Shapley's estimate of the diameter of the Milky Way were revised later. We now know that M31 is 700,000 parsecs away.

A major reason for the revisions in the distance estimates was the discovery that all Cepheids don't follow the period-luminosity plot given above. There actually are two *classes* of Cepheids: I and II.

Before we define what class I and II Cepheids are we must first define what population I and II stars are. All stars in a galaxy are broken into two *populations*: I and II. Population I stars (which include the Sun) are hotter, younger, concentrated in the disk of a galaxy, and have high metallicity (elements other than hydrogen and helium). Population II stars are older, less luminous, cooler, are found in globular clusters and the nucleus of a galaxy, and have lower metallicity.

Now we can define the two classes of Cepheids. Class I Cepheids are much more numerous, are population I stars, and most have a pulsation period between five and 10 days.[4] Class II Cepheids are population II stars and typically have pulsation periods of at least 10 days.[4]

The plot given above is for class I Cepheids. Delta Cepheid is class I, and W Virginis is an example of class II. Class II Cepheids have a slightly different period-luminosity relation.

A third class of Cepheid-like stars, the RR Lyrae stars (named after, no surprise, the star RR Lyrae) have their own period-luminosity relation, a lower luminosity than Cepheids, and a short period typically between 1.5 hours to one day.[4] In estimating the Milky Way's diameter, Shapley actually used RR Lyrae stars, not knowing they differed from Cepheids.[9]

What is the pulsation process in a Cepheid?

Both Cepheids and RR Lyrae stars have a small, dense core, and a large, low-density envelope enveloping the core. Inside the envelope and surrounding the core is a pulsation layer (PL). It is this layer that expands and contracts, while the core stays the same size and produces energy at a constant rate. The pulsations arise from variations in the rate at which radiation can escape from the star. [2,4]

Let's look at the details of this process. First, one must realize that deep inside a star radiation (starting off as gamma ray photons) from fusion takes millions of years to reach the surface (photosphere) of the star. Once photons reach the photosphere (now at visible and UV wavelengths) they stream away at the speed of light. After spending millions of years to reach the Sun's surface, light only takes about eight minutes to go from the Sun's surface to Earth. Why does light spend millions of years navigating the Sun's interior? Photons can stream unimpeded through a gas of neutral atoms, but are scattered by free electrons. The hydrogen (and some helium) in the core of a star is ionized. For hydrogen that just means the electron in the hydrogen atom is not bound to the proton. But for a helium atom, that means one or both of its two electrons are not bound to the nucleus (consisting of two protons and two neutrons for the most common isotope). The plasma (ions and free electrons) is so dense in the core of a star that the photons are constantly being scattered (in random directions) mainly by the vast number of electrons. The photons can barely make headway toward the surface of the star.

Now let us go back to looking specifically at a Cepheid variable. At the beginning of a pulsation cycle, the PL contains singly ionized helium. This allows radiation from the core to pass through the PL. But some of the radiation increases the temperature of the PL. The result: singly ionized helium atoms become doubly ionized. Radiation is now almost blocked from passing through the PL. The pressure from the blocked radiation causes the PL to expand outward, and the envelope of the star expands. But as expansion occurs the density and temperature of the PL drops.

Doubly ionized helium ions recapture an electron. Gravity and the reduced outward pressure cause the PL to subsequently contract to its former state, causing the star to contract. This process then begins all over again. Arthur Eddington was the first to characterize this activity in his 1926 book *The Internal Constitution of the Stars.*[6,7,8]

References

1. Gribbin, John (2002). *Science: A History 1543-2001.* Allen Lane. See pages 584-590 for a discussion of Cepheids. In general, a good source for the history of science from 1543 (the publication year of Copernicus' *De Revolutionibus Orbium Coelestium*—On the Revolutions of Celestial Bodies—and coincidentally the publication of Veslius' *De Humani Corporis Fabrica*—On the Structure of the Human Body) to 2001.
2. Ryden, Barbara. *Pulsating Stars.* Lecture 16, Astronomy 162, Ohio State University, Columbus. http://www.astronomy.ohio-state.edu/~ ryden/ast162_4/notes16.html
3. Croswell, Ken (October 1997). The first Cepheid. *Sky & Telescope,* p 90.
4. Pulsating Variable Stars. Commonwealth Scientific and Industrial Research Organization (CSIRO). Australian Telescope National Facility. https://www.atnf.csiro.au/outreach/education/senior/astrophysics/variable_pulsating.html
5. Cepheid Variable Stars and Distance Determination. Commonwealth Scientific and Industrial Research Organization (CSIRO). Australian Telescope National Facility. https://www.atnf.csiro.au/outreach/education/senior/astrophysics/variable_cepheids.html
6. Dorminey, Bruce (September 2009). The stars that revealed the vastness of the universe. *Astronomy Magazine,* p 48-53.
7. Percy, John (December 1984). Cepheids: cosmic yardsticks, celestial mysteries? *Sky & Telescope*, p 517-519. John Percy is Professor Emeritus of Astronomy and Astrophysics at the University of Toronto.
8. Carlisle, Camille M. (March 2019). Secrets of Polaris. *Sky & Telescope,* p 14-19.
9. Seronik, Gary (May 2022). Reply to a reader's letter. *Sky & Telescope,* p 6.

*Besides the god of commerce,
Mercury was also the patron
of shepherds, cheats, and thieves.*

—*Donald L. Wasson, historian*

TOPIC 25. Interesting characteristics of Mercury.

1. Mercury may be smaller in diameter than the two largest moons in the solar system, Ganymede and Titan, but Mercury's mass is greater than the *combined* masses of these two moons.

The diameters of Ganymede, Titan, and Mercury are respectively 5,268, 5,149, and 4,880 kilometers. The masses of Ganymede, Titan, and Mercury are respectively 1.48×10^{23}, 1.35×10^{23}, and 3.28×10^{23} kilograms.

The high mass of Mercury is due to its large, solid iron core. It is estimated that it is about 2,000 kilometers in diameter, about 40 percent of the diameter of the planet.

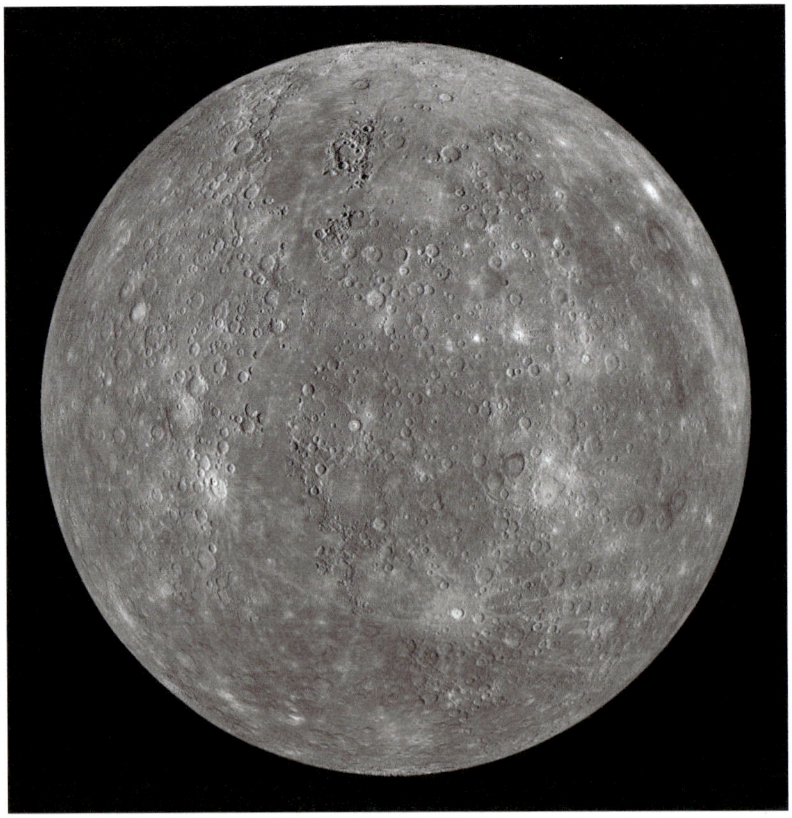

Figure 1. NASA composite image of Mercury taken by Messenger. The rayed crater Debussy can be seen toward the bottom.

2. Because Mercury's rotational axis is almost vertical, sunlight never reaches the bottoms of craters near the poles. Consequently it is permanently very cold there. NASA has discovered that a considerable volume of water ice is present.

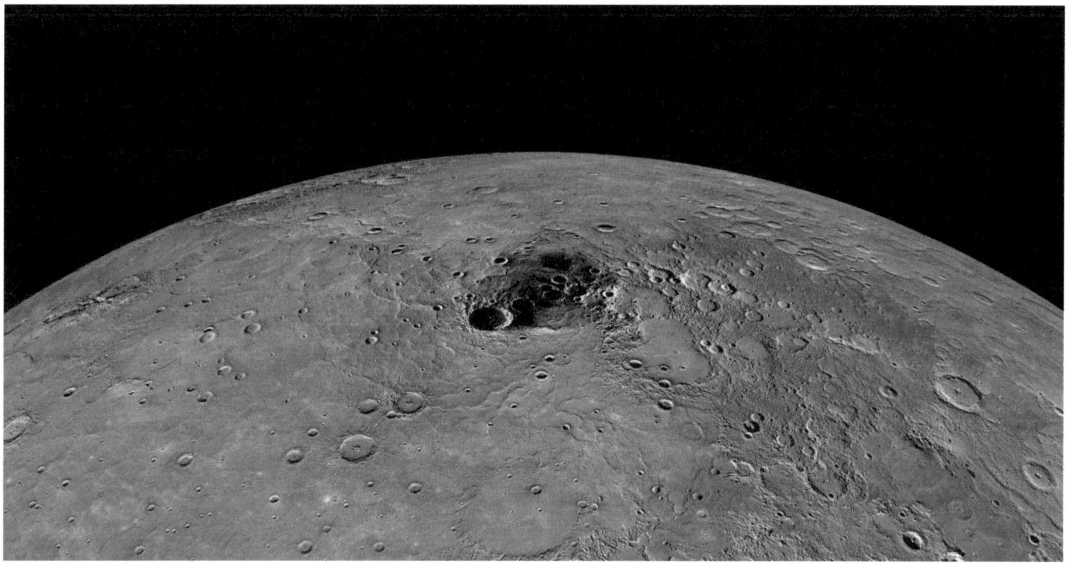

Figure 2. NASA composite image of the north pole region of Mercury taken by Messenger.

3. Mercury's youngest large impact basin and "Weird Terrain" are on directly opposite sides of Mercury. Is there a connection?

Caloris is Mercury's youngest large impact basin and is one of the largest impact basins in the solar system. The interior smooth plains have an area approximately that of Alaska.

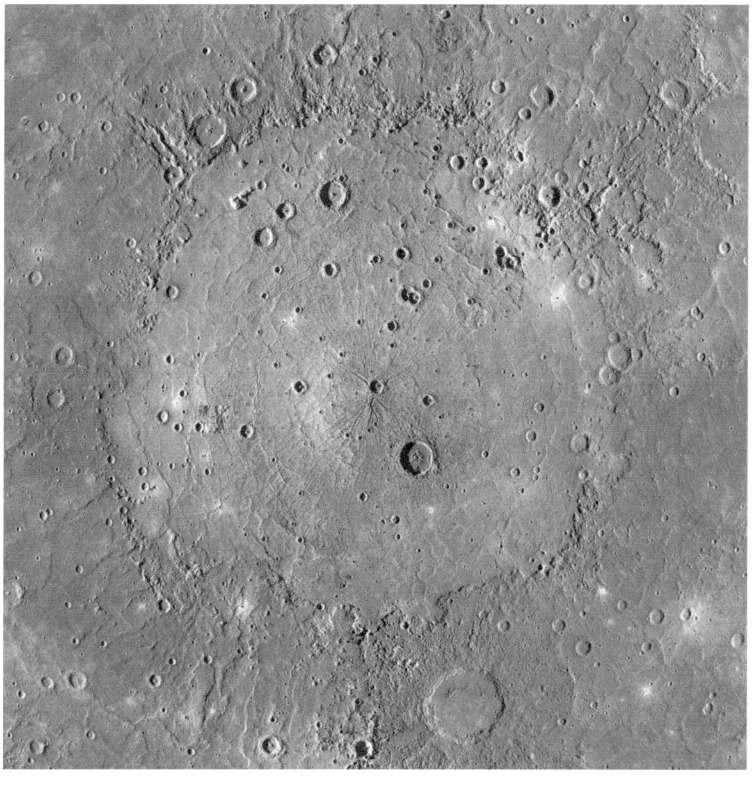

Figure 3. NASA composite image of the large Caloris basin on Mercury.

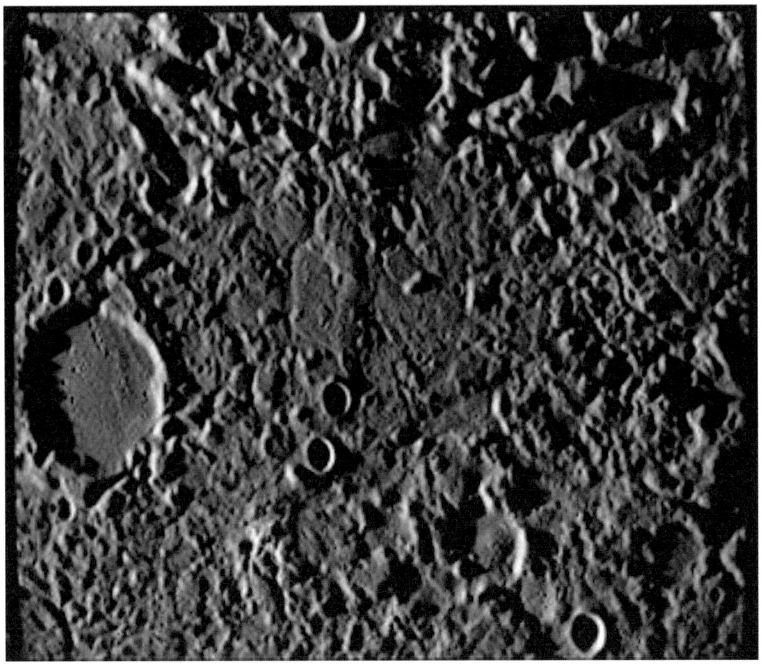

Figure 4. The "Weird Terrain" formed at the point antipodal to the Caloris Basin impact. Credit: NASA

One suggestion is that the impact that created the Caloris basin produced shock waves that traveled through the planet and ejected surface debris from the impact site. The debris then traveled halfway around the planet colliding at the antipodal point, jumbling the crust and breaking it into a series of complex blocks—the so-called Weird Terrain. But a more recent idea[1] from Alexis Rodriguez at the Planetary Science Institute and colleagues is that the chaotic terrain was formed by the collapse of the surface due to the escape of volatile material deeper underground. By counting craters one can get an idea of how old a surface is, with more craters indicating an older surface. Such a count indicates that the Weird Terrain is about 1.8 billion years old, which is about two billion years younger than the Caloris basin. Also, rays from more recent craters should extend over the Weird Terrain if the terrain were older. But they don't. In addition, some other areas of chaotic terrain have been discovered that aren't antipodal to a crater. Stay tuned for future updates from NASA.

4. How about the size and intensity of the Sun on Mercury?

At perihelion, when Mercury is closest to the Sun, the Sun's diameter is 3.3 times that as seen from Earth. At aphelion, when Mercury is farthest from the Sun, the Sun's diameter is 2.2 times that as seen from Earth. So during a Mercurian year, the Sun's diameter varies by a factor of 1.5. Compare that to the almost constant diameter of the Sun as seen from Earth throughout one year. The solar intensity (which is watts/square meter) varies from 4.7 to 10.6 times that on Earth. That would give one more than a suntan. See *Appendix 25.1* for details on how the above numbers were obtained.

References

1. Barbuzano, Javier (July 2020). Why have parts of Mercury's landscape collapsed? *Sky & Telescope,* p 8.

No matter what I do,
I'm always stuck with you.

—characterizing twins

TOPIC 26. Strange twins in the asteroid belt.

Many asteroids have at least one moon. But asteroid 90 Antiope is an intriguing case. It consists of two asteroids with almost identical diameters, 55 and 52 miles (88 and 84 kilometers),[1,2,3,4] both porous (i.e., rubble piles), both of carbonaceous composition, with very similar spectra, and orbiting each other. The distance between their surfaces is about the same as the diameter of one of the asteroids.[1,2,3,4] Imagine being on one of them and watching the show.

Figure 1. Image of 90 Antiope using adaptive optics at the Keck Telescope on Mauna Kea. Credit: W. Merline (SwRI), et al.

The asteroid was discovered October 1, 1866, by Robert Luther. But it wasn't until August 10, 2000, using adaptive optics at the Keck Telescope on Mauna Kea, that it was determined that the asteroid was really two asteroids. The twins orbit about a point in between them in 16.5 hours.[1,3,4,5] During their 5.6-year orbit around the Sun,[1,3] the orbital plane of the pair is seen edge-on from Earth for about half of the year; they partially eclipse each other every eight hours.[1,5]

They are in the outer third of the core region of the asteroid belt. Light-curve data analysis suggests one of them has an impact crater 68 kilometers in diameter (this is not a misprint).[6] Not surprisingly some theories of how they formed assume they were once part of the same body.

References

1. Berman, Bob (February 27, 2015).Weird object: Asteroid 90 Antiope. astronomy.com. https://astronomy.com/magazine/weirdest-objects/2015/02/44-asteroid-90-antiope. Bob Berman, as always, is witty in his writings on various astronomical subjects.
2. (March 2001). Twin asteroid discovered. *Astronomy Magazine,* p 22.
3. Johnston, W. Robert (November 23, 2008). (90) Antiope and S/2000 (90). Johnston's Archive. http://www.johnstonsarchive.net/astro/astmoons/am-00090.html
4. Marchis, F. (April 20, 2006-November 22, 2020). 90 Antiope A & B. Internet Archive Wayback Machine. https://web.archive.org/web/20080828002632/http:/astro.berkeley.edu/~fmarchis/Science/Asteroids/Antiope.html
5. Michałowski, T., et al. (September 2004). Eclipsing binary asteroid 90 Antiope. *Astronomy & Astrophysics, 423*(3), p 1159. https://www.aanda.org/articles/aa/abs/2004/33/aa0449-04/aa0449-04.html
6. Marchis, F., Enriquez, J. E., Emery, J. P., Berthier, J., & Descamps, P. (September 2009). The origin of the double main belt asteroid (90) Antiope by component-resolved spectroscopy. American Astronomical Society, DPS meeting #41, id.56.10 https://ui.adsabs.harvard.edu/abs/2009DPS....41.5610M/abstract

No one planet can tell us everything about the universe, but Neptune seems to hold more than its share of information about the formation of our own solar system

—Heidi Hammel, planetary astronomer

TOPIC 27. Galileo and the planet Neptune: He may have come close to detecting it 233 years before its actual discovery.

On December 28, 1612, while observing Jupiter and its moons, Galileo marked a "fixed star" near Jupiter on his drawing. Using computer simulations to calculate past positions of Neptune, it is now known that this was actually the planet Neptune (which was in conjunction with Jupiter). On January 28, 1613, he drew two stars near Jupiter, one being Neptune and the other a star with the current designation SAO 119234. Galileo also made a separate drawing showing just these two "stars." On the previous night Galileo had also seen Neptune and the star, but didn't record them in his notebook. From January 27 to 28 Neptune would have moved 2.5 Jovian radii closer to the star. In his notes for the 28th he wrote (translated from Latin, with star A being the real star and star B denoting what was actually Neptune):

> After fixed star A, following in the same line, is star B, which I saw in the preceding night, but they then seemed farther apart from one another.

So Galileo noticed movement. There are no known notes of a later date concerning this matter. It could be that this was due to weather, or that it was difficult to locate Neptune after it was no longer near Jupiter. We don't know the reason.

On January 4, 1613, Jupiter actually occulted Neptune. One might ask, how could Neptune be close to Jupiter for as long as a month (December 28 to January 28)? The answer is that the Earth in its orbit was passing the two of them and thus during this period they both went into retrograde motion, which would extend the time the two planets were visually near each other.

There is also an unlabeled black dot in an observation Galileo made on January 6, 1613, which doesn't appear in web posts of Galileo's notes. This dot is in the position where Neptune would have been that night. Professor David Jamieson, a physicist at the University of Melbourne, in 2009 suggested that perhaps Galileo, on January 28, retrospectively put this dot on the January 6 diagram, indicating that Galileo was looking for further evidence for movement of this "star." He has suggested testing the ink used to make this dot, comparing it with the ink used on the January 6 and 28 notes. But testing the ink used to make a single dot presents difficulties compared with testing the ink used on a whole page. Neptune's disk is only about 2.3 arc seconds in diameter at its closest distance, too small for it to have appeared as a disk in Galileo's telescope. And it is too faint to be seen with the unaided eye. No concrete evidence suggests that Galileo thought his "star" might be a new planet. He certainly would have been aware of the historic nature of such a sensational discovery.

It wasn't until September 23, 1846, *233 years later*, that Neptune was officially discovered by the German astronomer Johann Gottfried Galle. Two mathematicians, John Couch Adams (Great Britain) in 1843 and Urbain Le Verrier (France) in 1845-1846, separately predicted Neptune's position based on deviations in the predicted orbit of Uranus (which had been discovered by William Herschel on March 13, 1781). Galle used Le Verrier's charts to find Neptune. Galle, Adams, and Le Verrier are credited with the discovery, but only after conflicting claims between Adams and Le Verrier. If Galileo had discovered Neptune in the 17th century, the discovery of Uranus would have come after that of Neptune, turning history on its head. Barely visible to the unaided eye under a dark sky, Uranus' disk varies between 3.4 to 3.7 arc seconds in diameter and is brighter and larger than Neptune. Details of the above discussion are drawn from references [1] and [2].

Charles Kowal[1,3] is a noted American astronomer among whose achievements was the discovery of Chiron in 1977. Chiron is the first-identified member of a new class of objects now known as Centaurs—bodies orbiting between Jupiter and Neptune. Kowal and Galileo historian Stillman Drake discovered the connection between Galileo and Neptune.[1,2,3] (Note: In [1], Kowal is critical of the editing of his and Drake's paper that appeared in the periodical *Nature* [3].) In [4], Standish and Nobili discuss their discovery and conjectures concerning the dark ink dot appearing in Galileo's January 6, 1613, notes.

References

1. Kowal, Charles T. (December 2008). Galileo's observations of Neptune. *The International Journal of Scientific History, 15,* p 3-6. http://www.dioi.org/vols/wf0.pdf
2. Jamieson, David (May/June 2009). Galileo's miraculous year 1609 and the revolutionary telescope. *Australian Physics, 46*(3), p 72-76. https://www.ph.unimelb.edu.au/~dnj/AP-June-2009-Galileo-Neptune.pdf
3. Kowal, C. T., & Drake, S. (September 25, 1980). Galileos' observations of Neptune. *Nature, 287*, p 311-313. https://www.nature.com/articles/287311a0
4. Standish, E. M., & Nobili, A. M. (1997). Galileo's observations of Neptune. *Baltic Astronomy, 6,* p 97-104. http://articles.adsabs.harvard.edu/cgi-bin/nph-iarticle_query?1997BaltA...6...97S&defaultprint=YES&filetype=.pdf

TOPIC 28. A first of its kind: A stellar system of six stars, consisting of three pairs, where all three pairs (as seen from Earth) are eclipsing binaries.[1,2,3]

Other sextuple star systems are known (see Topic 29), but none consisting of three *eclipsing* binaries, i.e., pairs of stars that alternatively pass in front of one another as seen from Earth. As such the light from a combined pair diminishes in brightness twice for each orbit around their center of gravity. How much the light dims at any eclipse depends on the relative magnitudes and sizes of the two stars, the distance between them, and whether the brighter star is passing in front of or behind the fainter star.

The most well-known eclipsing binary star is Algol, the "Demon Eye," in Perseus. Algol is visible to the naked eye, and every 2.87 days it is in eclipse during a nine-hour period. This brightness change is noticeable to the unaided eye (and can be observed even better with binoculars) by comparing Algol's changing brightness to the fixed brightness of two neighboring stars. During most of its period, Algol has magnitude 2.1, the same as γ Andromedae, but at minimum it's no brighter than α Trianguli at magnitude 3.4.

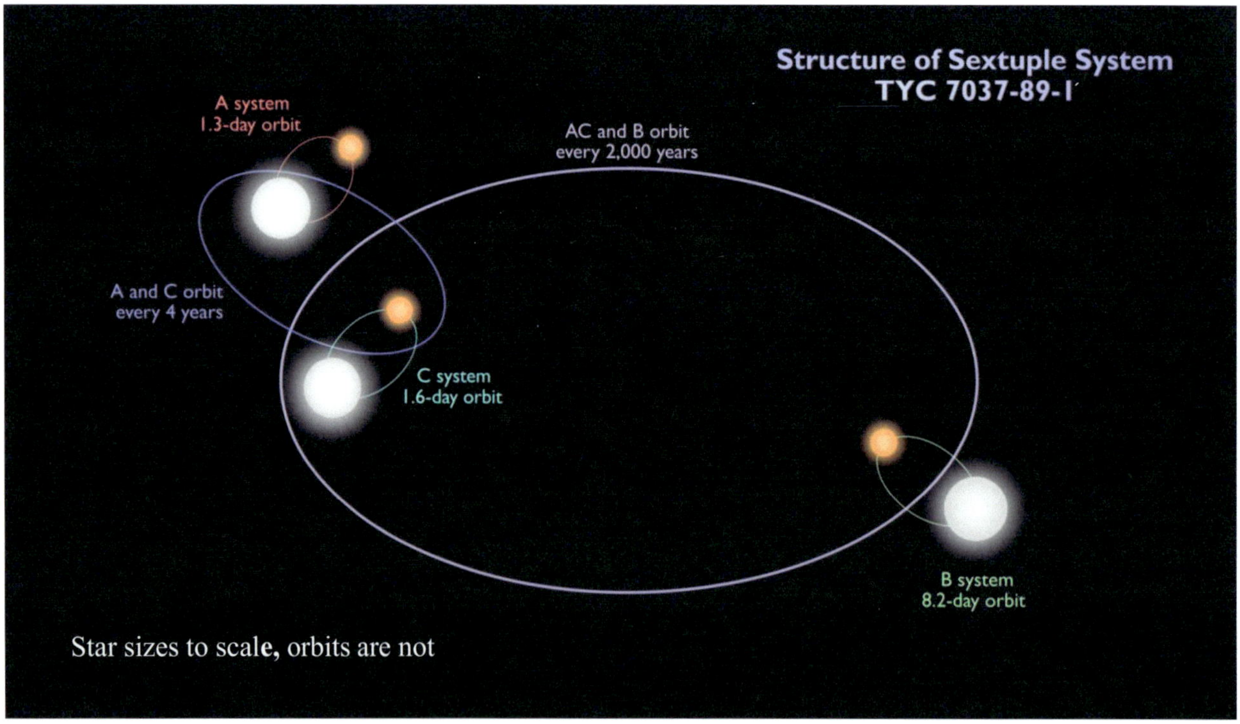

Figure 1. The sextuple star system. Credit: NASA's Goddard Space Flight Center (Text in lower left was altered to a larger size and bolded by author)

The sextuple system in the constellation Eridanus has the designation TYC 7037-89-1. Each binary has one star slightly larger, more massive, and about as hot as the Sun. The other star in each binary is about half the Sun's size and about a third as hot.

Here are the details.
The A system: The stars orbit a common center of gravity in 1.3 days.
The B system: The stars orbit a common center of gravity in 8.2 days.
The C system: The stars orbit a common center of gravity in 1.6 days.

The A and C systems are much closer to each other than to the B system.
The A and C systems orbit a common center of gravity in about four years.
The AC system (the combined A and C systems) and the B system orbit a common center of gravity about every 2,000 years.

The entire system is about 1,900 light years from Earth.

The interacting gravitational forces of the four stars on any planet in the AC system would be disruptive. It is possible, however, for the B system to have planets orbiting the two close stars, but at a safe distance from the pair. And what a sight that would be. In the first *Star Wars* movie we all were struck by the two suns in the sky as seen by Luke Skywalker on Tatooine. On a planet in the B system one would see two suns and in addition four very bright stars—the stars of the AC system.

The Discovery.
NASA's Transiting Exoplanet Survey Satellite (TESS) searches for exoplanets by looking for the light from a planet crossing the face of its sun and dimming the sun's light. This of course requires the orbital plane of the planet to be close to edge-on as seen from Earth. Planets not satisfying this condition (which would be the majority) are not detectable using this method. TESS's search began in 2018. Brian Powell, a data scientist, and Veselin Kostov, an astrophysicist, designed a neural network to identify eclipsing binary stars using TESS data. Nearly 80 million (yes, that's right) archived records of light-intensity changes were examined. Among them was TYC 7037-89-1, the proverbial needle-in-a-haystack.

It is interesting to note that TESS previously discovered that Thuban, the North Star at the time of the Pharaohs (see Topic 3), is an eclipsing variable.

References
1. Powell, Brian P., et al. (March 4, 2021). TIC 168789840: A sextuply-eclipsing sextuple star system. *Astronomical Journal, 161*(4), p 162. https://arxiv.org/abs/2101.03433
2. Kazmierczak, Jeanette (January 27, 2021). Discovery alert: First six-star system where all six stars undergo eclipses. NASA news. exoplanets.nasa.gov https://exoplanets.nasa.gov/news/1672/discovery-alert-first-six-star-system-where-all-six-stars-undergo-eclipses/
3. Andrews, Robin George (January 23, 2021, updated January 25, 2021). Six stars, six eclipses: "The fact that it exists blows my mind." nytimes.com. https://www.nytimes.com/2021/01/23/science/six-stars-eclipses.html

...six stars...in a complex dance
about a common center of mass

—Larry Sessions, earth/sky.org

TOPIC 29. The first-magnitude star Castor in the constellation Gemini is a six-star system, consisting of three pairs.

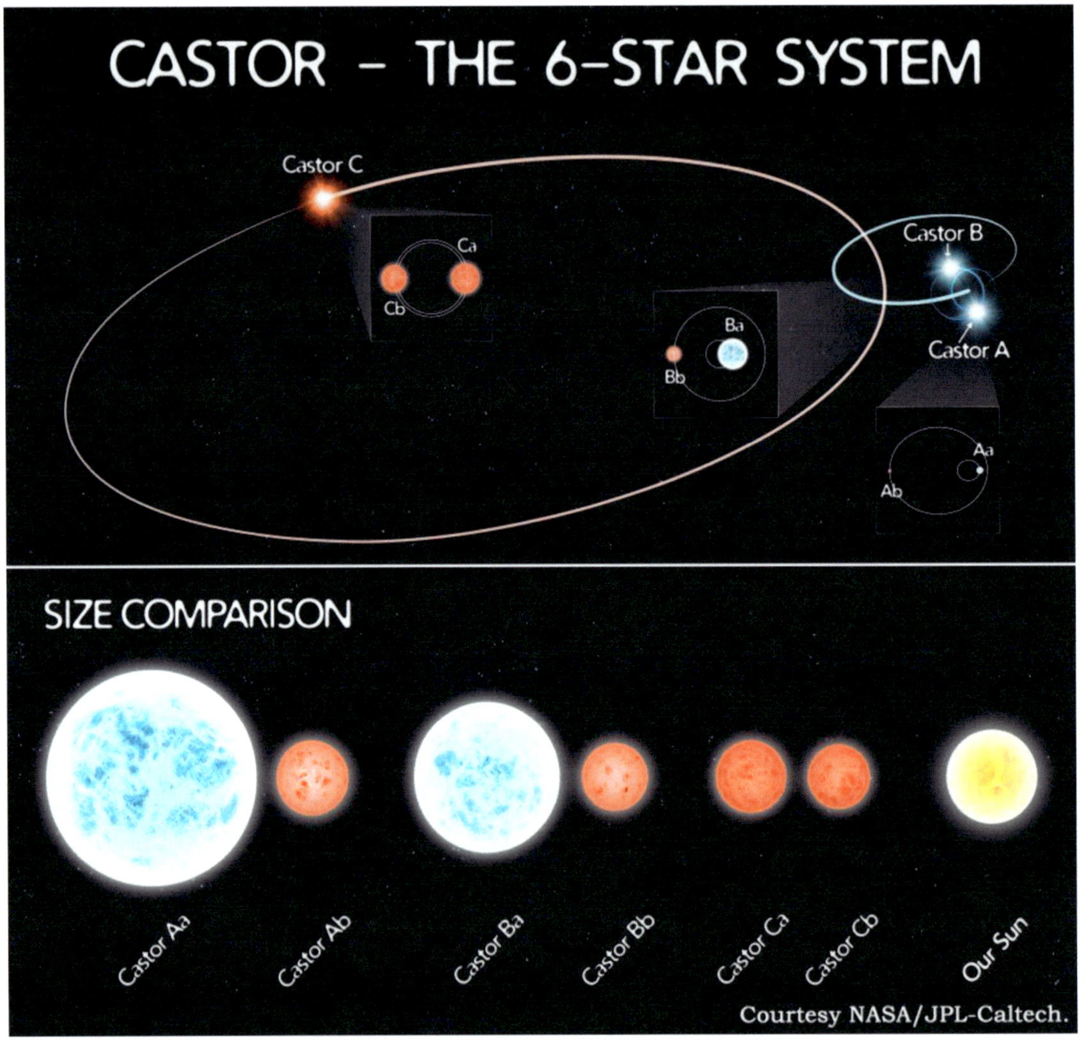

Figure 1. The Castor system of six stars. With each of the stars, Castor A, B, and C, is a small (faint) window revealing that each star is really a binary system and showing the orbits (as viewed from above). Also the relative star sizes are illustrated. Credit: NASA/JPL-Caltech; edited by Nicholas Beeson[1,2]

A small, amateur telescope reveals that Castor, one of the 20 brightest stars in the sky, is a triple. Castor A and B are separated by 3.9 arc seconds; Castor B and C by 71 arc seconds. But there are really six stars as each "star" is a binary. Currently the six-star system is 48.9 light years distant, with the AB and C systems separated by about 1,100 AU (about six light days and eight light hours).[1]

The orbital period of the C and the AB systems is unknown, but is thought to be at least 10,000 years.[1] The AB orbital period is about 467 years.[1]

The red dwarf pair Castor C is an eclipsing binary, with two eclipses occurring every 19.5 hours. At each eclipse the light is halved, dropping by about 0.7 in magnitude. The two components, with radii of 0.62 solar diameters, are only 3.9 solar diameters apart.[3] That's close.

The A system's stars' orbital period is 9.21 days; that of the B-system stars is 2.93 days.

While Castor has only one eclipsing binary out of its three binary systems, instead of all three as in TYC 7037-89-1 (see Topic 28), three of Castor's components are easily visible in a small telescope.

References

1. NASA/JPL. (August 10, 2012). Castor, the 6-star system - A3 poster. https://www.jpl.nasa.gov/infographics/castor-the-6-star-system-a3-poster
2. Letzter, Rafi (January 22, 2021). 'Sextuply-eclipsing sextuple star system' discovered whirling through the Milky Way. livescience.com https://www.livescience.com/sextuply-eclipsing-sextuple-star-system.html
3. Kaler, James B. (July 10, 2009). Castor. http://stars.astro.illinois.edu/sow/castor.html Kaler is Professor Emeritus of Astronomy at the University of Illinois.

Astronomy compels the soul to look upward,
and leads us from this world to another.
—Plato

TOPIC 30. Exoplanets.

We discuss four especially interesting exoplanetary systems, each having its own striking features.

I. The Trappist-1 system: seven planets.

This solar system consists of a red dwarf star called Trappist-1 and seven known planets named Trappist b, c, d, e, f, g, and h in order from closest to farthest from the star. The star is in the constellation Aquarius. See Figure 1.

The star's name is an acronym for TRAnsiting Planets and PlanetesImals Small Telescope, which is the corporate name for a pair of Belgian optic robotic telescopes, one at the La Silla Observatory in Chile and the other at the Oukaimeden Observatory in the Atlas Mountains of Morocco. The planets were revealed by Trappist and NASA's Spitzer space telescope.

From the Earth all the visible planets are just bright points of light to the naked eye. Our Moon, however, appears as a spherical body about 30 arc minutes in diameter. The Trappist-1 solar system, about 40 light years from Earth, is shockingly different. Its seven planets, five roughly Earth-sized and two only slightly smaller, are all closer to its sun than is Mercury to our Sun. In fact, much closer. Trappist h, the farthest planet, is at a distance from Trappist-1 that is only 16 percent of the distance of Mercury from the Sun. So from any of the planets, all the others would be visible as disks—some as quite large disks.

Apparent sizes of the planets as seen from a given planet.
The sizes, i.e., diameters, are given in terms of the diameter of our Moon (30 arc minutes), and all of the apparent sizes are determined when the two planets are closest to each other. (See *Appendix 30.1* for the calculations.)

b as seen from c: 2.6 times the size of our Moon—the largest.
h as seen from b: 15 percent the size of our Moon—the smallest.

Now consider the sizes of all the other planets as seen from e:

d and f would respectively be 10 percent and 12.5 percent *larger* than our Moon.
c would be 80 percent the size of our Moon.
g would be 64 percent the size of our Moon.
b would be 62 percent the size of our Moon.
h would be 23 percent the size of our Moon.

Three of the planets, Trappist e, f, and g, are in the star's *habitable zone*, that range of distances from the star where the temperature would allow water to exist in liquid form on a planet's surface given sufficient atmospheric pressure. Trappist c and e are believed to be entirely rocky, while b, d, f, and g have some type of volatile (easily evaporative) material covering their surfaces.[1] Such surface layers could be ice, water, or a thick atmosphere.

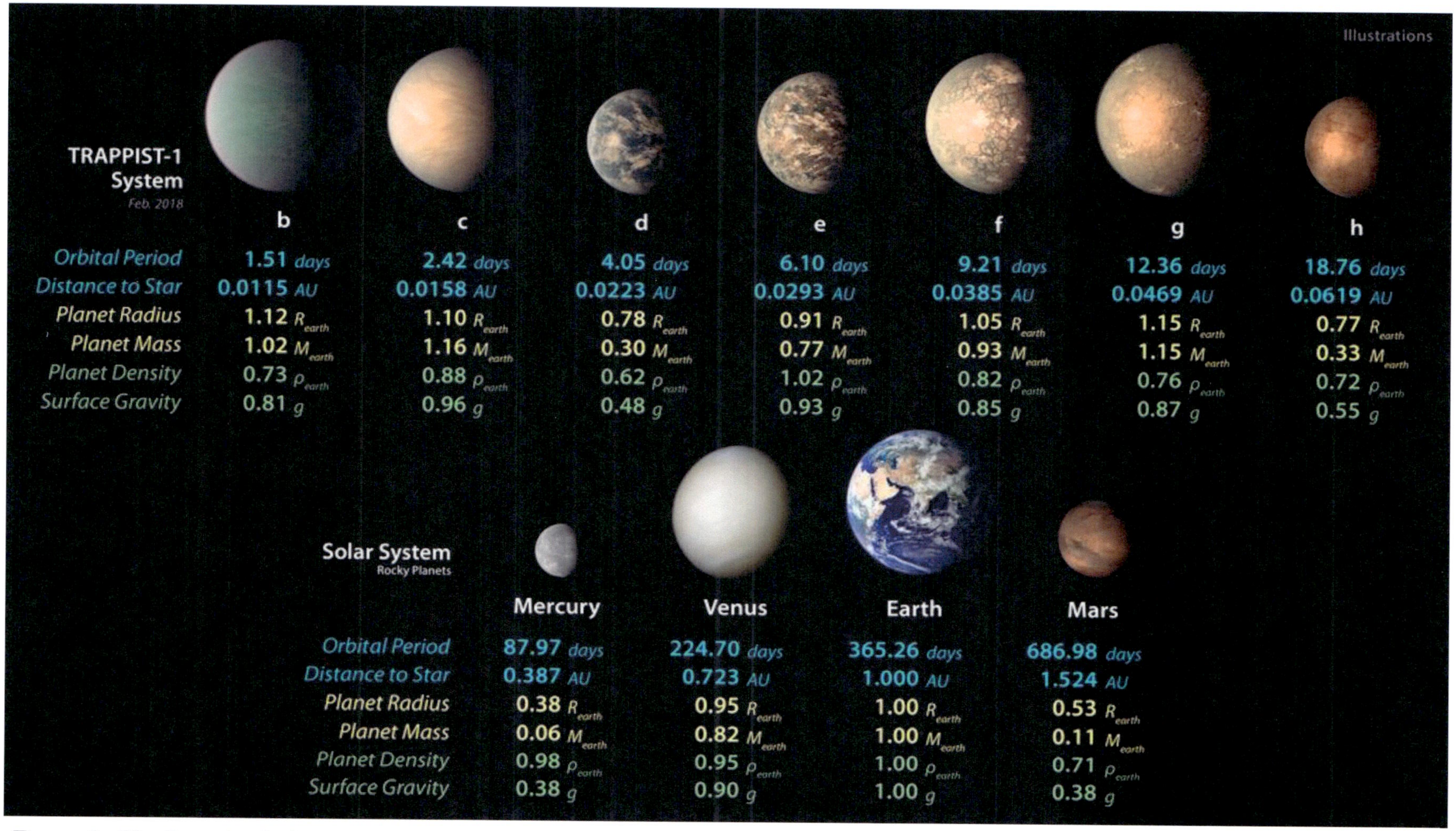

Figure 1. The Trappist-1 planetary system. The relative sizes of the planets are all shown to scale. Credit: NASA/JPL-Caltech/R. Hurt, T. Pyle (IPAC) Artwork, February 5, 2018, ssc2018-04c

Description of Figure 1:[2]
The top half of this descriptive chart shows artist concepts of the seven planets of TRAPPIST-1 with their orbital periods, distances from their star, radii R_{earth}, masses M_{earth}, , densities ρ_{earth}, and surface gravity as compared to those of Earth. Numbers are current as of February 2018.

Below, similar numbers are displayed for the bodies of our inner solar system: Mercury, Venus, Earth, and Mars. The TRAPPIST-1 planets orbit their star extremely closely, with periods ranging from only 1.5 to about 20 days. These periods are much shorter than the period of Mercury, which orbits our Sun in about 88 days.

The masses and densities of the TRAPPIST-1 planets were determined by careful measurements of slight variations in the timings of their orbits using extensive observations made by NASA's Spitzer and Kepler space telescopes, in combination with data from Hubble and a number of ground-based telescopes. These measurements are the most precise to date for any system of exoplanets.[2]

Figure 2. An artist's fantasy of the surface of TRAPPIST-1e, a stop on a tour of this seven-world system.

Take a trip with the Exoplanet Travel Bureau to the fourth planet in the TRAPPIST-1 system, TRAPPIST-1e, a world swimming in water and in perpetual twilight. Its sister planets gracefully light up the sky, promising another adventure just a hop away.

Credit: NASA-JPL/Caltech[3]

Regarding possible life on these planets and, in general, on any planet in orbit around a red dwarf star.
Co-discoverer Amaury Triaud of the UK's Institute of Astronomy suggests that planet f might be the most likely of the seven planets to have life.[10] Planet f has about 80 percent of the density of Earth with a diameter only five percent larger than Earth. He suggests it might be rich in liquid water or ice. It receives about the same amount of energy as does Mars. If Mars were the size of Earth and had a thick atmosphere it would not be the world it is now—dry and lacking a magnetic field. Mars is in the habitable zone of our Sun, but its small size is a big disadvantage for retaining life (at least on the surface). How would the star Trappist-1 appear in the sky from planet f? It would have a salmon-like color with a diameter three times that of our Sun. Quite stunning.

With the seven planets orbiting so close to each other it is likely that internal heat is generated in several of them through gravitational tidal stretching. So hydrothermal vents could exist at the bottoms of any oceans, providing an energy source for living organisms. Since the Trappist-1 planets are so close to each other, if life arose on one, meteorites from it could seed life on the others.

Like Spock's mythical home world Vulcan, the star Trappist-1 is a red dwarf. Its diameter is only about 12 percent of our Sun; its mass is about nine percent of our Sun. While our Sun is about 4.5 billion years old, Trappist-1 is about eight billion years old.[1] Our Sun will last about another 4.5 billion years before becoming a white dwarf—with life on Earth ending well before then. But like all red dwarfs, the most common type of star in the universe, Trappist-1's life span is in the trillions of years—well longer than the current age of the universe, about 13.7 billion years. That is good news for any life arising on a planet in orbit around such a star—especially a super Earth-type planet that could maintain a liquid iron core for a long time and thus keep its protective magnetic field.

Because they orbit so close to Trappist-1, several of the planets are probably tidally locked to their star. In particular any planet in the habitable zone of a red dwarf, being close in, is likely to be tidally locked. This situation had been thought to be a big negative for life, with any surface life having a greater chance of existing in the zone where the sun-lit and perpetually dark hemispheres meet. But a tidally locked planet with a substantial atmosphere should have strong winds blowing from the sunny side to the dark side, generated by the large temperature differences between these hemispheres. And studies have shown that an atmosphere would keep the temperatures on both sides roughly the same, and even if there were abundant ice on the far side, water could circulate beneath it.[4] If we ever visit such a planet, we might find that intelligent beings had built large windmills.

Recent results have introduced a new element into the issue of tidal locking of planets.[6,7,8] We all know that the hottest time of the day is not noon, but a few hours later. This is because sunlight takes some time to warm the Earth's surface, say at some point P, and then radiate heat into the atmosphere above the ground. During this period the Sun has moved to where it is noon at a different point P'. Thus the air at P is now warmer than at P' (assuming both points have the same latitude). Warmer air is less dense than colder air. So there is redistribution of the atmospheric mass with respect to the subsolar point (the point on the planet's surface where the sun is directly overhead). Research shows that the stronger gravitational attraction of a star on the more massive parts of the atmosphere produces a torque that tends to accelerate or decelerate the planet's rotation, depending on the direction of the solar motion. This is because friction couples the lower atmosphere to the ground, thus transferring angular momentum from the atmosphere to the ground. The result is that the planet's rotation can be changed, freeing the planet from the tidal lock to its star. Good news for the possibility of life.

In addition, gravitational interactions of a planet with other close-in planets could change the rotation rate of the planet, so that it rotates slightly slower or faster than the tidally locked rate. And perhaps the planet's orbital eccentricity would also be changed.[5]

One danger red dwarfs present is that they are very active in producing strong flares—massive bursts of charged particles. Such flares could strip away some of a planet's atmosphere. And it can take a long time (about a billion years) for a red dwarf to reach a quieter phase.[4,5] Before the flaring is reduced, a fledgling planet could be born, develop an atmosphere, and see it stripped away. So there is the issue of the likelihood that a planet orbiting a red dwarf can retain its atmosphere until its star quiets down.

In 2017 the Kepler space telescope watched Trappist-1 for 80 days and saw it flare 42 times. One of those flares was as strong as Earth's September 1-2, 1859, *Carrington Event*,[9] when a powerful solar coronal mass ejection hit Earth's magnetosphere. Ice core samples have revealed the Carrington Event to be twice the size of any other solar storm in the last 500 years.[11] The storm lit up skies all over the world with bright auroras. It completely disrupted telegraph lines. According to a 2008 report from the National Academy of Sciences, if such an event were to occur today it could cause "extensive social and economic disruptions" impacting power grids, satellite communications, and GPS systems.[11]

Of course, a planet could be born beyond the habitable zone, at a safer distance from the star's flares, and then later migrate into the habitable zone.

It is interesting to note that, while on Earth water ice *reflects* the shorter wave-lengths of sunlight, cooling the planet, on a planet around a red dwarf, the longer wavelengths would be *absorbed*, warming the ice.[5]

Another danger is the emission of high levels of X-rays and ultraviolet radiation. Such radiation can contribute over time to the loss of surface water on a planet. The good news is that a thick atmosphere can shield the surface from such radiation. Also, liquid water absorbs UV radiation and thus protects life just below its surface. Just three feet (one meter) of water would provide this protection.[1]

In speculating on the possibility of life on such worlds it is easy to overlook some important factors. With future telescopes, including the James Webb, we will learn more about the atmospheres, rotation rates, and other critical information. Stay tuned.

II. Proxima b—closest exoplanet—only 4.2 light years distant.

The Alpha Centauri triple-star system (discussed in Topic 17) consists of a binary system of two Sun-like stars, Alpha Centauri A and B, and the red dwarf Proxima. The distance between Alpha Centauri and Proxima varies between 4,300 AU and 13,000 AU, with the current separation being near maximum. Proxima's orbital period (around the center of gravity of the system) is about 547,000 years.[17] Alpha Centauri A and B orbit each other in about 80 years.

How bright would Alpha Centauri A (which is slightly brighter than Alpha Centauri B) be as seen from Proxima? The answer is about magnitude -6.68. Venus' magnitude variance as seen from Earth is -3.7 to -4.5, which means Alpha Centauri A as seen from Proxima is about 7.45 times as bright as Venus at its maximum as seen from Earth. In other words, brilliant. See *Appendix 30.2* for the calculation.

Of the three stars in this system, Proxima is the closest to Earth, 4.244 light years. The magnitude of Proxima Centauri as seen from Earth is a faint 11.1. First-magnitude Alpha Centauri is 4.37 light years from Earth.

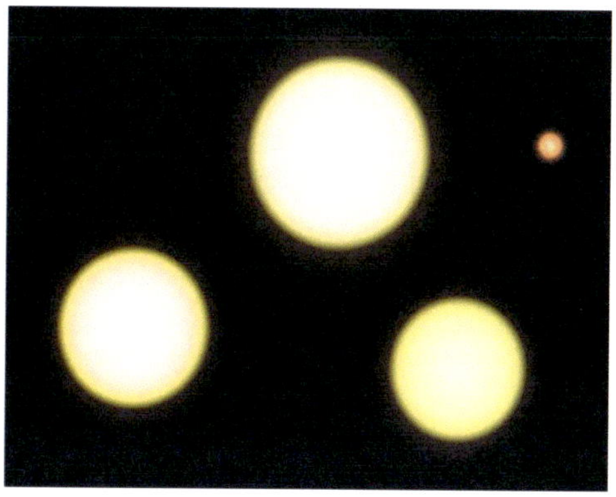

Figure 3. From left to right, the relative size of the Sun, α Centauri A, α Centauri B, and Proxima Centauri. Credit: Wikimedia Commons/RJHall (CC BY-SA 3.0)

At the time of this writing there are two *confirmed* planets in orbit around Proxima: Proxima b and Proxima c. Proxima b was discovered in August 2016 and Proxima c in April 2019.

The discovery in 2016 that the closest star to Earth had a planet certainly gained media attention. Proxima b is Earth-sized and is in its star's habitable zone. Its mass is at least 1.17 times that of Earth. If its density is the same Earth's, then its radius would have to be at least 1.05 times Earth's radius. It circles Proxima in 11.19 days, and, being tidally locked to Proxima, this is also the time it takes to rotate once on its axis. At 0.049 AU from its star, it is only about 13 percent of the distance of Mercury from our Sun.[17]

Proxima c has a mass about seven times that of Earth. It is about 1.49 AU from its star, far outside the habitable zone. It circles Proxima in about 5.3 years. Being very cold, it is unlikely to support life. But there is evidence that it has a large ring system.[17]

What about the chances of life on Proxima b? A big problem here is that Proxima, like so many red dwarfs, is a flare star. And Proxima seems to be quite active in that regard. A 2019 flare from Proxima was 100 times more powerful than any flare observed from our Sun, according to data from five telescopes.[18] That is certainly a negative. But consider Ross 128b, the next exoplanet in this topic. It orbits another red dwarf, Ross 128, that is very close to Earth. This exoplanet is in the habitable zone, and its star has a much quieter flare history.

III. Ross 128b—only 11 light years distant.

The red dwarf star Ross 128 was mentioned in Topic 17. As was stated there, currently it is the 12th-closest system to us. In about 80,000 years it will become the closest star to the Sun at about 6.2 light years. Now 10.9 light years away in Virgo, Ross 128 is magnitude 11.1; its mass is about 15 percent that of the Sun. It has high proper motion. And it has a planet, Ross 128b, orbiting it that potentially could support life. At present, Ross 128 is the second-closest star with a known exoplanet, after Proxima, to our solar system.

Ross 128b was discovered on November 15, 2017, at the inner edge of its star's habitable zone. Its mass is at least 1.35 times that of Earth, and it orbits its star in 9.87 days. It was located using the radial velocity method, not the transit method. As such, its radius, hence density, is not known, and it is not known if it has an atmosphere. But from other evidence its radius is thought to be about 10 percent greater than Earth's. You can bet Ross 128b will be a target of the James Webb space telescope.

In 2018, based on near-infrared, high-resolution spectra, astronomers determined the chemical abundances of several elements (C, O, Mg, Al, K, Ca, Ti, and Fe) present in the star Ross 128. They further concluded that its metallicity is similar to that of our Sun and that Ross 128b contains a mixture of rock and iron. By measuring the temperature of Ross 128 and estimating the radius of the planet, it was determined how much of the star's light should be reflecting off the planet's surface. From these calculations, they concluded that the planet has a temperate climate.[14,15]

Ross 128b, like Proxima b and perhaps all of the Trappist-1 planets, is tidally locked to its star. Furthermore, each of the stars is a red dwarf, with flares that can disrupt any planetary life. But...good news. Unlike Proxima and Trappist-1, the older star Ross 128 doesn't flare as often as its younger counterparts.[16]

IV. Kepler 64b.

In Topic 28, we discussed a sextuple star system (TYC 7037-89-1) consisting of three binaries. As was noted, in the first *Star Wars* movie where Luke Skywalker sees two suns in the sky on the fictional planet Tatooine, on any planet in the B binary system of this six-star system one would see two suns and in addition four very bright stars. Whether any planets exist in this system is unknown.

But the Kepler 64 system, consisting of two binary systems gravitationally bound, does have a large planet, Kepler 64b (also called PH1), in orbit around one of the binaries. The more remote binary is about 1,000 AU away. Even at that range, though, this second pair would be bright enough to be seen during the day, and at night the combined stars would shine about as bright as a full moon.[12] This is the first reported case of a planet orbiting a double-star that, in turn, is orbited by a second distant pair of stars. Kepler 64b was discovered October 15, 2012.

The Kepler 64 quadruple is several thousand light years away in the constellation Cygnus. The components of the first binary pair are called Aa and Ab; those of the second, more remote, pair are called Ba and Bb. Aa and Ab constitute an eclipsing binary, and they orbit a common center of gravity in 20 days. The planet Kepler 64b orbits this pair in about 138 days. Star Aa is a spectral type F star, hence it is brighter than our Sun (of type G). It has a radius about 1.7 times that of the Sun and a mass about 1.5 times that of the Sun. Star Ab is much smaller and dimmer. It is of type F, with a radius about 0.38 of the Sun and a mass about 0.4 of the Sun. The stars Ba and Bb are likely of type G and M, and are separated by about 60 AU. Based on stellar evolution models of Aa and Ab, the age of the planetary system is estimated to be about two billion years.[13]

The planet itself is about half the size of Jupiter and has about half its mass. Thus it is either a gas giant or, being only .65 astronomical units from the Aa/Ab pair—about the distance of Venus from the Sun—its surface is probably quite hot.[13]

Figure 4. Artist's illustration of Kepler 64b or PH1, a planet discovered by volunteers from the Planet Hunters citizen science project. Credit: Haven Giguere/Yale

References

1. Trefil, James, & Summers, Michael (2019). *Imagined life: a speculative scientific journey among the exoplanets in search of intelligent aliens, ice creatures, and supergravity animals.* Smithsonian Books. p 157-161.
2. NASA/JPL. (February 2018). *TRAPPIST-l planet data.* spitzer.caltech.edu https://www.spitzer.caltech.edu/image/ssc2018-04c-trappist-1-planet-data-feb-2018
3. NASA. (February 22, 2017). *Largest batch of Earth-size, habitable zone planets.* exoplanets.nasa.gov https://exoplanets.nasa.gov/trappist1/
4. Basri, Gibor (September 18, 2013). *The Kepler mission to detect Earth-like planets.* [Lecture]. Sonoma Astronomy Club, Proctor Terrace School, Santa Rosa, CA.
5. Palubski, Igor, & Shields, Aomaw (August 2019). Red dwarf habitability recipe. *Sky & Telescope,* p 34-40.
6. Gilster, Paul (February 11, 2015). *Overcoming tidal lock around lower mass stars.* centauri.dreams.org https://www.centauri-dreams.org/2015/02/11/overcoming-tidal-lock-around-low-mass-stars/
7. Leconte, Jeremy, et al. (February 6, 2015). Asynchronous rotation of Earth-mass planets in the habitable zone of lower-mass stars. *Science, 347*(6222), p 632-635. https://science.sciencemag.org/content/347/6222/632
8. Crockett, Christopher (February 7, 2015). Life might like cool stars' rocky planets. *Science News, 187*(3), p 8.
9. Grossman, Lisa (December 23, 2017 and January 6, 2018). Seven Earth-sized planets orbit the same ultracool star. *Science News,* p 25. Also see Grossman, Lisa (December 13, 2017). Year in review, Seven Earth-sized planets orbit the same ultracool star. *Science News*, https://www.sciencenews.org/article/trappist-1-earth-sized-planets-top-science-stories-2017-yir
10. Carlisle, Camille M. (June 2017). 7 Earth-size planets orbit dim star. *Sky & Telescope,* p 12-14.
11. Klein, Christropher (March 14, 2012, updated August 22, 2018). *A Perfect Solar Superstorm: The 1859 Carrington Event.* history.com https://www.history.com/news/a-perfect-solar-superstorm-the-1859-carrington-event.
12. Benningfield, Damond (May 7, 2021). *Tatooine-plus. May 7, 2021.* https://stardate.org/radio/program/2021-05-07
13. Schwamb, Megan E., et. al. (April 23, 2013). Planet hunters: a transiting circumbinary planet in a quadruple star system. *Astrophysical Journal, 768*(2), p 127. https://iopscience.iop.org/article/10.1088/0004-637X/768/2/127
14. Souto, Diogo, et al. (June 13, 2018). Stellar and planetary characterization of the Ross 128 exoplanetary system from APOGEE spectra. *Astrophysical Journal Letters, 860*(1). https://iopscience.iop.org/article/10.3847/2041-8213/aac896
15. American Association for the Advancement of Science. (July 10, 2018). *Rocky planet neighbor looks familiar, but is not Earth's twin—detailed chemical abundances of Ross 128 help us understand its exoplanet Ross 128b.* eurekalert.org. https://www.eurekalert.org/pub_releases/2018-07/cifs-rpn071018.php
16. Bochanski, John (March 2018). Ross 128b orbits quiet nearby star. *Sky & Telescope,* p 11.
17. Staff. (August 30, 2020). *Proxima Centauri.* star-facts.com https://www.star-facts.com/proxima-centauri/
18. Hot Bytes (August 2021). Record flash. *Astronomy Magazine,* p 7.

There are so many stars shining in the sky,
so many beautiful things winking at you,
but when Venus comes out, all the others
are waned, they are pushed to the background

—Mehmet Murat ildan, Turkish playwright

TOPIC 31. Venus vs Earth and some interesting facts about our "former twin."

Venus is now a planet of extremes. The surface temperature is about 870° F (466° C), hot enough to melt lead (which melts at 621.5° F or 327.5° C). The atmospheric pressure at ground level is 90 times that at sea level on Earth. This is approximately the same as the water pressure on Earth 3,050 feet (well over half a mile) under the ocean (see *Appendix 31.1* for this calculation). The planet rotates on its axis once in 243 Earth days, yet it has winds at high elevation that circle the globe in only four Earth days (a phenomenon called super-rotation). But close to the ground the wind speed is only about 10 kilometers per hour. Venusian clouds are composed mainly of sulfuric acid droplets.

And something else is rather strange: In 1995, radar reflecting off of the Venusian surface (which is opaque at visible-light wavelengths) from the Magellan spacecraft showed a brightening at higher elevations. It is thought that some volcanos and mountain tops are coated with a bright substance. And it certainly isn't snow or ice. Two suggestions are that it might be galena,[5] which is lead sulfide (PbS), or bismuth sulfide (Bi_2S_3). The element tellurium (element 52 in the periodic table which is a silver-white metalloid in crystalline form) had earlier been suggested. Galena is the primary ore for obtaining lead. The idea is that these substances would be present in trace amounts as vapor in the atmosphere (either from volcanic eruptions or vaporization at the hotter, lower elevations), rise up in the atmosphere, and then condense at higher, cooler elevations, coating the mountain tops. Is this what happens? Sherlock Holmes might say that we need more data.

In a PBS TV series,[6] astronomers discussed how Venus and the Earth were born similar, diverged, and will become similar again. They presented the following sequence of events:

A large, planet-sized object hit Venus reversing the direction it rotates. And the new, reversed rotation was slow. Thus Venus lost its dynamo and so had no magnetic field. Much granite is found on Venus, and granite requires water to form. So Venus once had abundant water. But the Sun got hotter over time, and the water became water vapor, a greenhouse gas. Because there was no magnetic field, the solar wind stripped away the water vapor, reducing its presence. But CO_2 increased because of volcanoes. So Venus became very hot and acquired a thick CO_2 atmosphere. Bismuth and lead melted and vaporized. Then the vaporized metals rained down on the mountain tops. The atmospheric pressure became 90 times that on Earth.

As the Sun gets hotter, Earth will heat up and produce more water vapor. But Earth's magnetic field will prevent the solar wind from stripping it away. The Earth will thus become hotter than Venus, and the surface pressure will become greater than that on Venus. The atmospheric pressure will become 270 times that of today and the temperature will reach 2,600° F. Earth will be a greater hell than Venus is today. If this suggested scenario is accurate, then perhaps we should enjoy going to the beach while we can. It should be mentioned that the claim that Venus contains much granite (and thus in the past must have had liquid water) is not settled.

The Mayans and the planet Venus.
The Pre-Columbian Mayan civilization of Mesoamerica realized some impressive accomplishments, including developing astronomical methods. They were aided in this by having a writing system. Currently some 6,000 languages are spoken in the world (about 1,000 of them on the island of New Guinea[4]), but worldwide, there exist only about 50 writing systems. Modern humans have been around about 150,000 years, but writing has been with us only about 6,000 years. Writing systems originated in different parts of the world. But starting from scratch is a difficult task, and very few writing systems developed completely independent of systems already in use. Most were inspired by, borrowed from, and/or adapted from others nearby. Two indisputably independent inventions of writing were achieved by the Sumerians of Mesopotamia somewhat before 3000 BCE and by the Maya in Mesoamerica before 600 BCE. Egyptian writing of 3000 BCE and Chinese writing by 1300 BCE may also have arisen independently.[1] The Maya also developed a positional numeral system (based on the number 20 and to some extent the number five) and a symbol for the number 0, again completely indigenous to Mesoamerica. They also used a sophisticated calendar to mark the passage of time.

The Maya were quite interested in the movement of the planet Venus and carefully tracked it. Their value for the synodic period of Venus (which equals the time between successive appearances of Venus as the evening star) was 583.9207 days, compared to the modern value of 583.93 (and compared to Ptolemy's value of 583.94267). They also estimated the lunar (synodic) month more accurately than Ptolemy[2] and the solar (synodic or tropical) year more accurately than did the Spanish when they first arrived in the Americas.[3]

References

1. Diamond, Jared (1999). *Guns, Germs, and Steel.* W. W. Norton & Co. p 217-219.
2. Finley, Michael John (n.d.), *Mayan astronomy.*
http://ircamera.as.arizona.edu/NatSci102/NatSci102/text/extmayaastronomy.htm
3. Leon-Portilla, Miguel (September 1, 1990). *Time and reality in the thought of the Maya.* University of Oklahoma Press. ISBN 9780806123080
https://books.google.com/books?id=wjFOQ8U4qKcC&dq=maya+length+tropical+year+spanish.
4. Diamond, Jared (1999). *Guns, Germs, and Steel.* W. W. Norton & Co. p 27.
5. Otten, Carolyn Jones (February 10, 2004). *'Heavy metal' snow on Venus is lead sulfide.* Newsroom. Washington University in St. Louis. https://source.wustl.edu/2004/02/heavy-metal-snow-on-venus-is-lead-sulfide/
6. How the Universe Works, PBS TV series season 4 (2015), episode 2, "Earth, Venus's Evil Twin."

The world is little more than a Rorschach blot in which we see our own desired systems projected.

—Prem Prakash, Indian actor & producer

TOPIC 32. The Dark Side—the Great Emu, the Coal Sack, the Dark Doodad, the Field of Nothing, and the Ink Spot.

All of the following are dark clouds of dust visually surrounded by stars that frame them. Though they look like holes in the sky, hidden behind the dust clouds are stars whose light is blocked from reaching us.

1. The Great Emu.

While Greek astronomy focused on star patterns, Australian aboriginal astronomy took notice of dark patches between stars in the Milky Way. An outstanding example is the great Emu in the sky.

Figure 1. The Emu, endemic to Australia, is the second-tallest living bird after its relative, the ostrich.

Figure 2. The Great Emu in the night sky over Orford, Tasmania. Credit: James Garlick and Australian Broadcasting News

Figure 3. Image on a T-shirt given in 2007 to participants in the Deepest South Texas Star Safari—a star party in Coonabarabran, Australia, organized by Texans and Australians.

The Great Emu runs from the constellation Crux (the Southern Cross) down (i.e., north!) through the pointers (a line from Alpha through Beta Centauri points to Crux) to the main body in Scorpius and then into Sagittarius. It is the dark dust clouds in the Milky Way that constitute the Emu. The dark region called the Coalsack (see below), which is in Crux, marks the head of the Emu.

2. The Coal Sack.

The Coal Sack covers nearly seven degrees by five degrees of the sky. The Southern Cross looks more like a diamond as there is no star marking the intersection of the vertical and horizontal pieces of a cross. Crux itself has the smallest area of the 88 constellations. Alpha Crucis is the bottom star in the

Figure 4. The Coal Sack to the lower left of the Southern Cross. Credit: ESO/S. Brunier

diamond, Beta Crucis is the left star in the "cross piece," Gamma is at the top, and Delta is on the right. Between Delta and Alpha is the fainter star, Epsilon.

The flag of New Zealand shows the four stars in Crux that form the diamond. The Australian flag shows six stars: five in Crux (including Epsilon) and, below the union jack, the first-magnitude star Beta Centauri. Whether Epsilon appears on the left or right side of the diamond in the Australian flag depends on which side of the flag you are looking at.

Figure 5. The Australian (left) and New Zealand flags, both depicting the Southern Cross.

3. The Dark Doodad.

Figure 6. The Dark Doodad's vertical streak with globular cluster NGC 4372 at its southern end. Credit: CC-BY-SA-3.0, author: Naskies

4. The Field of Nothing.
The *Field of Nothing* is the colorful name given to the dark cloud *Bernes 157* (Be 157) in the constellation Corona Australis. It is virtually devoid of stars and is about a degree in length, quite long. Other interesting objects are visible to its northwest (moving up and to the right in Figure 7).

Figure 7. The Field of Nothing or Bernes 157 is the dark cloud extending from the middle of the image down and to the left. Credit: Marco Lorenzi, lorenzi70@gmail.com (slightly cropped by author)

Just at its NW boundary is NGC 6729, a tiny, comet-shaped emission and reflection nebula—an object which, like the more famous Hubble Variable Nebula in Monoceros, changes unpredictably in brightness, size, and shape. In its "head" is the variable star R CrA, which fluctuates erratically between magnitudes 9.5 and 13. A fainter variable, T CrA, lies in the "tail."

Just northwest of NGC 6729 is NGC 6726/27, a reflection nebula (the upper blue region) about nine by seven arc minutes in size and containing two bright stars about 1.2 arc minutes apart. One is magnitude 7.3 and the other a variable with a magnitude range of 9.4-9.8.

Just to the southwest (diagonally down and to the right) of NGC 6726/27 is the double star BSO 14 (also known as Brs 14). Its two blue-white stars of magnitudes 6.6 and 6.8 are separated by 12.2 arc seconds—not visible in the image as a double but easily split with a small telescope. BSO 14 is embedded in a faint nebulosity, the reflection nebula IC 4812 (the lower blue region).

In the upper right corner, in the neighboring constellation Sagittarius, is the *Chandlier Globular Cluster,* NGC 6723. It is seventh magnitude and about 11 arc minutes in diameter. A visual treat.

5. The Ink Spot, Barnard 86, embedded in the Milky Way in Sagittarius.
Barnard 86 (B86), which resembles an ink drop spilled onto the Milky Way, and the open cluster NGC 6520 (on its left in Figure 8) are almost the same distance from Earth, about 6,200 light years distant.

Figure 8. The Ink Spot, Barnard 86, with the open cluster NGC 6520 on its left. Image is from the Wide Field Imager on the MPG/ESO 2.2-meter telescope at ESO's La Silla Observatory in Chile. Credit:4.0 International (CC BY 4.0)

On the right edge of B86 is a bright (magnitude 6.7) orange star about 4,300 light years distant.[1] B86 is a cloud of mainly hydrogen gas and thick dust, the dust blocking the light from stars behind it. It is thought that gas in this cloud collapsed to form the stars in the open cluster. Such a relatively small, dust-thick gas cloud from which stars are formed is called a *Bok globule*. Bart Bok (1906-1983) was a Dutch-American astronomy professor at Harvard University who was director of the Harvard Observatory and who discovered such objects in the 1940s. His wife, Priscilla Fairfield, worked closely with him on research and is the co-author of several of his papers.

References

1. The Set of Identifications, Measurements, and Bibliography for Astronomical Data (SIMBAD). http://simbad.u-strasbg.fr/simbad/. SIMBAD is an astronomical database of objects beyond the solar system. SIMBAD gives a parallax for this star (HD 164562) of .7663 milli-arc seconds. Using the formula D = 1/P for the distance of the star in parsecs when P is in arc seconds gives 1,310 parsecs or about 4,300 light years.

Zeus, the father of the Olympic Gods,
turned mid-day into night,
hiding the light of the dazzling Sun;
and sore fear came upon men.

—*Archilochus, Greek poet*
Referring to the total solar eclipse of April 6, 648 BCE

TOPIC 33. Total solar eclipses, the saros period, and saros series.

We pose a question: What is the time period, called the *saros period (or cycle)*, between successive identical total solar eclipses?

To answer we must first clarify what we mean by "identical eclipses." We mean that the Moon and the Sun have the same apparent sizes and the same respective distances from Earth at the two eclipses, and that the duration of the eclipses is the same. We do not require that the eclipses can be seen from the same points on the Earth, i.e., that the center lines of the eclipses are the same. Four conditions (given below) are needed.

For a total solar eclipse to occur, the Earth, Moon, and Sun must be in a line in that order. The Moon's orbit is not circular, but elliptical. Currently it is inclined at an *average* of 5.15 degrees to the Earth's orbital plane (with a variation from 4.99 to 5.3 degrees over less than a year).[4] The two points where the Moon's orbit around the Earth and the Earth's orbit around the Sun cross each other are called the *lunar nodes*. The *ascending node* is where the Moon crosses to the north of Earth's orbit. The *descending node* is where it crosses to the south. The line joining these nodes is called the *nodal line*. The Moon must be at one of these nodes for a solar eclipse to occur. The time it takes for the Moon to go from one node back to the same node is 27.21222 days. It is called the *nodal period* (also called the *draconic or draconitic period*).

1. *The saros period must be (close to) an integer multiple of the nodal period.*

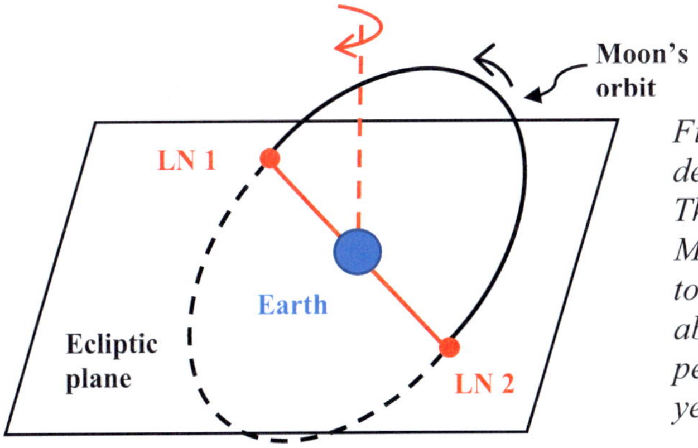

Figure 1. The Lunar Orbit. LN 1 is the descending node; LN 2 is the ascending node. The sold, red line is the nodal line. The Moon's orbital plane is inclined 5.15 degrees to the ecliptic plane. The nodal line rotates about the red, dashed line (which is perpendicular to the eclipitic plane) in 18.6 years—see nodal precession below.

A solar eclipse can only occur at new moon (i.e., when the Moon is between the Earth and the Sun). The period of time between new moons (or between any specific phase of the Moon back to that same phase) is called the *synodic period*, which is 29.53059 days. If the Moon has gone from say, lunar node 1 back to lunar node 1, in order to have the Earth, Moon, and Sun in a line (i.e., have the Sun behind the Moon as seen from Earth) it must be new moon.

2. *The saros period must also be (close to) an integer multiple of the synodic period.*

Note that the synodic period is not the same as the *sidereal period* (which is the Moon's orbital period around the Earth with respect to the fixed stars). This is due to the fact that while the Moon makes one revolution about the Earth with respect to the stars, the Earth has traveled a small distance along its orbit around the Sun (about 1/12 of the way around). So the Moon has to travel a little farther around the Earth to have the Earth, Moon, and Sun line up (i.e., be at new moon).

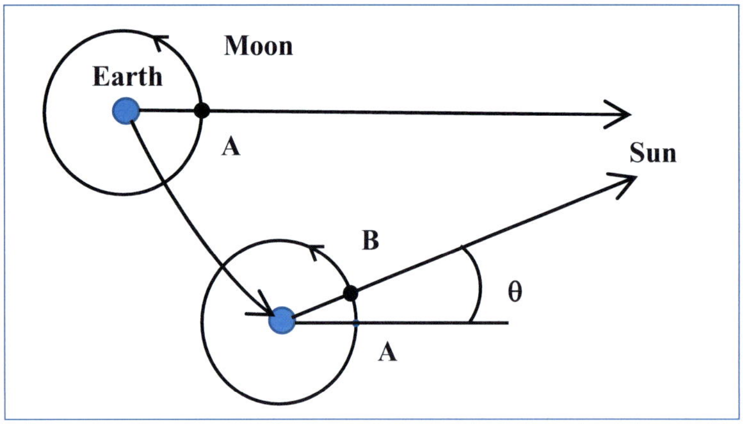

Figure 2. The Moon must revolve by angle 360 + θ degrees (from A to A and then to B) to go from new moon to new moon. So the synodic period is the sidereal period plus the extra time it takes for the Moon to revolve θ degrees. Distances are not to scale.

We want the Earth-Moon distance to be the same at successive identical eclipses, so that the Moon appears to be the same size. The *anomalistic period*, which is 27.55455 days, is defined as the period between successive *perigees*, or closest approaches of the Earth and Moon. This is the same as the time for the Moon to return from any fixed point on its orbit (relative to the perigee point) back to the same point. To have the Moon again have the same size as seen from Earth, condition 3 below is required.

3. *The saros period must be (close to) an integer multiple of the anomalistic period.*

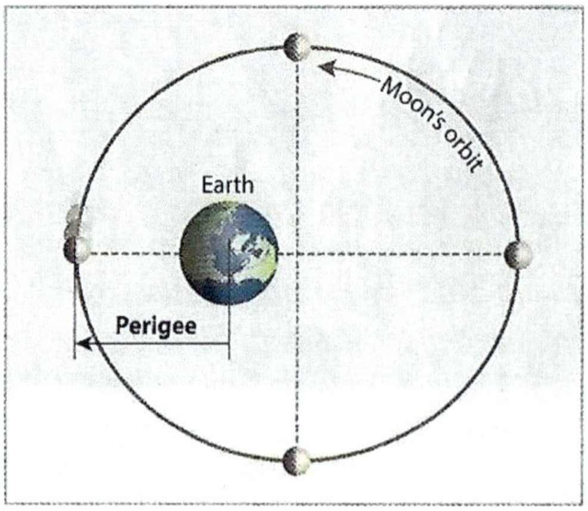

Figure 3. Perigee (illustration from reference [3]).

At this point you might ask, "Why isn't the anomalistic period the same as the nodal period?" It is because the nodal line actually precesses around the ecliptic. Put another way, the nodal line, which is always in the ecliptic plane, rotates about an axis perpendicular to the ecliptic plane at the Earth. This is called *nodal precession*. One complete rotation of the nodal line takes about 18.6 years and projects a yearly arc of approximately 19.4 degrees in the westward direction, opposite to the eastward movement of the Moon in its orbit around Earth. While the angle between the Moon's orbital plane and the ecliptic plane is currently always close to 5.15 degrees, the angular elevation of the Moon itself about the ecliptic varies from -5.15 to +5.15 degrees over a lunar month. Every 18.6 years, the Moon's declination (the angle between the Moon's orbital plane and the celestial equator) reaches a maximum of 23.45 + 5.15 = 28.6 degrees. This is called a *major lunar standstill.* Around this time, the Moon's declination will vary from −28.6 to +28.6 degrees in one lunar month. Then, 9.3 years later, the Moon's declination reaches its minimum of 23.45 - 5.15 = 18.3 degrees. This is called a *minor lunar standstill.* Around this time, the Moon's declination will vary from −18.3 to +18.3 degrees in one lunar month. For a satellite in low Earth orbit, nodal precession of the satellite's orbit is due to torque on the satellite produced by the Earth's equatorial bulge, which causes a precession of the satellite's orbital plane about Earth's rotational axis. The reasons for the Moon's nodal precession are more complicated, but the Earth's equatorial bulge plays a role.

We need to know about one other period of the Moon, its *sidereal period.* This is the orbital period of the Moon with respect to the background stars and is 27.32166 days.

4. *We want the saros period to be (close to) an integer multiple of the Moon's sidereal period.*

But why is this desired? With the Sun, Moon, and Earth in a line, having the Moon against the same background stars as before means the Sun is also against the same background stars as before, and so the Earth is at the same point in its orbit as before. So the axial tilt of the Earth and thus the season is the same as before, and the Earth-Sun distance is the same. Thus the size of the Sun doesn't vary from one eclipse to the eclipse one saros period later. This and condition 3 above mean the durations of the two eclipses are the same. It is true, however, that in fact the size of the Sun as seen from Earth doesn't vary much over a year as the Earth's orbit is almost a circle. Actually, most articles (e.g., [3]) discussing the saros period don't include requirement (4), i.e., don't consider the Moon's sidereal period in determining the saros period—they only consider the other three periods.

Finally, we note that the sidereal period is not the same as the anomalistic period due to another complicating detail, namely the fact that the axis joining the perigee and apogee points on the Moon's orbit (called the major axis of the ellipse) itself rotates with respect to the Earth (and the fixed stars). This rotation is eastward, and one rotation takes about 8.85 years. The perigee and apogee points are called the *apsides,* and the rotation is called *apsidal precession (or perihelion precession)*. Do not confuse nodal precession with apsidal precession or with the Moon's axial precession (which is the lunar counterpart to the Earth's axial precession).

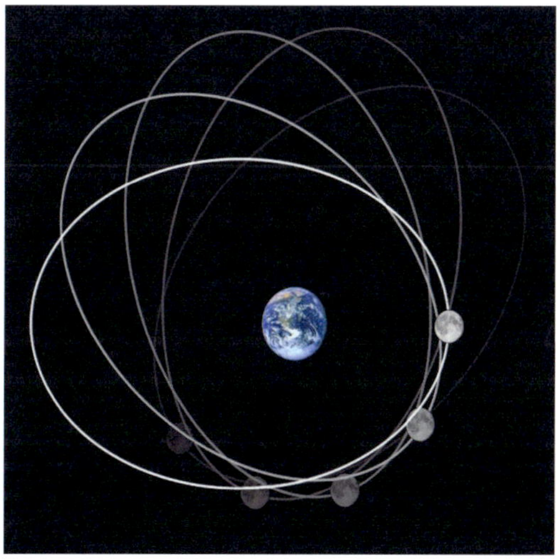

Figure 4. Moon's apsidal (or perihelion) precession. Credit: Rfassbind https://commons.wikimedia.org/wiki/File:Moon_apsidal_precession.png

From the above we therefore need to find the smallest number that is (close to) an integer multiple of all four periods. This will be the saros period. In other words, we would like to find integers a, b, c, d such that if

AP = the anomalistic period, NP = the nodal period, SP = the sidereal period, and SNP = the synodic period, then

$$a \times AP = b \times NP = c \times SP = d \times SNP ,$$

with the requirement that this choice of integers makes this common number as small as possible. We then would have the saros period as this common number. Note that each of

$$239 \times AP,\ \ 242 \times NP,\ \ 241 \times SP, \text{ and } 223 \times SNP$$

is a close approximation to 6,585.3211, which is the saros period accurate to four decimal places.

For those interested, the four products respectively are 6,585.5375, 6,585.3572, 6,584.5201, and 6,585.3216. Not equal, but close. So the saros period is close to an integer multiple of each of these periods. (And, as stated above, often the sidereal period is not considered, as the size of the Sun doesn't vary that much, and that corresponds to leaving out the product 6,584.5201, the number that is farthest from the actual saros period.)

Let's try to say a little about the relationship between the locations on the surface of the Earth where one can see total solar eclipses separated by one saros period. A saros period is about 11 days longer than 18 years. The second eclipse will occur at a different place on the Earth: The saros period is not an integer. It differs from being an integer by 0.3211 days, which is seven hours, 42 minutes, and 23 seconds. So the Earth will have rotated for a period of time that much longer, causing it to be that much later in the day when the second eclipse occurs. So every spot on the Earth at which the first eclipse was visible will shift 115.6 degrees to the west. Another way to express this is to say that due to nodal precession westward, the Sun is farther west at the next eclipse.

There is also a northward (respectively southward) shift of the path of totality about 300 kilometers on average when the Moon is at its descending (respectively ascending) node. This is due to the fact that the Moon's orbit is tilted 5.15 degrees to the ecliptic, so the descending node (which through nodal precession moves in the opposite direction to the Moon's orbital motion) will be farther north at the next eclipse. Likewise, the ascending node will be farther south at the next eclipse at that node.

The exeligmos cycle.
But there is more. After three saros periods, a period of 54 years and 33 days, the points of visibility shift 3 x 115.6 degrees which is just 13.2 degrees less than a complete circle of 360 degrees. So the second eclipse will occur at about the same location within less than one hour of the same time of day (as 13.2 degrees is less the 15 degrees that the Earth rotates in one hour) as the eclipse of about 54 years earlier. This three-saros interval (19,755.96 days) is known as a *triple saros or exeligmos* (Greek: turn of the wheel) cycle.

Central eclipses and their three types: total, annular, and hybrid.
In the above we have only considered total eclipses. Now we will have a discussion where we include two other types of solar eclipses, namely annular and hybrid eclipses. These three types of eclipses are

all *central eclipses*, meaning that they produce a (narrow) track on the surface of the Earth. If one is inside this track the eclipse can be seen. For viewers outside this track the eclipse is only partial.

An annular eclipse is an eclipse in which the Moon is too far from the Earth to completely cover the face of the Sun, but from the eclipse track on the surface of the Earth one sees the Moon surrounded by a blazing ring of light—the Moon is completely inside the face of the Sun and a ring of sunlight surrounds it.

A hybrid eclipse is one in which the eclipse alternates between annular and total along different sections of its path on the Earth's surface. In most cases, hybrid eclipses begin as annular, transform into total, and then revert to annular before the end of their track. In rare instances, a hybrid eclipse may begin annular and end total, or vice versa.

Saros series.
A *saros series* is a sequence of eclipses, each separated from the next by one saros period (about 18 years). There are about 238 total solar eclipses a century, so approximately 42 occur during one 18-year saros period.[3] None of these 42 are identical (as per our definition of identical) as they are ipso facto not separated by an integer multiple of a saros period. Thus each of these approximately 42 total eclipses is in a different saros series. A saros series need not involve only total eclipses. Let us explain what is going on.

Every saros series begins with a sequence of partial eclipses near one of Earth's polar regions. To be specific, let us say the saros series begins near the north pole. The sequence of partial eclipses is followed by a sequence of central eclipses, the track of each succeeding central eclipse moving south by about 300 kilometers. After the sequence of central eclipses, and near the end of the saros period, is a sequence of partial eclipses near the south pole. The partial eclipses at the beginning and at the end of each saros series occur when the Sun is on the horizon. Due to the ellipticity of the orbits of the Earth and the Moon, the exact number of eclipses in a Saros series is not constant and neither is the duration (i.e., how long the series lasts). A series may last from 1,226 to 1,550 years and is comprised of 69 to 87 eclipses, of which about 40 to 60 are central. There exists a numbering system for the saros series, but we will not discuss it here.

Examples of saros series.[1]
In the following, T, A, H, and P stand respectively for total, annular, hybrid, and partial.

Saros series 137:
Began May 25, 1389; ends June 28, 2633. Duration: 1,244.08 years. 70 eclipses.
Sequence: 8P, 10T, 6H, 4A, 3H, 32A, 7P
(i.e., the sequence started with eight partial eclipses, was followed by 10 total eclipses, then six hybrid eclipses, four annular eclipses...and ends with seven partial eclipses).

Saros series 151:
Began August 14, 1776; ends October 1, 3056. Duration: 1,280.1 years. 72 eclipses.
Sequence: 18P, 6A, 1H, 39T, 8P

Saros series 136 (see Figure 5 below)**:**
Began June 14, 1360; ends July 30, 2622. Duration: 1,262.11 years. 71 eclipses.
Sequence: 8P, 6A, 6H, 44T, 7P
Saros series 136 is currently producing the longest total solar eclipses of the 20th and 21st centuries.

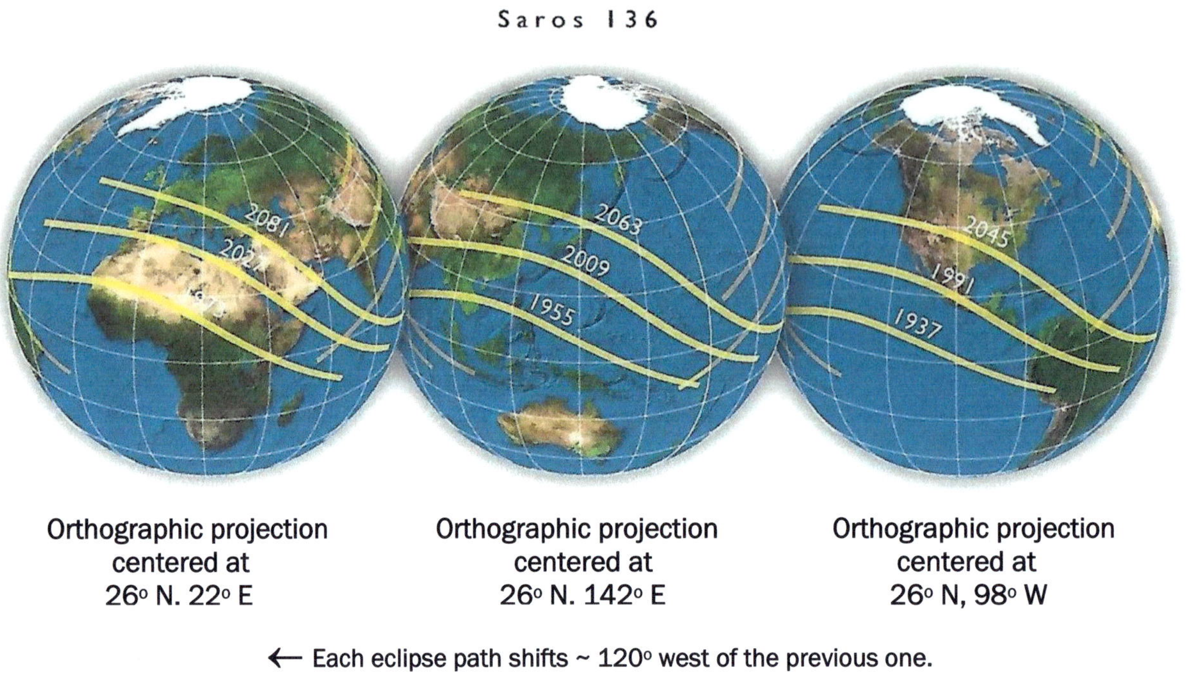

Figure 5. Path 1937 (right globe) moves left by ~120 degrees and northward to become path 1955 (middle globe) which does the same to become path 1973 (left globe), which then becomes path 1991(right globe), etc. to path 2081(left globe). Credits: Map illustration by Michael Zeiler, Paths of totality from eclipse calculator by Xavier Jubier, Eclipse predictions by Fred Espenak, NASA Goddard Space Flight Center. See [2]. (Text enlarged and bolded by author)

References

1. Espenak, Fred. *Predictions for Solar and Lunar Eclipses, Catalog of Solar Eclipses of Saros Series—including Saros Series 33 to 190.* EclipseWise.com. http://www.eclipsewise.com/solar/SEsaros/SEsaroscat.html. The following website also includes a glossary of solar eclipse terms. http://www.eclipsewise.com/solar/SEhelp/SEglossary.html.

2. *Illustration of Saros Series 136.* https://eclipse.gsfc.nasa.gov/SEsaros/image/Saros136-large.pdf http://eclipse.gsfc.nasa.gov/SEsaros/SEperiodicity.html

3. Bakich, Michael E. (August 2016). Decipher the eclipse pattern. *Astronomy Magazine,* p 58-59. The third figure is from this excellent article on the subject of Topic 33.

4. Meeus, Jean (2017). *Mathematical Astronomy Morsels.* Willmann-Bell, p 12.

Time and tide wait for no man.

—*Chaucer c. 1395*

TOPIC 34. How does the Moon raise tides on (both sides of) the Earth? The role of the Sun. Some factors affecting the heights of high and low tides, including spring tides and neap tides (which occur seven days after spring tides).

I. The tidal effect of the Moon (the role of the Sun is considered in section II).

The gravitational attraction of the Moon at the point on the Earth's surface that is directly below the Moon is upward toward the Moon. On the opposite side of the Earth from this point the Moon's gravitational attraction is also toward the Moon (although somewhat less in strength as this point is about 8,000 miles farther from the Moon) and thus downward toward Earth's center. See Figure 1. This causes some people to say that while they can understand why there is an upward bulging of the ocean at the point on Earth that is opposite the Moon, they don't see why this also is true at the antipode of this point. Yet water does bulge upward at this antipode. Let us examine what is really going on.

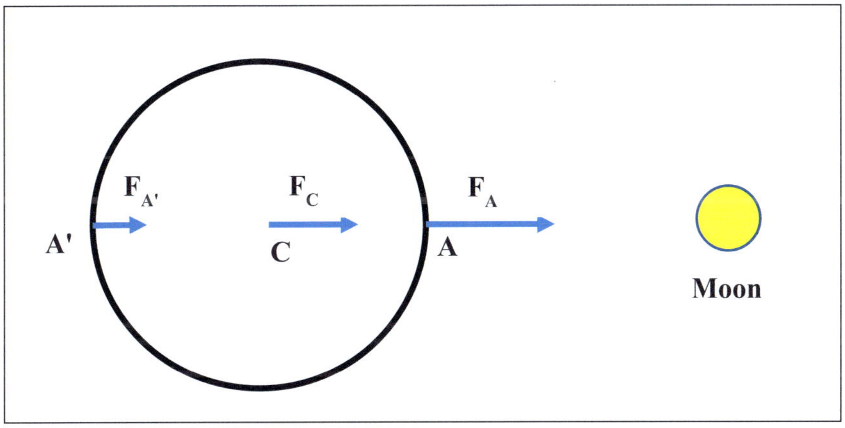

Figure 1. The Moon's gravitational force at point A, at its antipode A', and at the Earth's center C.

The easiest way to look at this is to first imagine that the Earth is perfectly spherical and consists entirely of water. The point opposite the Moon, call it A, is closer to the Moon than the center of the Earth, call this C, and that is closer to the Moon than the antipode of A, call this A'. Thus the gravitational force of the Moon at A is greater than at C which is greater than at A'. Force vectors F_A, F_C, and $F_{A'}$, respectively at A, C, and A' are indicated in Figure 1.

The *key idea* here is that we are really interested in the *differential or relative force vectors, with the force vector at C being set to zero*. That is, our frame of reference is the Earth, with a zero force at C. So if the force vectors at A, C, and A' are respectively F_A, F_C, and $F_{A'}$, we wish to consider the relative force vectors, F_1, F_0, and F'_1, called *tidal forces*, defined by

$$F_1 = F_A - F_C, \qquad F_0 = F_C - F_C = 0, \qquad \text{and} \qquad F'_1 = F_{A'} - F_C. \qquad (1)$$

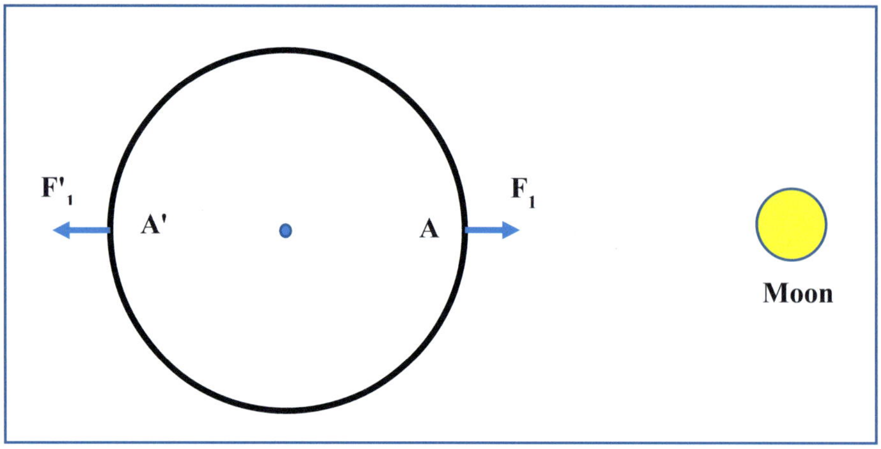

Figure 2. The tidal forces at point A and its antipode A' (relative to a tidal force of zero at the center of the Earth).

With the positive direction being toward the Moon (to the right in our figures) and the negative direction to the left, we see that F_1 points to the right (as expected) and F'_1 points to the *left* since F'_1 is negative (as $F_{A'} < F_C$). So we get a bulge not only on the right, but also on the left. While the gravitation force at A (and A') is inversely proportional to the *square* of the distance from the (center of the) Moon, the tidal force is actually inversely proportional to the *cube* of the distance. The equation for the (magnitude of the) tidal force at A (and A') is the following:

$$F_1 = (2GMm/R^3)r\ , \qquad (2)$$

where M is the mass of the Earth, m is the mass of the Moon, R is the distance between the centers of the Earth and the Moon, and r is the radius of the Earth. See *Appendix 34.1* for a proof. The above equation corresponds to equation (3) in the Appendix.

Tidal forces F_1, F'_1, and the Moon's attraction at all the other surface and internal points on the spherical body of water cause the spherical shape to change to that of a prolate ellipsoid as seen in Figure 3.

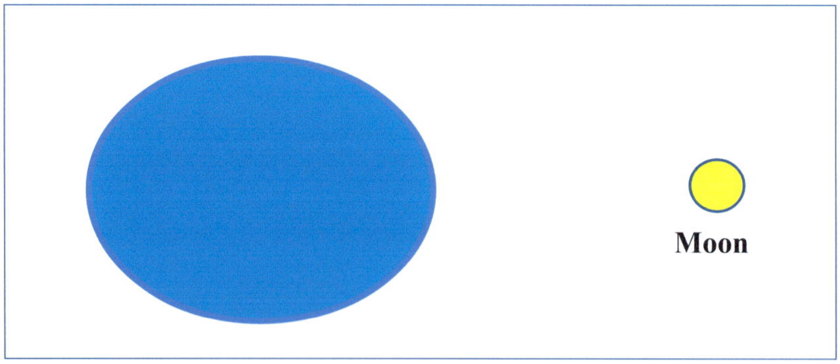

Figure 3. The spherical ball of water becomes a prolate ellipsoid due to the Moon's gravitational force at all points.

Now consider the Earth as a solid sphere covered by water of uniform thickness. Both the solid Earth and its water are stretched into a prolate ellipsoid, but the Earth, being much more rigid, is not stretched very much. It's a different story for the water. We obtain Figure 4.

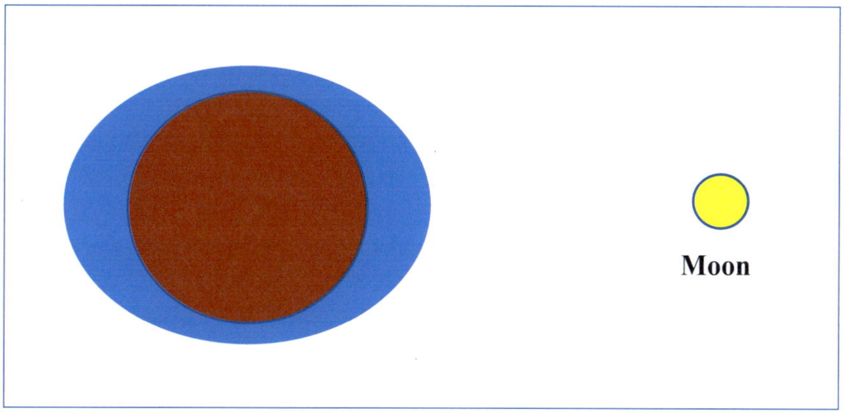

Figure 4. The spherical Earth with an ellipsoidal shell of surface water.

It is to be noted that the tidal force vector at any point on the Earth's surface can be broken into a tangential component and a vertical component. The actual movement of water on the surface of the Earth, due to the Moon's gravitation, is tangential to the Earth's surface. So it is the tangential component of the tidal vector at a point on the surface that determines the direction of water's flow at that point. Notice that this component is zero at the two vertical points in Figure 5. (And these are not the two poles of the Earth as the orbit of the Moon is tilted with respect to the Earth's equatorial plane. In fact this tilt varies in 18.6 years from 18.3 degrees to 28.6 degrees and back to 18.3 degrees. See nodal precession in Topic 33.)

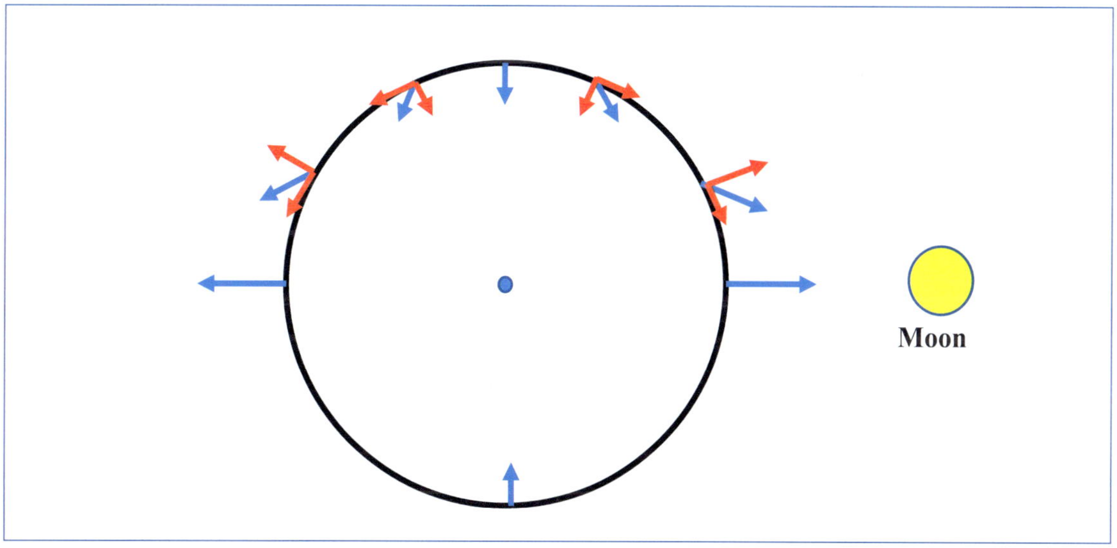

Figure 5. Tidal forces (blue arrows) are indicated at a few points with their vertical and tangential components indicated with red arrows.

The Earth is not a perfect sphere—its equatorial diameter is somewhat larger than its polar diameter due to centrifugal force from its rotation. But here we are only interested in the principal issues, not secondary ones.

Here is an interesting fact:[2] Any two objects which have the same density and, from the viewpoint of an Earthly observer, *appear* to have the same diameter (regardless of their actual diameter) produce the same tidal forces on the Earth (assuming both are far away compared to the diameter of the Earth). This situation is true regardless of how much farther away one is from the Earth than the other. Why? With a fixed perceived diameter of a body, its actual diameter increases in direct proportion to its distance from Earth. This means its volume increases as the cube of its distance. And since its mass is its density times its volume, its mass must also increase as the cube of its distance. But the tidal force decreases as the cube of the body's distance—see equation (2) above. So the two effects cancel each other out and the tidal force remains the same.

Of course our Moon and the Sun have different densities, so this doesn't apply to them. But one can say that since the Sun and Moon do have the same perceived diameters (1/2 degree) in the sky, and the Moon has a much higher density, the Moon must have the larger tidal effect.

II. The role of the Sun in raising tides. Some factors affecting the heights of high and low tides, including spring and neap tides.

The Sun has about 46 percent of the effect of the Moon on raising tides on Earth. See *Appendix 34.1* for a proof. To put it another way, the Sun's force is responsible for almost a *third* of the total tidal rise. (If the Moon's contribution is 1 and the Sun's is .46, then the total contribution is 1.46, and .46/1.46 = .315, almost one third).

We now will look at some factors affecting the heights of high and low tides. We will only consider factors *other* than the location and shape of land masses. So for our purposes consider the Earth to be completely covered by water.

When the Sun and Moon are in a line with Earth, which is at every new moon and full moon, the combined tidal forces of the Sun and Moon reinforce each other producing higher high tides and lower low tides. Such tides are called *spring tides*.

When the Sun and the Moon are 90 degrees apart, which is every first- and third-quarter Moon, the tidal forces of the two bodies interfere with each other producing lower high tides and higher low tides. Such tides are called *neap tides*.

During a 24-hour high-tide period, there are two high tides, but one of them is lower than the other. Why?

The short answer is because the Earth is tilted on its axis from the plane of the ecliptic (the orbital plane of the Earth around the Sun) and (to a smaller extent) because the Moon can be anywhere from -5.15 to +5.15 degrees above the ecliptic plane. The Earth's tilt is 23.5 degrees from a perpendicular to the ecliptic plane. Let's look at the details.

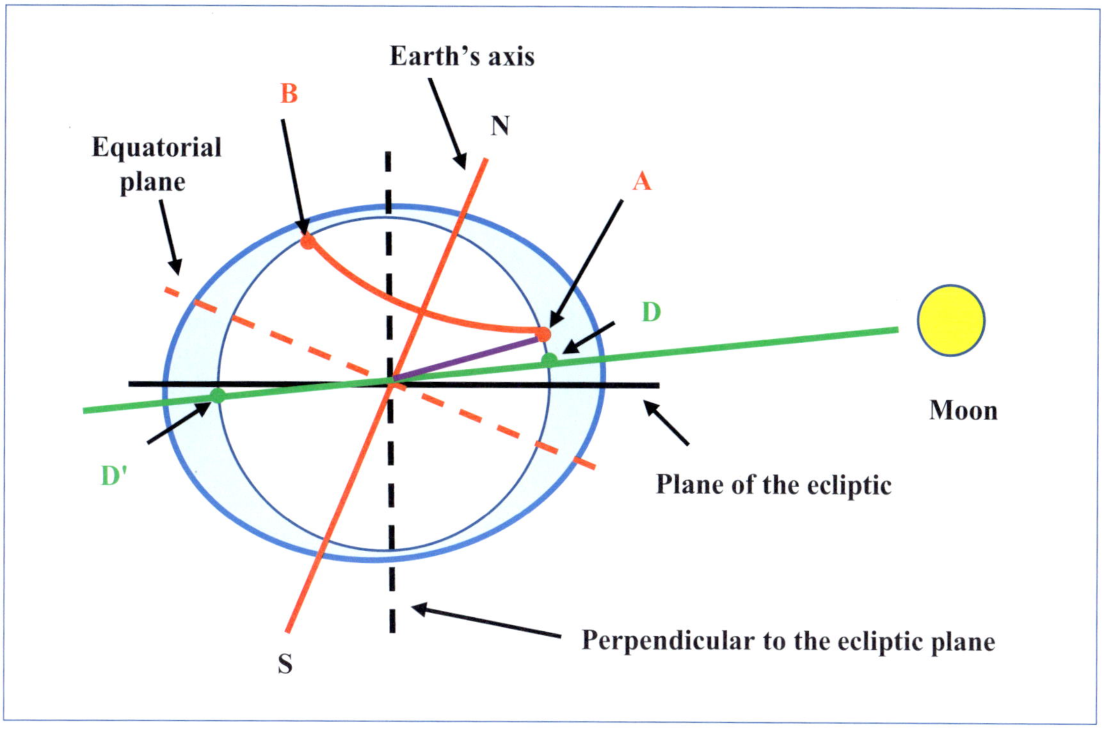

Figure 6. The high tide for a person at location A is higher than the high tide for that same person approximately 12 hours later when he/she has rotated to location B. For purposes of illustration we assume the Moon is four degrees above the ecliptic plane (it can be anywhere from -5.15 to +5.15 degrees). So four degrees is the angle between the green line and the horizontal black line. (Equivalently, the Moon's declination is 23.45 + 4 = 27.45 degrees.) Note: While the green line is in the Moon's orbital plane, the orbital plane is NOT perpendicular to our cross section of the Earth, i.e., to the page in this book.

The tilt of the Earth's axis to the ecliptic, which is the angle between the dashed black line and the solid red line, is 23.5 degrees. We assume the Moon is four degrees above the ecliptic plane, i.e., four degrees is the angle between the green line and the horizontal black line.

Suppose you are at point A in Figure 6 whose latitude is 40 degrees N. The solid red arc joining A and B in Figure 6 is part of the 40-degree circle of latitude. Point D on the surface of the Earth is directly opposite the Moon, so the tidal bulges are at D and its antipode D'. D can be shown to be 12.5 degrees south of A. About 12 hours later, because of the rotation of the Earth, you are now located at point B (halfway around the globe) in the figure. (Since the Earth rotates CCW as viewed from above the north pole of the Earth, you actually rotate from A to B "behind the Earth" in Figure 6, but that is irrelevant here.) The Moon is almost in the same position as when you were at A, so the bulges are still approximately at D and D'. B can be shown to be 67.5 degrees from D' (and B is 112.5 degrees from D). *So A is only 12.5 degrees from the bulge while B is 67.5 degrees from the bulge.* (See *Appendix 34.2* for a detailed argument showing this.) *Thus the high tide at B will be lower than at A.*

Of course, rather than the first high tide being the highest, if you had initially been at point B instead of A (both being 40 degrees latitude), then it would be the second high tide that would be the highest.

If the Earth's axis were not tilted and the Moon's orbital plane were the same as the elliptical plane, then A and B would both be 40 degrees from their respective tidal bulges, so the tidal heights would be the same. So it's the tilt of the Earth's axis and the Moon's declination (or equivalently, its angle above or below the ecliptic plane) that makes the two tidal heights different in an approximately 24-hour period.

In the above situation you might be concerned about the fact that during the 12 hours that the Earth rotated, the Moon has been moving. It is easy to see that the Moon will have moved about 6.6 degrees in its orbit around the Earth during these 12 hours. So it will actually take about 12.44 hours rather than 12 hours before the second high tide occurs. See *Appendix 34.3* for a proof.

The Earth's orbit around the Sun and the Moon's orbit around the Earth are not circles, but ellipses. So the distance of each object from the Earth varies. The Moon's gravitational attraction is strongest at *perigee*, its closest point to Earth, and weakest at *apogee*, its farthest point from Earth. Similarly the Sun's gravitational attraction is strongest at *perihelion*, its closest point to Earth, and weakest at *aphelion*, its farthest point from Earth. Aphelion is about July 2, and perihelion is about January 2.

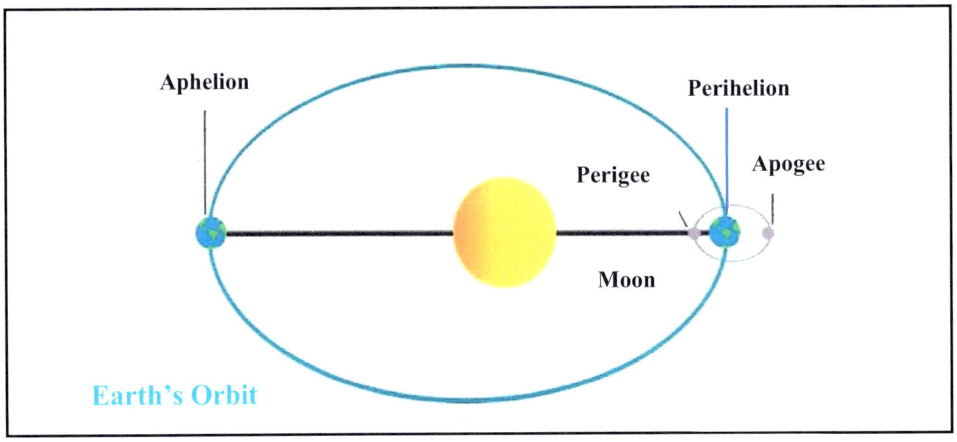

Figure 7. Important points on the Earth's and Moon's orbits. Credit: NASA

From the above, one can see that height of the tide will be greatest (a king tide) when there is a confluence of the following events:

1. The Moon is at perigee.

2. It is about January 2 (so the Sun is at perihelion).

3. It is full moon or new moon (so the Earth, Moon, and Sun are in a line)—so it is a spring tide. (A spring tide with the Moon at perigee is called a *perigean spring tide*.)

Finally, the 18.6-year lunar nodal precession has a large effect on the Moon's declination (its angle below or above the equator). Changes in the Moon's declination, in turn, have a large effect on the character of lunar tidal forcing. With rising sea levels this effect can contribute to flooding in coastal areas.

References

1. Barger, V., & Olssen, M. (1973). *Classical Mechanics: a modern perspective*; Section 7.2. The tides. p 265-274. McGraw-Hill. (Referenced in the Appendix.)
2. Editors. (March 2011). A tidal secret. Q&A section, *Sky & Telescope*, p 71. This fact was pointed out in an answer to a question by Richard Kilgallen.

...[most] people don't get to see 100 billion tonnes of seawater flowing back and forth in front of their kitchen windows every day.

—Terri, Bay of Fundy blogger

TOPIC 35. Tides continued: The Bay of Fundy.

In Topic 34 we discussed the basic reason that there are tides on the Earth. We ignored local effects due to land masses. Here we mention a particularly striking local effect.

The highest tides in the world are at the Bay of Fundy which is between Nova Scotia and New Brunswick in Canada. In the open Atlantic Ocean the tidal variation (the height between low and high tides) is about two meters. At the mouth of the Bay of Fundy, the variance is six to 10 meters, and farther up the Bay it is about 15 meters (over 50 feet). The highest average variance in the Bay is about 17 meters at Burntcoat Head, on the shore of the Minas Basin (see map). The highest tide was set in the Minas Basin during the tropical cyclone of October 4-5, 1869, called the Saxby Gale. That height was 21.6 meters (70.9 feet).

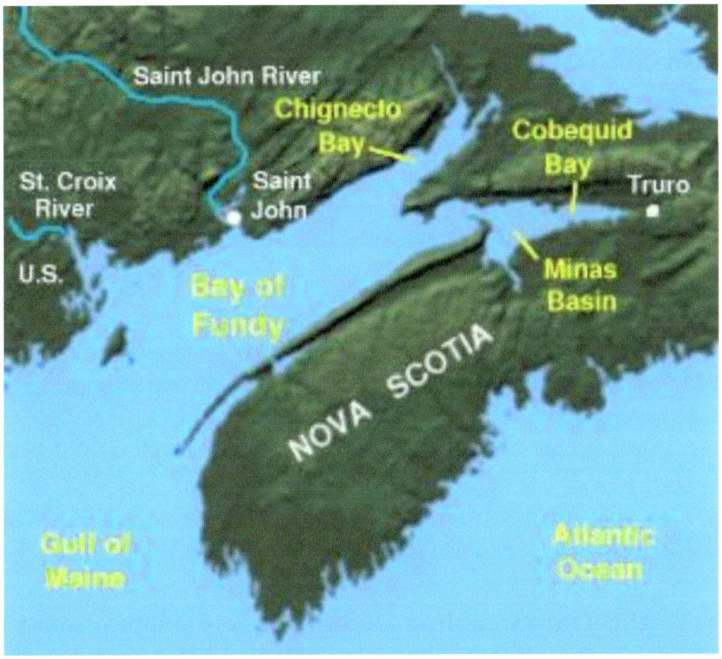

Figure 1. Bay of Fundy. Credit: Decumanus at en.wikipedia, CC BY-SA 3.0

What is the cause of such high tides at the Bay of Fundy? Looking at a map one is struck by how the bay narrows as one proceeds into it. So there is a natural funneling of water to a greater depth due to this narrowing. But this is actually only a minor reason for the high tides. The major reason is what is called a *resonance*. Let us elaborate. The time for tidal water to move from the entrance of the bay, reach the far end, and then move back to the entrance is about 12.4 hours. That number should seem familiar from Section II of Topic 34. It is the time between successive high tides. So when the water returns to the entrance to the Bay it is met by water moving inward because of the next high tide. It is this coincidence of timing that is the main reason for the unusually high tides in the bay. If the bay were a different shape, say, longer, wider, or narrower, the tide's one-way travel time would not almost exactly coincide with half the time between high tides, and tidal water returning to the ocean would not meet head on the next tide's water entering the bay. A good analogy is that of a child on a swing. Every time the swing moves back, the child's parent pushes on the swing at just the appropriate moment to give the swing an extra kick along its path forward. The timing of the push is critical.

The Bay of Fundy is about 270 kilometers long. It takes about six hours for the tide to go from its entrance to the far end. That means the average rate of flow of the tide is about 45 kilometers an hour or 28 miles per hour. Would you like to try to outrun it?

We might say a little more about the Saxby Gale of 1869 which produced a record-setting high tide in the Minas Basin (see above map) in the Bay of Fundy. In addition to the alignment of the Moon and Sun with respect to the Earth, what factors contribute to a tide as high as 71 feet? One ingredient is low pressure over a body of water, such as occurs during a storm. Low pressure allows the ocean water to rise. Also wind can push the water in a given direction. The two combine to give us what is called a *storm surge*.

Nine thousand to 10,000 years ago the Moon was closer to the Earth and the tides were up to four times higher.[1] And native Americans were living in the area. Can you imagine seeing monthly tides 200 feet high? At that time the sea level was lower and the continental shelf was exposed. This meant that bays and channels were more steeply sided, resulting in more funneling of the tidal water.[1]

With the Moon exerting such a strong tidal effect on the Earth, just think about the tidal effect of the Earth on the Moon. While the Moon doesn't have any body of water, the stresses in the interior of the Moon caused by the gravitational tug of the Earth result in some of the Moon quakes that have been recorded by seismometers astronauts have left on the Moon.[1,2]

Another interesting phenomenon occurs in rivers and streams flowing into the Bay of Fundy. It is called a *tidal bore*.

A tidal bore is a wavefront of the incoming tide that moves upstream in a river. It is common for it to have a height of two to three meters. So one could surf against the flow of the river.

In the Bay of Fundy, the Salmon River at Truro is one of a number of places to observe a tidal bore (as the author has done). It is striking to be next to a dry creek bed leading into the river and then see a small wall of water surging up the creek and in about 10 minutes see the water's depth reach about five feet.

Tidal bores occur at several places in the world. One example is Morecambe Bay, a large estuary in NW England, just to the south of the Lake District National Park. (See Figure 2 below.)

Figure 2. Tidal Bore in Morecambe Bay in the UK.
Credit: Arnold Price/Tidal Bore/CC BY-SA 2.0

Some other tidal bore sites[1]:
(a) Ungava Bay in northern Quebec. Here (in an area called Leaf Basin) the maximum tidal range is 16.8 meters (55.1 feet).

(b) The Qiantang tidal bore on China's Fuchun River. This is the largest tidal bore in the world. Here the wave can move at about 40 kilometers/hour (25 mph) and reach up to nine meters (30 feet) high. Dangerous.

(c) The River Severn's estuary, along the west coast of England, which empties into the Bristol Channel. A 15-meter (49 feet) tidal range produces a wave which can carry surfers up to 10 kilometers (6.2 miles) inland, making the river a famous surfing site.

(d) The Pororoca tidal bore on the Amazon River and its estuaries in Brazil. Here the bore can run 13 kilometers (eight miles) inland; the surfing record is a 36-minute ride.

References
1. Wysession, Michael. The world's greatest geological wonders: 36 spectacular sites. Great Courses lecture series. thegreatcourses.com. (Lecture 23 of 36 highly interesting lectures in this course.)
2. Klesman, Alison (December 2019). The Apollo missions left seismometers on the Moon. How long did they work? What conclusions can we make about moonquakes? Ask Astro, *Astronomy Magazine*, p 69.

Synchronicity: A meaningful coincidence of two or more events where something other than the probability of chance is involved.

—Carl Jung

TOPIC 36. What causes the Moon to be tidally locked to the Earth, i.e., have its axial rotation period the same as its orbital period, and thus always have the same side facing the Earth?

It is because the Earth pulls the Moon slightly out of round, i.e., it creates slight tidal bulges on the facing and opposite sides of the Moon. The bulges on the Moon cause internal friction that slows the Moon's rotation until it spins at the same rate that it orbits the Earth, in which case there is no internal friction.

People don't notice whether it's winter or summer when they're happy.

—Anton Chekhov

TOPIC 37. Why is the northern summer on average (93.65 days[1]) longer than the average northern winter (88.99 days[1])?

First remember that it is the tilt of the Earth's axis that gives us the seasons, not how far away the Sun is. In fact, in the northern hemisphere, the Earth is closer to the Sun in winter than in summer.

The reason for the different averages is that the Earth's orbit is not a circle with the Sun at the center, but an ellipse with the Sun at one focus. The ellipse doesn't depart much from being a circle. Technically the *eccentricity* of an ellipse (the distance between the center to either focus divided by the length of the semi-major axis) is a measure of how squashed the ellipse is, with a circle having eccentricity 0, and a straight line (a degenerate ellipse) having eccentricity 1. The Earth's orbit has an eccentricity of .0167, so it is almost a circle, but not quite. The Earth travels faster in its orbit when closer to the Sun, and in northern winter the Earth is close to *perihelion* (the closest point in the Earth's orbit to the Sun, about 91.5 million miles). So the Earth stays close to the Sun for less time than it stays far from the Sun (*aphelion* being the farthest point from the Sun, about 94.5 million miles). Thus the northern hemisphere is tilted away from the Sun (i.e., winter time) for less time than it is tilted toward the Sun (i.e., summer time). This gives us the different average times for the two seasons.

References

1. Gleason, Edward (June 2001). Ask Astro, *Astronomy Magazine,* p 81.

The Moon continues to spin away from the Earth at the rate of 3.78cm (1.48in) per year... about the same speed at which our fingernails grow.

—BBC

TOPIC 38. The Moon's distance from Earth is increasing over time. Why? And what is happening with some of the other moons in the solar system? E.g., the moons Phobos and Triton are moving closer to their planets Mars and Neptune respectively and in the future will break apart and form rings around them.

In general, tidal bulges should occur on opposite sides of a planet's surface and be in a line with its moon. But this is not what actually takes place. Friction between the rock under the two tidal bulges on the planet and the bulges' material (water in the case of the Earth) will cause the bulges to slightly move in the forward direction of the planet's rotation and thus slightly *ahead* of the two (opposite) spots on the surface of the planet directly below the moon. This is illustrated in Figures 1 and 2 below. (In case 3a below, the bulges are slightly *behind* the two spots on the surface. See Figure 3.)

Case 1. A moon orbits its planet *slower* than the planet rotates on its axis. Such is the situation with the Earth (one day) and the Moon (27.6 days—sidereal time). We consider two subcases, 1a and 1b.

1a. The moon's orbital direction and the planet's axial rotation direction are the *same* (as is the case with the Earth and the Moon). This is called a *prograde orbit*.

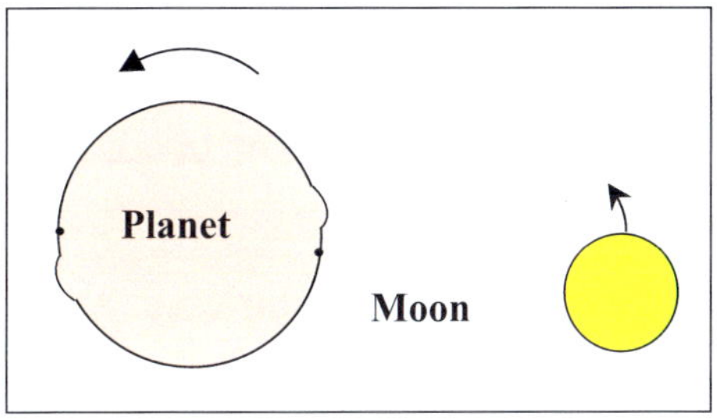

Figure 1. Prograde orbit for a moon (looking down from above). The two black dots on the surface of the planet are directly "below" the moon, meaning they are in a line with the (center of the) moon.

In the situation of the Earth there is friction between the ocean floors and the bulges consisting of water, and this gradually slows the Earth's axial rotation. In addition, the moon exerts a gravitational torque on the planet's bulge, also slowing the planet's axial rotation. Nine hundred million years ago our day was only about 18 hours long. In general, the bulges on the planet need not be of water. As the planet's spin slows, its axial angular momentum decreases. But by conservation of angular momentum, the moon must increase its orbital angular momentum, thus increasing its distance from the planet. How is this accomplished? The planet's bulge exerts a gravitational torque on the moon, increasing the moon's orbital velocity and hence its angular momentum. By the way, the increased velocity of the moon decreases as the moon recedes from the planet. This continues until the moon reaches a distance from the planet where Kepler's third law holds, namely that the square of the period of the moon in its orbit around the planet is a constant times the cube of the moon's average distance from the planet (where the constant is independent of which of the planet's moons is being discussed).

Note that here we are considering only the total angular momentum of the planet-moon system *about the planet's axis*, which equals the sum of the angular momentum of the rotation of the planet on its axis plus the (component of the) orbital angular momentum of the moon about the planet's axis. The rotation of the moon about its own axis plays no role in the angular momentum currently being considered.

1b. Here the moon's orbital direction is *opposite* that of the planet's axial rotation, which is not the case for our moon, but is the case for Triton (see below). This is called a *retrograde orbit*.

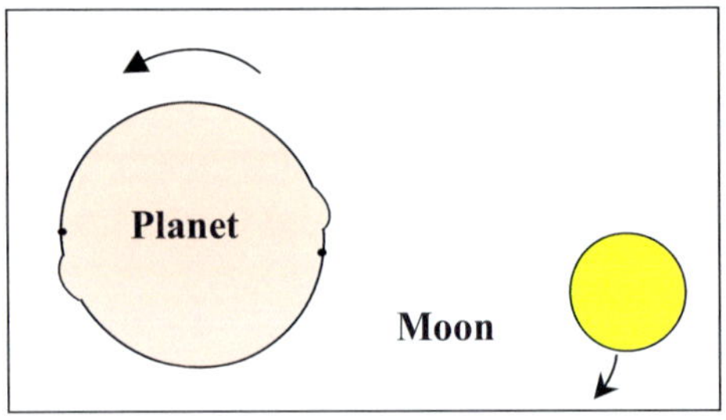

Figure 2. Retrograde orbit for a moon (looking down from above).

The tidal bulge is moving around the planet in the opposite direction to the planet's rotation. Friction and the moon's torque on the planet's tidal bulge gradually slow the planet's rotation, decreasing the planet's axial angular momentum. As before, the bulges are slightly *ahead* of the two (opposite) spots on the surface of the planet directly below the moon. But now these bulges will exert torque on the moon so as to *decrease* the moon's orbital speed. Since the moon's orbital speed decreases, the moon moves closer to the planet.

At this point you might be puzzled, and ask: Conservation of angular momentum requires the moon's orbital angular momentum to increase. How can this increase if the moon's orbital speed decreases? The answer is that angular momentum also has a direction. In the case of a retrograde orbit, the angular momentums of the planet and its moon have *opposite* signs. So if the angular momentum of the planet is positive, the angular momentum of the moon is negative. And a decrease in the *magnitude* of a negative number (the moon's angular momentum) results in a *larger* number. So the moon's angular momentum does increase, and we have no violation of the law of conservation of angular momentum.

Case 2. Both the planet and the moon are tidally locked to each other. In this case the distance of the moon will not change. This is the case for Pluto and its large moon Charon.

Case 3. The moon's orbital revolution about the planet is *faster* than the axial rotation of the planet. We give arguments similar to those for Case 1 to establish Case 3.

3a. The moon has a *prograde orbit* about its planet. Then the bulge on the planet rotates around the planet faster than the planet rotates on its axis. Friction causes the rotation of the planet to increase. Also, friction causes the bulges to lag *behind* the two (opposite) points on the planet below the fast-moving moon.

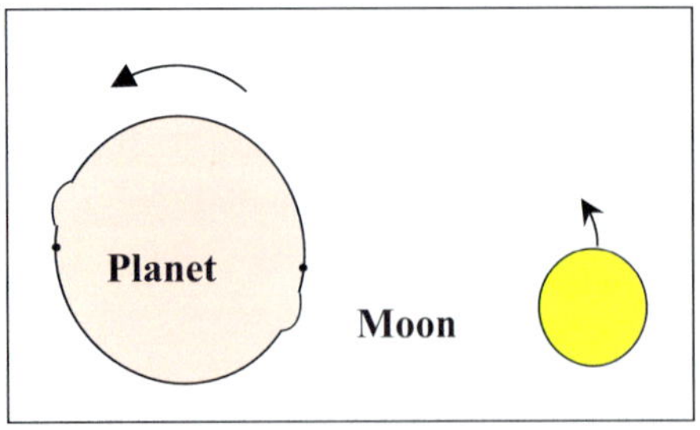

Figure 3. Prograde orbit for a moon (looking down from above). Here the moon orbits the planet faster than the planet rotates on its axis.

Hence the moon's torque on the bulge also increases the rotation rate of the planet. So the angular momentum of the planet increases. Because of conservation of angular momentum, the moon's orbital angular momentum, and hence its speed, decreases. The moon will thus move closer to the planet. The decrease in the angular momentum of the moon is caused by torque on the moon due to the planet's bulge.

3b. The moon has a *retrograde orbit* about its planet. The result is the same as in case 1b. The argument is identical to that given for case 1b, with Figure 2 needing no changes.

Summary.
Let M stand for the moon, and P for the planet.

1. If M's orbit is retrograde, then regardless of whether M orbits P slower or faster than P rotates on its axis, P will rotate slower and M will move closer.

2. Assume M's orbit is prograde. Then if M orbits P slower than P rotates on its axis, P will rotate slower and M will recede. If M orbits P faster than P rotates on its axis, then the opposite is true, namely P will rotate faster and M will move closer.

If a moon approaches its planet, it will either eventually impact the planet or will break apart because of tidal effects of the planet on the moon.

See *Appendix 38.1* for some other mathematical results, in particular that the tangential velocity of the Moon in its orbit about the Earth is inversely proportional to the square root of its distance from the (center of the) Earth.

Further remarks about our Moon.
The outward movement of our Moon means its orbital period around the Earth increases over time. Since the Moon is tidally locked to the Earth, this means the Moon will spin more slowly on its axis. Also, as the Moon recedes from the Earth, the tidal bulges it creates on Earth are diminished, and so the Moon's rate of recession decreases over time.

Will the Moon ever stop moving away from the Earth? The answer is yes. When the Earth's rotation rate is slowed to the point that the Earth is tidally locked to the Moon (so the Earth will present the same face toward the Moon), the distance between the Earth and the Moon will stop changing. But that would be in about 50 billion years[1]—and the universe is only 13.7 billion years old.

It is thought (but it is not known for sure) that when the Moon was formed in a collision of a Mars-sized object with the Earth (which is the leading theory) that the Moon initially was between 15,000 and 20,000 miles from Earth. Today it is 238,000 miles away. How much larger would the Moon appear in the sky at that time than it appears today? The answer is about 12 times larger in diameter. (See *Appendix 38.2* for the easy proof.) So instead of the half-degree size of the Moon today, its size at the earlier time would have been about six degrees. That would be some sight—even though conditions on Earth then would not have been pleasant for us humans.

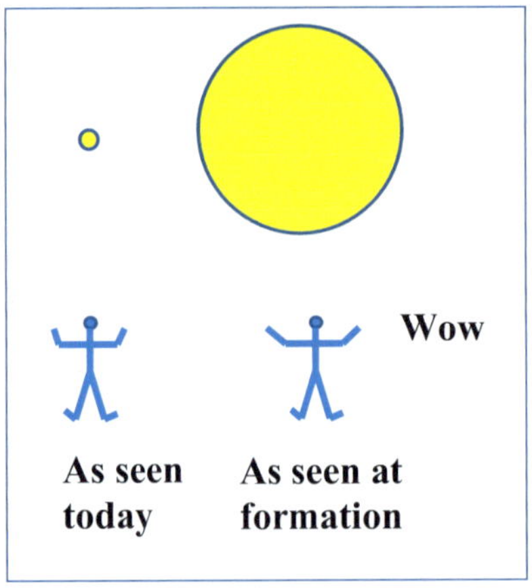

Figure 4. Comparative sizes of the Moon as seen from Earth today and when it formed.

For a discussion on tides see Topics 34 and 35 above.

Let's look at some other examples, using the above summary.

Phobos.
Phobos revolves around Mars faster than Mars rotates on its axis and the motion is prograde. Thus Phobos is getting closer to Mars.

Deimos.
Deimos, the smaller moon of Mars, takes longer than a Martian day to revolve in a prograde orbit around Mars. Thus, like our moon, it is receding from Mars over time.

A possible reason Venus and Mercury currently have no moons.
If Venus had a moon which orbited it (in a prograde orbit) faster than Venus' very slow axial rotation (243 days), the moon would approach Venus. Maybe Venus had a moon long ago, but has no moon now because its very slow axial rotation rate resulted in the moon moving closer to Venus and breaking up. Exactly the same thing can be said about Mercury, as its axial rotation period is 59 days—quite slow.

Triton.
Triton is by far the largest moon of Neptune, with a diameter of about 78 percent of our moon. It is the seventh-largest moon in the solar system and is slightly larger than Pluto. Triton is the only large moon in a retrograde orbit, though some of the smaller moons in the solar system do have retrograde orbits.

Triton is believed to be a Kuiper belt object (the largest example being Pluto) that was captured by Neptune's gravity when it approached too close. The retrograde orbit supports this contention because as such Triton could not have formed from orbiting material around Neptune.

Triton takes about 141 hours to orbit Neptune, and Neptune takes about 16 hours to rotate on its axis. Because of its retrograde orbit Triton is moving closer to Neptune. Currently Triton is closer to Neptune than the Earth-Moon distance. In millions of years it will break apart to create a ring around Neptune.

References
1. Klesman, Alison (December 2019). The Moon is slowly moving away from Earth. When will the Moon leave Earth's orbit? Ask Astro, *Astronomy Magazine,* p 69.

If a moon has a moon,
is its moon called a moonmoon?

—science writer Jason Daley
Smithsonian Magazine, October 2018

TOPIC 39. Could a moon itself have a moon?

The discussion in Topic 38 is very relevant. We consider the common case where a moon M is tidally locked to its planet. With this assumption the axial rotation period of M is the same as its orbital period around its planet, so M rotates “slowly” on its axis. Thus any moon m orbiting M would assuredly orbit M much faster than M rotates on its axis. Assuming m has a prograde orbit around M, from Topic 38 we see that over a long period of time m would spiral inward toward M and eventually break apart through tidal forces (perhaps forming a ring around M). So while such a moon M could itself have a moon, the situation would be unstable over a long period of time.

There are stars in the southern sky
Southward as you go...

—The Eagles. 'Seven Bridges Road' 1980

TOPIC 40. A brief discussion of a few stunning objects to be seen from the southern hemisphere through amateur-sized telescopes.

Everyone interested in astronomy should at least once in their lifetime observe from a dark-sky site in the southern hemisphere—my choice: Australia. This discussion is given to whet your appetite for a visit to the southern half of our world.

1. The great barred spiral galaxy NGC 1365—The Mark of Zorro.

Without a doubt, NGC 1365 is the most prominent barred spiral galaxy that can be seen from Earth through an amateur telescope. It is an in-your-face "Mark of Zorro"—big and bright with an obvious bar. The two best barred spirals visible in the northern hemisphere are probably NGC 7479 in Pegasus and NGC 1300 in Eridanus.

Figure 1. NGC 1365. A combination of images using three different filters (B, V, R) taken with the 1.5-meter Danish telescope at the ESO La Silla Observatory in Chile. Credit: ESO, IDA, Danish 1.5 m, R. Gendler, J-E. Ovaldsen, C. Thöne, and C. Feron

NGC 1365 is 56 million light years distant in the constellation Fornax. It is about 200,000 light years across (compared with 100,000 light years for the Milky Way galaxy). Visually its size is about 50 arc seconds x 40 arc seconds.

Its central black hole of two million solar masses has been found to be spinning relativistically (at least 84 percent the speed of light) by the Nuclear Spectroscopic Telescope Array (NuSTAR) and the European Space Agency's XMM-Newton X-ray satellites.[1,2]

2. The massive star Eta Carinae, at the center of the Homunculus, which is embedded in the huge Eta Carina Nebula.

The star Eta Carinae is about 7,600 light years distant in the large star-forming region known as the Eta Carina nebula (NGC 3372). Surrounding the star is the *Homunculus* (Latin: little man). It mainly consists of two large bubbles of gas expelled on opposite sides of the star whose axis is aimed about 60 degrees away from us. This is my favorite deep-sky object in the entire sky. There is nothing like it to be seen anywhere else—certainly if one is using a small telescope. The Homunculus is 16 arc seconds x 12 arc seconds. Using high power in an eight-inch telescope it appears as an orange bar. In larger amateur telescopes, dark lines and irregularities are visible within the orange bulges.

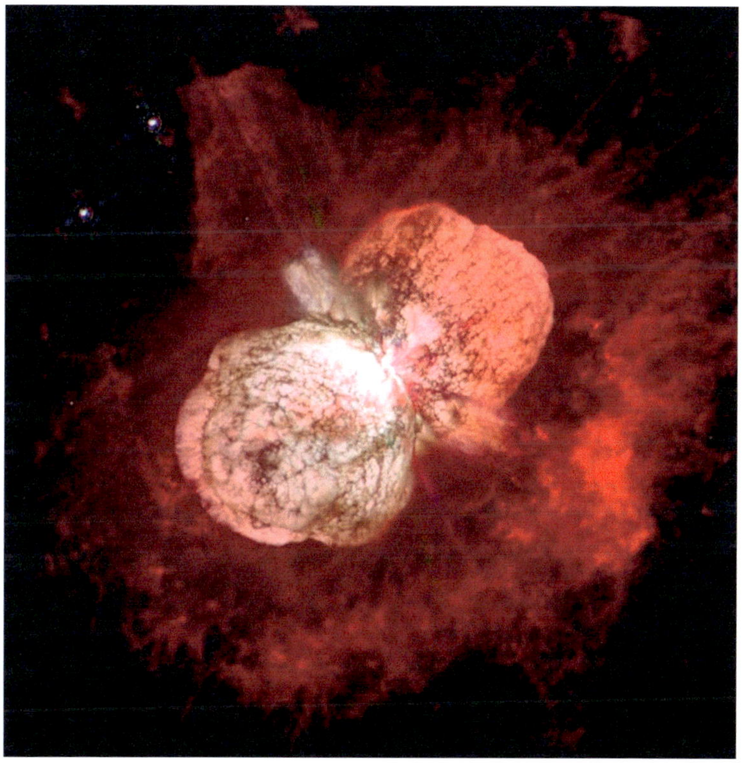

Figure 2. Optical image of Eta Carinae's Homunculus Nebula made by the Hubble Space Telescope. Credit: NASA/HST/J. Morse/K. Davidson http://chandra.harvard.edu/photo/1999/0099/

The image shows two spectacular bubbles of gas expanding in opposite directions away from a bright, central region at speeds in excess of a million miles per hour. The inner region visible in the Chandra image appears to be associated with a central disk of high-velocity gas rushing out at much higher speeds perpendicular to the bipolar optical nebula.

Figure 3. The greater Carina Nebula (NGC 3372). Eta Carinae is the bright star in the upper center of the image, slightly left of a faint, dark, S-shaped area called the Keyhole, and above the dark, downward-pointing wedge in the center. The image was captured by the VLT Survey Telescope at ESO's Paranal Observatory on June 5, 2012. Credit: ESO. Acknowledgement: VPHAS+ Consortium/Cambridge Astronomical Survey Unit http://www.eso.org/public/images/eso1250a/

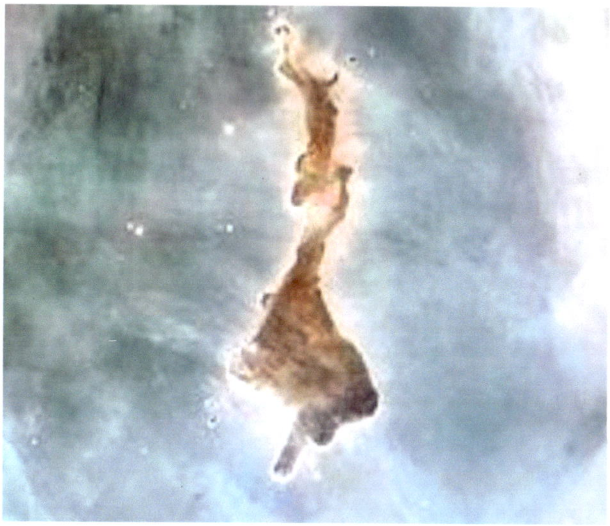

Figure 4. The Keyhole, about two light years in length, is a dark nebula located within the Carina Nebula. In Figure 3 it is just to the right of the star Eta Carinae. Credit: NASA, ESA, N. Smith (University of California, Berkeley), and The Hubble Heritage Team (STScI/AURA); credit for CTIO Image: N. Smith (University of California, Berkeley) and NOAO/AURA/NSF https://upload.wikimedia.org/wikipedia/commons/8/82/%22Finger_of_God%22_Bok_globule_in_the_Carina_Nebula.jpg

Astronomers have discovered that Eta Carinae has a companion star. The larger star is called Eta Carinae A; the smaller is Eta Carinae B. They orbit each other over a period of about 5.54 years. Both stars are inside the Homunculus, so it is difficult to determine their exact physical properties. Rough estimates based on models indicate the A star currently has a mass of between 100 to 120 solar masses and the B star between 30-60 solar masses. Both stars are thought to be about three million years old.[3]

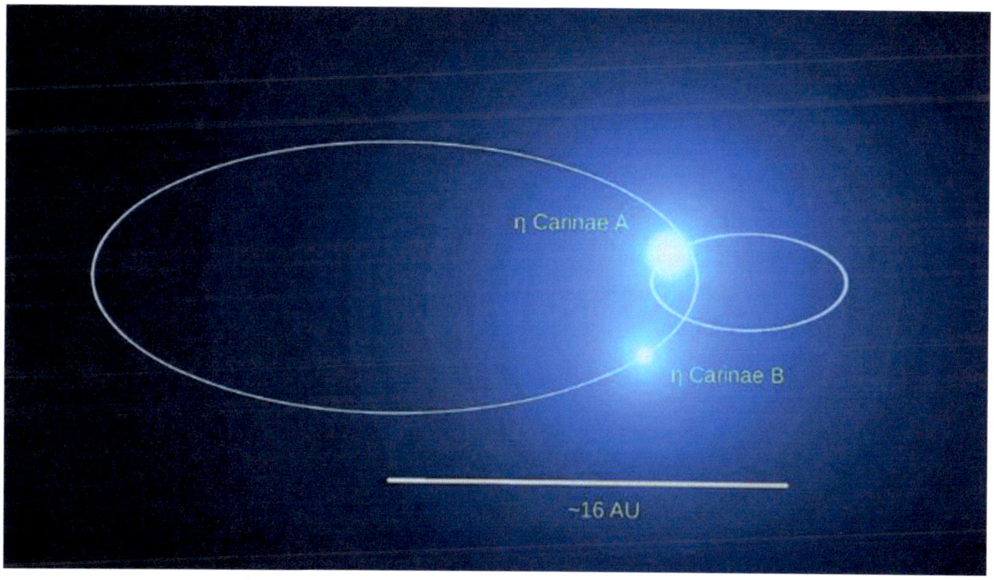

Figure 5. Orbits of Eta Carinae A and B. Credit: https://upload.wikimedia.org/wikipedia/commons/thumb/a/a8/Eta_carinae_orbit.png/640px-Eta_carinae_orbit.png

A little history of brightness changes in Eta Carinae.[3,4]

Edmond Halley made one of the earliest recordings of Eta Carinae in 1677. It was about fourth magnitude. In 1751, the French astronomer Abbe de Lacaille recorded it as second magnitude in his catalogue of southern stars observed at the Cape of Good Hope. It was also listed as second magnitude in several star catalogues between 1822 and 1826. The English botanist and traveler W. J. Burchell on

February 1, 1827, noticed it to be first magnitude, as bright as Alpha Crucis. He was the first to realize it was a variable star, as he had no recollection of having seen a bright star in this area during his earlier travels in the south. John Herschel also observed Eta Carina while in the Cape Province of South Africa. He noticed its brightness changes and, on January 2, 1838, that its brightness was almost equal to that of Alpha Centauri. Subsequently it faded until a final rise from second magnitude in 1842 to a magnitude of about -1 in April 1843, perhaps due to the two stars coming close to each other. Thus it became the second-brightest star in the sky. Only Sirius, a thousand times closer to Earth, was brighter. This period in the early 1840s is known as *The Great Eruption.* It is when the Homunculus was created.

The brightness decreased from 1845 to 1856 until in 1857 it was below naked-eye visibility. Then what is called *The Lesser Eruption* occurred in 1887, peaking at magnitude 6.2. In 2003, using Hubble, a miniature version of the Homunculus was discovered inside the Homunculus. It is believed this is due to the Lesser Eruption.[5] By 1895 Eta Carinae had faded to magnitude 7.5. From 1900 to 1940 it was at a constant magnitude of about 7.8. But by 1953 it had increased to 6.5. From then it has more or less steadily risen until in 2021 it is about 4.5. Since 1996 a 5.54-year variation in brightness was noticed, after which star B was confirmed to exist. But the five-year periodic brightness change does not always take place.

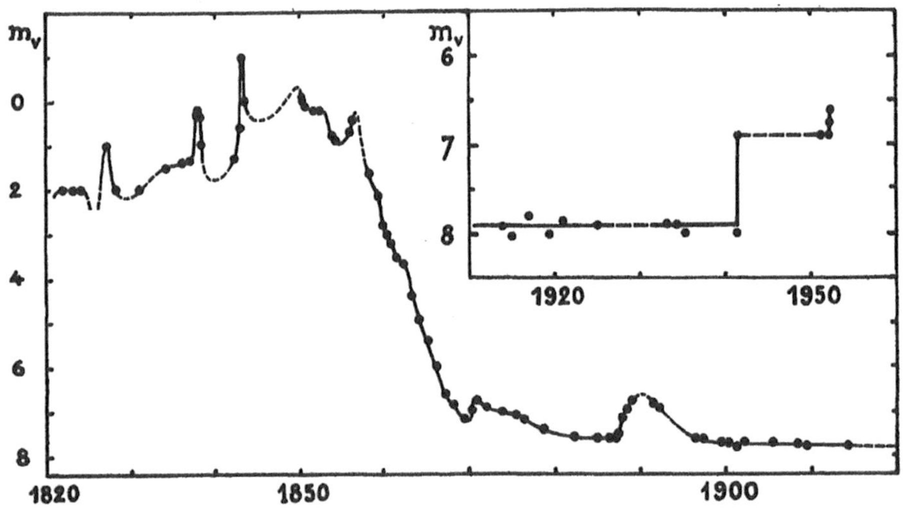

Figure 6. Light-curve of Eta Carinae 1882-1952. A plot of visual magnitude against time. Credit: Chart from p 246 in [4].

What does the future hold? Star A is one of the most massive stars known. Perhaps the answer is "supernova." If so, at only 7,600 light years away we will have ring-side seats—bring your sunglasses.

3. Omega Centauri (NGC 5139).
Omega Centauri (NGC 5139) appears as a magnitude 3.65 star to the naked eye and is so prominent that it was labeled as a star on early star charts. It is included in Ptolemy's mid second-century CE star catalogue. Johannes Bayer, assigning Greek letters to the brighter stars, labeled it Omega Centauri.

Figure 7. Globular cluster Omega Centauri, NGC 5139, with as many as 10 million stars. The image was taken with the WFI camera from ESO's La Silla Observatory. Its angular size is 36 arc minutes (about the size of the full Moon). North is up; east is to the left. This color image is a composite of B, V, and I filtered images. Credit: ESO https://www.eso.org/public/images/eso0844a/

Edmond Halley in 1677 was the first person to record it as a cluster of stars. The stars in Omega Centauri are bright enough that a view through a large amateur telescope provides enough light to activate color vision. It is the most massive and largest-known globular in the Milky Way galaxy, one of the closest (about 17,000 light years) to Earth, and is one of only eight globulars of less than 6.0 magnitude. Its diameter is about 150 light years. Usually all the stars in a globular have about the same metallicity, but in this globular some stars have three times the metallicity of others. This peculiarity has been cited as evidence that Omega Centauri might be the core of a dwarf galaxy that has been stripped of its outer layers. Or it could be that Omega is a globular whose heavy metals, created by early stars in the globular's history and not all blown out because of the globular's mass, were incorporated into younger stars in the globular. Larry Mitchell, a well-known amateur astronomer and a member of the Houston Astronomical Society, gave a talk at the 2019 Texas Star Party in which he said that Omega Centauri is 12 billion years old and has four-, six-, and 10-billion-year-old stars, which is evidence for it to be the core of a dwarf galaxy. In 2008, research indicated that Omega Centauri might contain an intermediate-mass black hole at its center. If true this would support the idea that it is not really a globular. Whatever its true nature, its sheer size and perhaps 10 million stars is a riveting sight to behold through a small telescope.

4. The Tarantula Nebula, NGC 2070, in the Large Magellanic Cloud.
The Tarantula Nebula is the most spectacular feature of the Large Magellanic Cloud (LMC), the largest satellite galaxy of our Milky Way galaxy. Also known as 30 Doradus (after a bright star near its center), the Tarantula Nebula owes its name to the numerous loops that emanate from the central region which are suggestive of spider legs. And a big spider at that. It is 160,000 light years distant. At its core, in a region known as R136 (see Figure 10), are a number of very luminous and massive stars—including many stars of spectral type O3 and several Wolf-Rayet stars. Wolf-Rayet stars are unstable (prime supernova candidates), supermassive, hot, young stars that are rapidly evolving and ejecting material (gas) at high speeds.

To the naked-eye the Tarantula appears as a faint, fuzzy star. It is more than 100 times as far away as the Great Orion Nebula M42 (whose distance is about 1,500 light years). Estimating the Tarantula's angular size as 20 arc minutes, at a distance of 160,000 light years, we glean two interesting facts:

1. The diameter of the Tarantula is on the order of 1,000 light years. Compare that with the Orion nebula's approximately 25 light-year diameter.

2. If the Tarantula were as close as the Orion Nebula, it would cover an area of the sky with a diameter of about 36 degrees, which would fill the whole constellation of Orion. (I note that the angular distance between Betelgeuse and Rigel is only 17 degrees.) Now imagine being on a world around a star in the LMC, and looking up at the Tarantula Nebula in the night sky. A beautiful sight for all except those who suffer from arachnophobia.

For an easy proof of items 1 and 2 see *Appendix 40.1.*

Figure 8. This first-light image from the TRAPPIST national telescope at La Silla shows the Tarantula Nebula, NGC 2070. The image was made from data obtained through B, V, and R filters, and the field of view is about 20 arc minutes across. Credit: TRAPPIST/E. Jehin/ESO - http://www.eso.org/public/images/eso1023a/ (direct link)

The Tarantula Nebula and its surroundings are absolutely awe inspiring. It is the second brightest and largest gaseous nebula in our local group of at least 80 galaxies. NGC 604 in M33 (2.6 million light years away) barely beats out the Tarantula in this regard.

Figure 9. This image from the VLT Survey Telescope at ESO's Paranal Observatory in Chile shows the region around the Tarantula Nebula (upper center) and its rich, surrounding star clusters, glowing gas clouds, and the scattered remains of supernova explosions. Credit: European Southern Observatory. https://www.eso.org/public/images/eso1816a/ , CC BY 4.0

Figure 10. A close-in view of the R136 star-forming region in the Tarantula Nebula, as seen by the Hubble Space Telescope. Credit: NASA/ESA/P. Crowther (University of Sheffield) https://earthsky.org/space/tarantula-nebula-spitzer-telescope-image/

5. The Large and Small Magellanic Clouds (LMC and SMC)—and 47 Tucanae.
The LMC and SMC are separated in the sky by about 22 degrees, and to the eye look as if they are detached pieces of the Milky Way. But they both are dwarf satellite galaxies of the Milky Way galaxy, being respectively about 161,000[6] and 204,000[7] light years distant. They are separated from each other by about 80,000 light years.[9]

Figure 11. The Magellanic Clouds low over the horizon as seen from Paranal Observatory in the Chilean Andes. Credit: ESO/Yuri Beletsky

The angular diameter of the LMC is about 11.5 degrees, giving it a physical diameter of about 32,140 light years. For the SMC, the angular diameter is about 5.5 degrees, for a diameter of about 18,800 light years.[8] But for both galaxies, these numbers include the faint, outer regions.

It is known that there is a bridge of gas and stars stretching between the LMC and SMC. This suggests they had a close encounter in the past. More recently, in a series of papers to be published in the journal *Astronomy & Astrophysics*, the European Space Agency's Gaia mission collaboration will show that while stars in the LMC are revolving, many stars in the SMC appear to be flowing toward and into the LMC along the bridge.[10] Gaia is a space observatory of the European Space Agency (ESA) launched in 2013 and expected to operate until around 2025. The Gaia mission will create a precise, three-dimensional map of astronomical objects throughout the Milky Way and plot their motions, which will allow the discovery of the past and subsequent evolution of the Milky Way.

The LMC is a virtual astronomical treasure trove—a bright, close galaxy to peer into. Its myriad globular clusters, open clusters, planetary nebulae, and H II regions can be seen in detail with amateur telescopes. And it contains the enormous Tarantula Nebula (see item 4 above and Figure 12 below). The LMC is also the site of 1987a, the first supernova visible to the naked eye in almost 400 years.

The SMC is farther south than the LMC—about 17 degrees from the south celestial pole—and also contains many objects accessible to amateur telescopes. Of note is NGC 346, an open cluster and emission nebula. It is the largest H II region in the SMC and easily seen through binoculars.

Figure 12. The Large Magellanic Cloud with its prominent Tarantula Nebula on the left. Credit: Anglo-Australian Observatory/David Malin Images

Figure 13. The Small Magellanic Cloud and the globular clusters 47 Tucanae, upper right, and NGC 362, upper left. Neither globular is in the SMC; both are local to our galaxy. Credit: courtesy of Sky-Map.org. Image obtained from https://cosmicpursuits.com/2481/

47 Tucanae (NGC 104) is the only globular cluster other than Omega Centauri that also has a star designation. It is the second-brightest (magnitude 4.0) globular and is noted for having a very dense, concentrated core. 47 Tucanae is about 120 light years across and about 16,700 light years from Earth. Omega Centauri and 47 Tucanae are the two most impressive globular clusters that can be seen. Even through a small telescope each takes your breath away.

Figure 14. A closer look at the globular cluster 47 Tucanae (NGC 104) in Figure 13. Image taken by ESO's VISTA (Visible and Infrared Survey Telescope for Astronomy) from the Paranal Observatory in Chile. Credit: ESO/M.-R. Cioni/VISTA Magellanic Cloud survey

A few further comments.

In addition to all the deep-sky objects to be seen from the southern hemisphere, a few other surprises are in store for northern observers who venture south. First, the familiar, northern constellations will all be upside-down. Are you a northern viewer accustomed to seeing Scorpius rise head-first with its tail eventually coming into view? In the southern hemisphere be prepared to see Scorpius rise horizontally. What is the best latitude, north or south, from which to see the greatest wonders in the night sky? My answer: About 35 degrees *south.*

I have observed nine times from Australia. For someone who has never been there (and for those who have), where does one go to observe the night sky in convenience, with access to large telescopes, and in the company of friendly fellow amateur astronomers?

The answer: Go to https://ozsky.org/. This group of Australians throws star parties, and folks from all over the world are invited. Check it out. I have. You might meet some Aussies who will become life-long friends. But be careful—Australia itself is addictive.

References

1. Risaliti, G., Harrison, F. A., Madsen, K., et al. Unambiguous determination of the spin of the black hole in NGC 1365. https://arxiv.org/ftp/arxiv/papers/1302/1302.7002.pdf . This is a pdf version of the entire article from reference 2.

2. Risaliti, G., Harrison, F. A., Madsen, K., et al. (February 27, 2013). A rapidly spinning supermassive black hole at the centre of NGC 1365. *Nature, 494*, p 449-451. https://www.nature.com/articles/nature11938 (This URL only gives an abstract.)

3. The Nine Planets. (January 4, 2021). *Eta Carinae Facts.* https://nineplanets.org/eta-carinae-%CE%B7-carinae/

4. de Vaucouleurs, G. (September 1952). The wonder-star: Eta Carinae. *Astronomical Society of the Pacific Leaflets, 6*(281), p 244. http://adsabs.harvard.edu/full/1952ASPL....6..244D

5. Reddy, Francis (June 2007). The supernova next door. *Astronomy Magazine,* p 36.

6. Pietrzyński, G., Graczyk, D., Gallenne, A., et al. (2019). A distance to the Large Magellanic Cloud that is precise to one percent. *Nature, 567,* p 200–203. https://doi.org/10.1038/s41586-019-0999-4

7. Graczyk, Dariusz, et al. (November 18, 2020). A distance determination to the Small Magellanic Cloud with an accuracy of better than two percent based on late-type eclipsing binary stars. *Astrophysical Journal, 904*(1), p 13. https://iopscience.iop.org/article/10.3847/1538-4357/abbb2b

8. NED (NASA/IPAC Extragalactic Database). http://ned.ipac.caltech.edu/ Data were acquired from this source.

9. Cooper, I., Kay, J., & Kepple, G. R. (2008). *The night sky observers guide. Volume 3, The southern skies.* Willmann-Bell. p 181.

10. Schilling, Govert (April 2021). New multidimensional map of the Milky Way. *Sky & Telescope,* p 8-9.

a minuscule push on a small body in space,
imparted by nothing more than sunlight

—*earth/sky.org*

TOPIC 41. The Yarkovsky effect.

Sunlight strikes the surface of a body, such as a small asteroid, heating its surface. The subsequent infrared radiation emitted by the surface acts as a jet exerting a small force on the asteroid. Over a long time period, the jet action is capable of altering the asteroid's orbit. This is the Yarkovsky effect,[1,2] and it is an important consideration in determining the orbit of a small asteroid. Furthermore, the effect can be employed in one method of altering the course of an asteroid that could hit Earth. In addition, the Yarkovsky effect has a role in explaining why so many asteroids have orbits which cross Earth's orbit. (See Topic 42.)

Just as the hottest time of the day on Earth is not at noon, but later in the afternoon, so the hottest part of the surface of an asteroid is where the local time would be p.m. during the day. A spot on the surface takes a while to heat up once sunlight first reaches it. So later in the day it will be at its warmest. Thus there is differential emission of infrared radiation from the body's surface. The result is a force acting in the opposite direction to the direction of strongest emission. You might ask, how can light (infrared or otherwise) exert a force when photons have no mass? They do not have mass, but they do have momentum. (A photon's momentum is Planck's constant divided by the wavelength of the photon—thank you quantum mechanics.) So photons emitted by the asteroid can impart momentum to the asteroid.

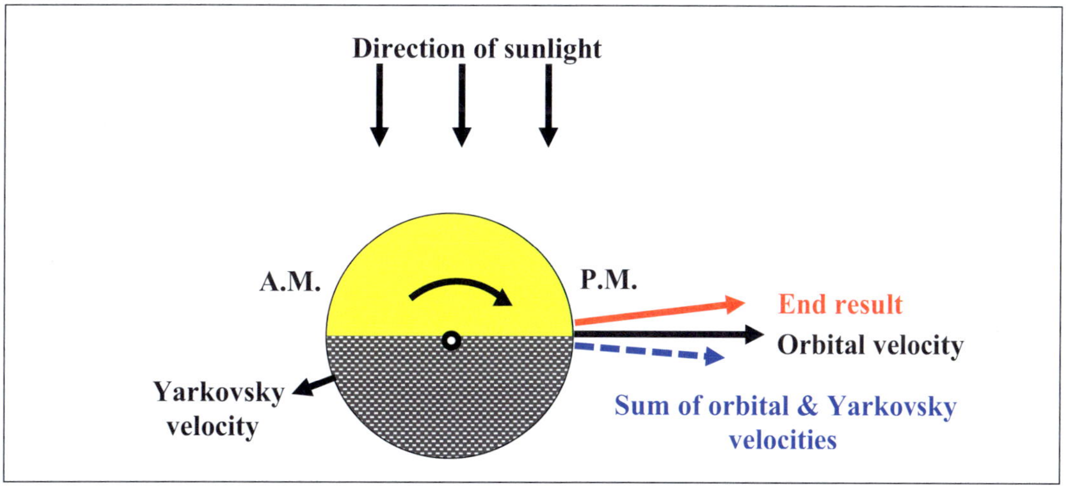

Figure 1. Yarkovsky effect, with retrograde rotation of the asteroid. The Yarkovsky velocity vector is the result of the Yarkovsky force. The Yarkovsky force can be thought of as similar to mounting a rocket on the asteroid's surface at about 2 o'clock and firing it straight up, pushing the asteroid in the direction of the Yarkovsky velocity. The blue, dashed arrow is the sum of the orbital velocity vector and the Yarkovsky velocity vector. The end result (see the explanation below) slows the asteroid's orbital velocity causing it to move inward (in the direction of the End result arrow) into a closer orbit around the Sun.

Consider Figure 1. Here the rotation of the asteroid is retrograde, i.e., it is rotating on its axis in the opposite direction to that of its motion around the Sun (in this case, clockwise on its axis and counter clockwise around the Sun). The Yarkovsky thrust acts in a direction to *slow* the asteroid's orbital velocity. The result is that the asteroid moves into an orbit *closer* to the Sun. In Figure 1, the Yarkovsky velocity vector is exaggerated in length (or magnitude) for purposes of clarity. The blue, dashed arrow is the sum of the orbital velocity vector and the Yarkovsky velocity vector. You might wonder why the asteroid doesn't move in that direction. If the asteroid were moving in a straight line, not under the influence of the Sun's gravity, and a force equal to that of the Yarkovsky thrust were applied, the asteroid would indeed change its direction to that of the blue, dashed arrow. But the asteroid is in orbit around the Sun. So what is significant here is that the Yarkovsky force *decreases* the *magnitude* of the orbital velocity. Under the Newtonian law of gravity this slowdown results in the asteroid moving *closer* toward the Sun, as indicated by the solid red End result arrow.

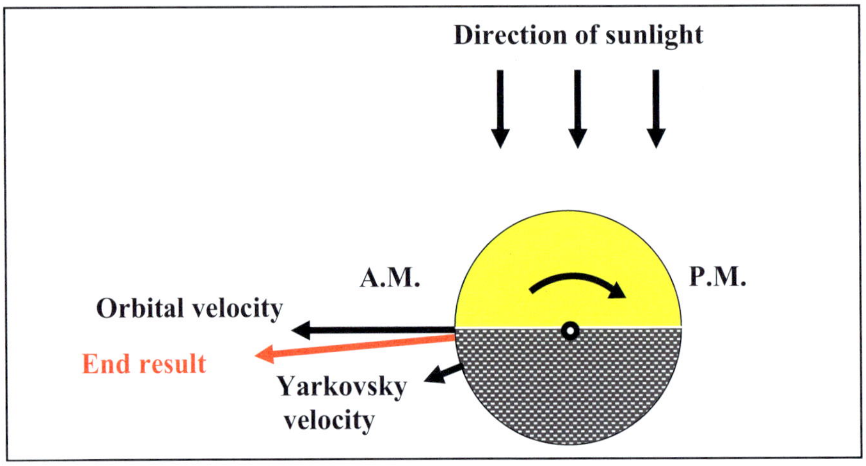

Figure 2. Yarkovsky effect, with prograde rotation of the asteroid. The net result is that the asteroid moves outward into a wider orbit around the Sun.

Now consider Figure 2 where the asteroid's axial rotation is prograde, as this rotation and the asteroid's orbital revolution are both clockwise. Here the Yarkovsky force *increases* the magnitude of the orbital velocity, and, according to the Newtonian law of gravity, the asteroid will move away from the Sun, as the solid red End result arrow indicates.

It should be noted that in the figures we assumed that the asteroid's rotational axis is perpendicular to the direction of the incoming rays of light. Of course the axis could be at any angle. The Yarkovsky effect is also affected by the (most likely non-spherical) shape of the asteroid and the varying texture of its surface material.

The effect is named after the Polish-Russian civil engineer Ivan Osipovich Yarkovsky (1844-1902) who first suggested it.

References

1. Tytell, David (April 2004). Seeing Yarkovsky at last. *Sky & Telescope,* p 22.
2. Chandler, David L. (December 2002). Dodging an asteroid. *Astronomy Magazine,* p 44-46.

There are only three ways to increase our chances against an asteroid aimed at Earth: Find it early; find it early; find it early.

—Don Yeomans, NASA Near-Earth Object Program

TOPIC 42. Asteroid/comet collisions with Earth—past events; possible remedies.

Most asteroids are in the asteroid belt between Mars and Jupiter. Asteroids in a resonance with Jupiter can have their orbits perturbed by Jupiter's strong gravitational field.

Resonances are discussed in Topics 53 and 54. Objects A and B are in *resonance* if there are integers, e.g., two and five, so that A makes two trips around its orbit in the same time, call it T, that B makes five trips around its orbit. This implies that if they come into resonance when close to each other, they will be close to each other again in T years, 2T years, 3T years, etc. This creates the possibility of either a future collision or, if A were much larger than B (e.g., A is Jupiter and B is an asteroid), B's orbit being modified by A's gravity.

An asteroid's orbit can be elongated by Jupiter to such an extent that its new orbit brings it close to Earth. Such a resonance with Jupiter is an example of an *escape hatch* from which asteroids can endanger the Earth. Another example of an escape hatch is in the main asteroid belt near Mars. While numerous, these escape hatches are extremely narrow. So how can so many asteroids end up in the inner solar system?

One idea is that the Yarkovsky effect (see Topic 41) can cause asteroids to change their orbits and, after a long period of time, enter a resonance with, say, Jupiter or Mars. It's a two-step process: First an asteroid enters an escape hatch in part due to the Yarkovsky effect, and then gravitational effects from a larger body extend the length of the asteroid's orbit so that it crosses Earth's orbit.[1,2,3]

Near-Earth asteroids (NEAs) are those whose orbits cross the Earth's and thus potentially could collide with our planet. There is a program to discover all NEAs over 140 meters (1.5 football fields) in diameter and accurately determine their orbits.[5] Earth has sustained many NEA collisions in the past; several are discussed below.

How about comets? Both short-term and long-term comets orbit the Sun. Long-term comets are believed to originate in the Oort cloud (mentioned in Topics 17 and 18), a sphere of billions (perhaps trillions) of comets encircling the Sun. This sphere is thought to extend from between 2,000 and 5,000 AU out to perhaps 100,000 AU from the Sun.[4] Sometimes, perturbed by a nearby passing star, the orbit of an Oort cloud comet can be altered, sending the comet inward toward the Sun. Except for comets whose orbits are known, the only warning that a comet might be on a collision course with Earth is when it is spotted heading in. So, unlike asteroids, where the warning time could be decades or longer, for Oort comets the warning time could be less than one year. And...more bad news. Many comets are more massive than NEAs and travel at much higher speeds. On impact, the energy released (the kinetic energy of the object) is directly proportional to the object's mass and to the square of its speed. So a comet with say, twice the mass and three times the speed of an asteroid, would release 2×3^2 or 18 times the energy of the asteroid. Moreover, all asteroids circle the Sun in the same direction as does the Earth, while comets can circle the Sun in the same or the opposite direction as Earth. Therefore a head-on collision with Earth is possible for a comet but not for an asteroid. Fortunately far fewer comets than asteroids come close to Earth.

One must not forget, while the effort spent trying to calculate the probability of a future collision with an asteroid/comet is worthwhile, it is less meaningful because the human race cannot afford even one major collision. Such an event must not be allowed to happen. Preparations are mandatory. No dinosaurs remain to give us advice on the subject. And as Benjamin Franklin said, "If you fail to plan, you are planning to fail."

The following table gives a rough idea of how high the collision danger is relative to an asteroid's size:

Asteroid size	Result	TNT explosion equivalent	Frequency
16 feet (5m)	Bright fireball	10 kilotons	3 years
82 feet (25m)	Airburst event	1 megaton	200 years
164 feet (50m)	Local devastation	10 megatons	2,000 years
460 feet (140m)	Regional devastation	300 megatons	20,000 years
985 feet (300m)	Continental devastation	2,000 megatons	70,000 years
1,970 feet (600m)	Widespread devastation	20,000 megatons	200,000 years
0.6 mile (1km)	Global catastrophe	100,000 megatons	700,000 years
3 miles (5km)	Global catastrophe	10 million megatons	30 million years
6 miles (10km)	Mass extinction	100 million megatons	100 million years

Source: Asteroid Day expert panel

Figure 1. Asteroid danger level and frequency. Credit: Asteroid Day expert panel (July 2015 Astronomy Magazine, p 35.)

Methods of preventing an asteroid collision with Earth.

1. Use a "gravity tractor." Here a heavy spacecraft would position itself close to a small asteroid and have its gravitational attraction nudge the asteroid into a different orbit. When an asteroid is very far from Earth, only a slight change in its orbit is necessary to result in the asteroid just missing the Earth years later. This simple method has the advantage in that it does not require an analysis of the asteroid's composition which would be needed for more aggressive techniques. Ed Lu and Stan Love of NASA's Johnson Space Center in Houston have stated in the periodical *Nature* that, with a lead time of at least 20 years, it would take just a year for a 20-ton spacecraft to drag a 200-meter-wide asteroid away from Earth's path.[21]

2. Paint part of the asteroid's surface so that the Yarkovsky effect can be used to alter its orbit over a long period of time. Again, the asteroid would have to be small. Here the rotation of the asteroid would have to be ascertained, and calculating the Yarkovsky effect would be no easy feat.

3. From a close distance, use a mirror to redirect strongly focused sunlight onto the object or make use of a laser. This heats a part of the surface, breaking it apart. The resulting ejected material gives the object a rocket-like push.

4. Attach rockets to its surface to change the trajectory.

5. Set off an explosion on its surface, propelling material away, giving the object a push.

6. Detonate an atomic weapon at a short distance from the asteroid. A pulse of neutrons would cause portions of the asteroid's surface to be blasted away, giving the asteroid a jolting push.[19]

7. Direct a heavy, solid object to slam into the asteroid. A NASA mission named DART (double asteroid redirection test) was launched November 24, 2021, as a test of this method. The asteroid Didymos, a half a mile in diameter, has a small moon, Dimorphos, orbiting it. The plan is to hit the 4.8-billion-kilogram moonlet with the 610-kilogram spacecraft at a speed of 6.58 kilometers per second, changing the orbital period by about 10 minutes.[6] The moonlet, 160 meters in diameter,[20] is about the size of an asteroid that could endanger a city. Scientists will gauge how well the collision will transfer momentum to the moonlet. It should be noted that the direct impulse of the spacecraft hitting the moonlet is a minor factor in how much momentum is imparted to the moonlet. The major contribution is from material ejected from the moonlet's surface due to the impact.[19] Why was a moonlet chosen for the test instead of a single asteroid? The answer is because it is simple to measure how much the moonlet's orbit changes by observing it pass in front of its parent body. It is expected that the 12-hour orbital period of the moonlet will be shortened by about 10 to 12 minutes.[20] Impact is scheduled to occur between September 26 and October 2, 2022.

In methods 3 through 7 the object would have to be reasonably solid, in some cases more so than others. With an asteroid that is a loosely held rubble pile, there is the danger of having the object split into pieces, creating a buckshot problem. Also, its rotation might have to be determined. An advantage for methods 4 through 7 is that a change in the orbit could be almost immediate.

Major asteroid hits on Earth and cometary fragments impacting Jupiter.

1. Barringer Meteor Crater (commonly referred to as Meteor Crater).
Meteor Crater in Arizona was the first crater confirmed to have resulted from an asteroid strike. The major research resulting in this confirmation was carried out by geologist Eugene M. Shoemaker (1928–1997), one of the founders of planetary science. In the early 1960s Shoemaker and Edward C.T. Chao came upon the "smoking gun"—shocked quartz (SiO_2) inside the crater. Coesite and stishovite,[7] the two specific forms of quartz they found, both require high temperature and instantaneous extreme pressure to form. Only an impact event or a nuclear explosion will suffice.

Figure 2. Meteor Crater in Arizona. Credit: Planetary Society

Daniel Moreau Barringer (1860-1929), lawyer, geologist, and mining entrepreneur, in 1903 bought the land where the crater is located. It is still privately held by the Barringer Crater Company, dedicated to preserving the crater and furthering the science of meteoritics. The history of Daniel Barringer and the crater is a colorful one, which the reader might want to investigate. I have visited this extraordinary site and highly recommend a stopover.

The crater is believed to have been formed roughly 60,000 years ago by an iron meteorite (containing about seven percent nickel) about 160 feet across.[7] The crater is 3,900 feet in diameter and 560 feet deep, with a rim about 148 feet above the surrounding desert plain.[7] The rim is believed to have lost some 50-65 feet of its height due to erosion.[8] Also, since the impact, an ancient lake has deposited about 100 feet of sediment on the crater floor.[7,8]

2. The K-Pg boundary, Chicxulub Crater, and the end of the dinosaurs.

An amazing and historic discovery starts with something called the K-Pg boundary (formerly called the K-T boundary). Everywhere on Earth where various geological strata of the crust are exposed on the surface, at the level corresponding to 65.5 million years ago is found a layer of limestone (calcium carbonate) and just above it is a thin layer of red clay, and just above that is another layer of limestone.

The *K-Pg boundary* is the layer of clay just above the lower layer of limestone, called the *Cretaceous limestone*. The layer of limestone just above the red clay is called the Tertiary limestone or the *Paleogene limestone*.

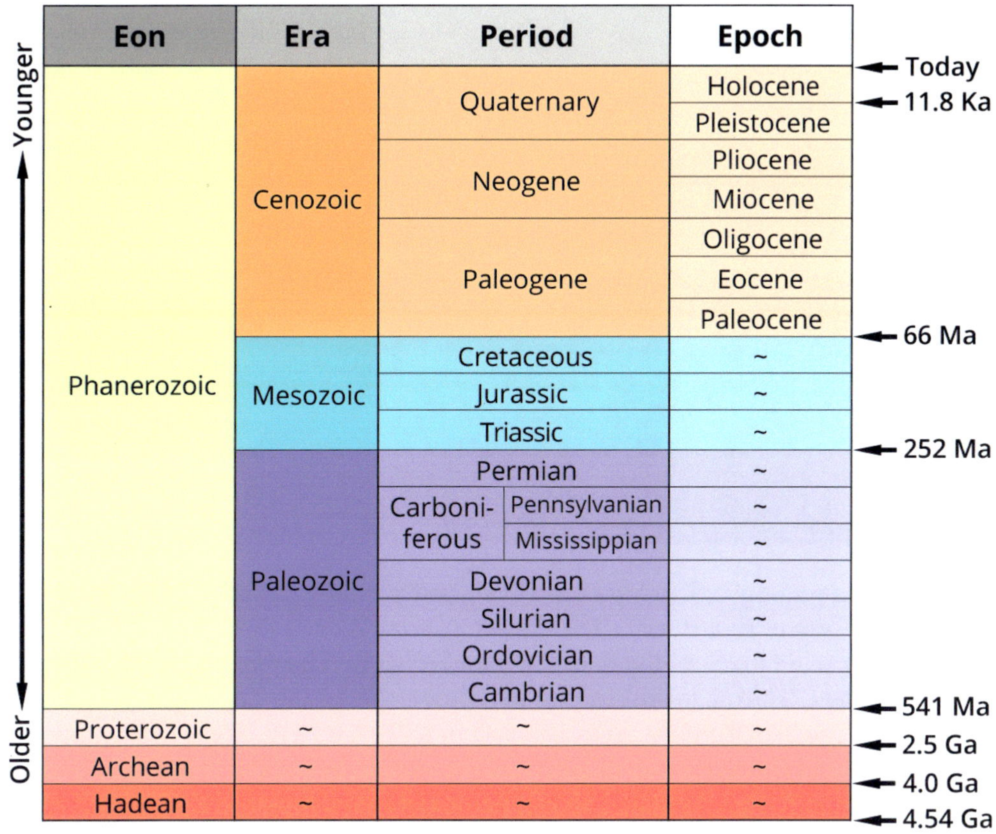

Figure 3. The geological time scale. Ka, Ma, and Ga stand respectively for "thousands, millions, and billions" of years ago. The K-Pg boundary is at the 66 Ma level (see the right-hand column). Credit: Jonathan R. Hendricks, CC BY-SA 4.0

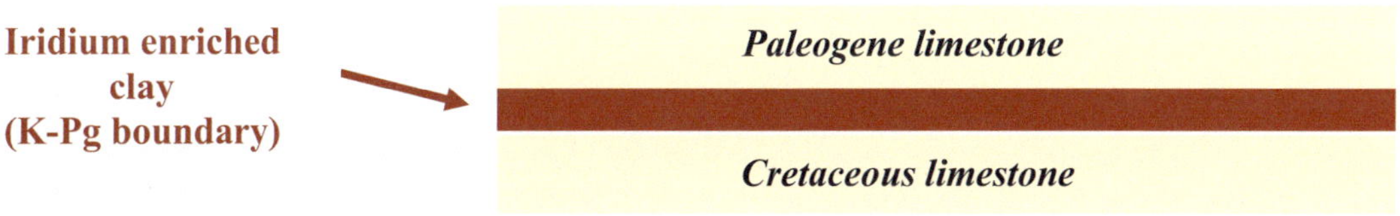

Figure 4. The K-Pg boundary consists of a clay layer containing a rich source of iridium. This was created by an asteroid hit 65.5 million years ago. Below this is a layer of limestone formed during the Cretaceous geological period (145.5-65.5 million years ago). Above the K-Pg boundary a new limestone layer was formed during the Paleogene (65.5-23 million years ago).

Figure 5. The finger is pointing at the inch-thick Cretaceous limestone layer (white). Just above it is the dark K-Pg boundary of clay. And above that is the Paleogene limestone layer, which in Trinidad lies under a thick sandstone layer.

First some geological terminology: The Mesozoic Era's last period is called the Cretaceous (Latin: creta, meaning chalk, a form of limestone). The dinosaurs were then at their peak. Chalk in German is Kreide, and the K in this word gives us the K in K-Pg and in K-T. The era after the Mesozoic is called the Cenozoic period. The Cenozoic is broken into two time periods: the Tertiary (not labeled in Figure 3) and the Quaternary. The T in Tertiary gives us the T in K-T. The Tertiary is broken into two time periods, the first of which is called the Paleogene (and the second called the Neogene). The P and g in this word gives us the Pg in K-Pg. As stated above, geologists have replaced K-T with K-Pg.

Essentially all dinosaur fossils are found below the K-Pg boundary. Also, nearly 75 percent of all existing species of life on Earth (plants, animals, and microorganisms) disappear from the geological record above this boundary. This suggests the boundary is quite important. What is going on?

The amazing answer. Walter Alvarez, a professor of geology at the University of California, Berkeley, and son of the 1968 Nobel Prize-winning physicist Luis Alvarez, also a UC Berkeley professor, was doing field work in 1977 in Gubbio, Italy. He found that the Cretaceous limestone (below a layer of red clay) contained the fossils of many different types of tiny marine creatures called foraminifera (or foram). But the Paleogene limestone (above the layer of red clay) contained only one species of foram. And the clay contained no foram.[9] Returning to Berkeley with samples of the clay, he showed them to his father who suggested they be analyzed by Lawrence Berkeley Lab nuclear chemists Frank Asaro and Helen Michel. They were stunned when the results showed that the clay contained about 600 times as much iridium as the surrounding limestone.[9] Iridium, a silvery-white metal related to platinum, is almost non-existent in the Earth's crust. When the Earth was molten, iridium, being very heavy, would have sunk deep into the planet's core. But high concentrations of iridium are found in asteroids and comets.

Then they looked elsewhere. In Denmark, New Zealand, and many other places all over the globe a high concentration of iridium appeared in the clay above the Cretaceous limestone. Comparisons between the amount of iridium and other key elements in the clay were essentially independent of the region from which the samples were taken,[9] indicating a common source. In 1980 Walter and Luis Alvarez wrote their seminal paper, Extraterrestrial Cause for the Cretaceous-Tertiary Extinction,[10] published in the journal *Science*.

It was later discovered that there was shocked quartz, microtektites, and an abundance of soot in the clay layer from samples taken from all parts of the world.[9] Only extremely high pressure and high temperatures can produce shocked quartz and tektites—more supporting evidence for their claim. (Tektites, gravel-sized bodies composed of black, green, brown, or gray natural glass fused from *terrestrial* matter, are a tell-tale sign of a meteorite impact.) In addition it was verified that the chemical composition of the clay *between* the layers of limestone is different from that of clay *mixed in* with the limestone. And the clay mixed with the two limestone layers is similar.[10] More evidence that the boundary-layer clay is from dust thrown up into the stratosphere by the impact.

Figure 6. Location of Chicxulub Crater. Credit: thetimes.co.uk

Still the question persisted: Where was the crater? The final piece of the puzzle fell into place in 1981. Geophysicists Antonio Camargo and Glen Penfield, working for PEMEX, Mexico's state oil company, had been looking for petroleum in the Yucatán Peninsula during the late 1970s. Their investigations traced out the features of a huge, round shape partly under the land of the Yucatan and partly under water off the coast. The crater was named Chicxulub (Mayan, Chick-shoo-loob, place of the deer's antler) for the small Mayan village northeast of Merida which lies almost at its geographic center. The tale of its discovery and its verification as the location of the impact cited in the Alvarez paper is a story in itself. The crater's size is about 261 by 162 kilometers (162 by 100 miles)[12] and lies deep in the continental crust.

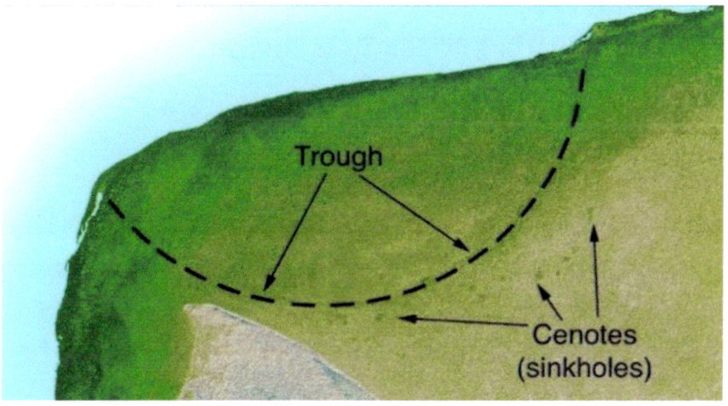

Figure 7. Chicxulub Crater. Credit: NASA/JPL https://photojournal.jpl.nasa.gov/catalog/PIA03379

Along the edge of the crater is a trough, only about three to five meters (10 to 15 feet) deep and about five kilometers (three miles) wide.[12] *Cenotes* or water-filled sinkholes[13] cluster along the trough, suggesting that, after the impact, water filled the crater's basin. Think of the Yucatan Peninsula as a giant slab of limestone. Groundwater from the crater basin dissolved the limestone creating caves and cenotes—Mayan cultural sites and tourist delights to this day.

Figure 8. The cenote Hubiku, at Temozon, Yucatan, Mexico. Credit: Son of Groucho, CC BY 2.0, via Wikimedia Commons, https://www.flickr.com/photos/sonofgroucho/15238545692/

Based on geochemical evidence, it is now widely believed the impactor was a carbonaceous chondritic asteroid[11] with a diameter estimated at 10 ± 4 kilometers (6.2 miles).[10] The total global mass of iridium in the K-Pg clay layer is estimated to be about 2.0-2.8 x 10^{11} grams. A 10-kilometer-wide asteroid would deliver approximately 2.3x10^{11} grams. In contrast a seven-kilometer-wide comet would deposit only about 0.1x10^{11} grams of iridium as comets contain much ice.[11] This virtually rules out a comet as the impactor. A comet and asteroid of the above sizes would produce about the same size crater as comets travel at higher speeds.

Walter Alvarez was awarded the 2008 Vetlesen Prize, considered the equivalent of the Nobel prize for geology. (Geology is not a field in which the Nobel is awarded.)

The Effects. The strike resulted in impact-spawned mega tsunamis, earthquakes, and a blast of hot, rocky material ejected into the atmosphere which covered the globe. As it fell to Earth this substance ignited forest fires world-wide and acidified rain. The little sunlight reaching the surface was reduced by dust and sulfate aerosols. Photosynthesis was suppressed, causing a breakdown in the ocean food chain and the death of most marine creatures. Plant life perished, as did animals dependent on plants and other animals for food.

3. Tunguska—the biggest collision with Earth in recorded history.

In remote Siberia it is the morning of June 30, 1908. There is a massive explosion. Eyewitnesses some distance away report a fireball as bright as the Sun crossing the sky. In locations around the world seismometers and sensitive barometers record seismic and pressure/shock waves. For the next few days, in places across Eurasia, people observe bright lights in the sky. People report being able to read newspapers at midnight.

Some of these lights we now describe as noctilucent clouds or night-shining clouds—very small ice crystals that form around micro meteoric dust in the cold, upper mesosphere (the atmospheric layer above the stratosphere and below the thermosphere extending from about 50 to 85 kilometers (31 to 53 miles) high). While it is dark on the ground, sunlight can still strike these clouds, so they are usually seen after sunset or before sunrise.

Figure 9. Location of the 1908 Tunguska event. Credit: Wikipedia

Figure 10. Flattened trees from the Tunguska event.

The explosion killed hundreds of reindeer, and 80 million fallen trees formed a radial orientation away from ground zero. Though the devastation covered about 2,150 square kilometers (826 square miles), no crater marks the center point from which the fallen trees radiate and no "smoking gun" meteoritic material has ever been found.[27,29] At first, no one correlated the seismic records, barometric records, and the lights in the sky. Only later were they all tied to the Tunguska event.

In 1921, Leonid Kulik, chief curator of the Mineralogical Museum in St. Petersburg, led an expedition toward Tunguska. But harsh Siberian weather forced him to turn back short of his goal. It wasn't until 1927, when he led a second expedition, that details of the devastation became known.

Why did it take 19 years to reach ground zero? The site was remote and the weather unpredictable. Also, in the meantime, the First World War and the Russian Revolution had intervened, focusing attention elsewhere.[28]

It is believed that the Tunguska event was caused by a comet or asteroid exploding in the air. One prominent theory points to a stony asteroid, as no ferrous meteoritic material has been discovered.

4. Recent, strong evidence indicates the city of Tall el-Hammam (near Jericho), in what is now Jordan, was obliterated by an asteroid about 3,600 years ago.[30]

The asteroid is believed to have been about the size of the Tunguska asteroid. All of the 8,000 people and animals perished. And about a minute after the explosion, winds from the blast flattened the walls of Jericho and the city burned. Scientists have been excavating the site of Tall el-Hammam for 15 years. At the site is a dark, roughly five-foot-thick jumbled layer of charcoal, ash, melted mudbricks, and melted pottery. The result of a firestorm. Found at the site is shocked quartz that only forms at a pressure of 725,000 pounds per square inch. Wood and plants in the area were instantly turned into nanodiamonds by the high pressures and temperatures. Experiments with laboratory furnaces showed that bubbled pottery and mudbricks liquefied at above 2,700° F. Tiny balls of melted material the size of dust particles (called spherules) made of vaporized iron and sand are present. Spherules melt at about 2,900° F. Surfaces of pottery and meltglass contain tiny melted grains of iridium, platinum, and zirconium silicate. These melt respectively at 4,435°, 3,215°, and 2,800° F. The only natural process that can result in these conditions is a cosmic impact. The city and over 100 other area settlements were abandoned for several centuries. It is thought that this abandonment could have been caused by vaporized or splashed toxic salt water from the Dead Sea, only about seven miles southwest of the city.[30]

Figure 11. Tall el-Hammam. Credit: (left) Phil Silvia, CC BY-ND; (right) NASA/CC BY-ND

5. Pingualuit Crater in Canada.

Pingualluit (an Inuktitut term for skin blemishes caused by cold weather) Crater, 3.44 kilometers (2.1 miles) in diameter and about 400 meters (1,300 feet) deep, was formed by a meteorite 1.4 million years ago in the Pleistocene. At 267 meters (876 feet), the lake is Quebec's deepest. The lake is slow to freeze and thaw. The crater's rim rises more than 100 meters above the surrounding surface and is pitched 40-45 degrees down to the water. The surrounding uplifted area formed by the impact has a radius almost twice that of the crater itself.[14]

Figure 12. Pingualuit Crater in Nunavik, northern Quebec, Canada, looking west. Lake Pingualuk lies within the crater. Photographed on October 12, 2007. Credit: NASA. Courtesy of Denis Sarrazin

6. Jupiter is hit with kilometer-sized chunks from the breakup of a comet that ventured too close to the planet.

Comet Shoemaker-Levy 9 was discovered by Carolyn and Gene Shoemaker and David Levy in a photograph taken on March 18, 1993, with the Mt. Palomar 0.4-meter Schmidt telescope. It was the trio's ninth short-period comet discovery. This one grabbed the attention of not only the world of professional and amateur astronomers, but public news networks and members of Congress. These dramatic collisions also vividly displayed a potential threat to our home planet. At the time of its discovery the comet had already broken into more than 20 pieces, and they were in a two-year periodic orbit around Jupiter. It was shortly realized that these pieces were headed for a collision with Jupiter. For the first time in recorded history a major collision of solar system objects was observed.[15] (There is some evidence that in 1973 two amateur astronomers imaged clouds on Mars resulting from a cometary collision.[31])

Figure 13. This composite was assembled from separate images of Jupiter and comet Shoemaker-Levy 9 taken by the NASA/ESA Hubble Space Telescope in 1994. Jupiter's moon Io is transiting Jupiter: Its shadow is the black dot. (Io itself is an almost invisible yellowish dot just to the upper right of the black dot.) The "pearls" lined up below Jupiter are the cometary fragments. Credit: NASA, ESA, H. Weaver and E. Smith (STSCI) and J. Trauger and R. Evans (NASA's Jet Propulsion Laboratory)

Astronomers discovered that the comet, estimated to originally have had a diameter of 1.5 to two kilometers (0.9 to 1.2 miles), had made a close approach to Jupiter in July 1992 and Jupiter's powerful gravity had tidally torn it apart. It is thought that the intact comet had been circling Jupiter for about a decade, perhaps longer.[15]

The pieces, one after the other, collided with Jupiter from July 16 to July 22, 1994. NASA's Galileo orbiter, which was still on its way to Jupiter, captured spectacular direct views. In addition, many Earth-based observatories and orbiting spacecraft, including the Hubble Space Telescope, Ulysses (on its way to study the Sun), and Voyager 2 (in the distant, outer reaches of the solar system) were turned to watch the collisions and the aftermath.[15]

Figure 14. String of Pearls. This Hubble Space Telescope image, courtesy of H. Weaver, shows the "string of pearls" comet Shoemaker-Levy 9 in March 1994, four months before its collision with the planet Jupiter. Jovian tides fragmented the comet into this series of pieces in July 1992, the first such occurrence observed by mankind.

Figure 15. Shoemaker-Levy 9 Fragment D and G impact scars on Jupiter. Credit: Data, H. Hammel, MIT, and NASA; processing, Judy Schmidt

In Figure 15, fragments of comet Shoemaker-Levy 9 crashed into Jupiter over several days in July 1994. Fragments D and G struck Jupiter on July 17 at 11:45 and July 18 at 07:28 UTC, respectively. The relatively fresh fragment G impact has produced a concentric set of scars: an inner dark circle, an outer thin ring, and an outermost diffuse ring. Fragment D is responsible for the small, dark circle at one o'clock from G. The three photos used to make this color composite were taken at 09:19, 09:22, and 09:25, nearly two hours after the impact.

Figure 16. Ultraviolet image of Jupiter taken by the Hubble Space Telescope's Wide Field Camera. Credit: NASA/Hubble Space Telescope Comet Team - NASA: http://www2.jpl.nasa.gov/sl9/image 129.html

In Figure 16, Jupiter's atmosphere is shown at a wavelength of 2,550 angstroms after it suffered many impacts from fragments of comet Shoemaker-Levy 9. The most recent impactor is fragment R, third from right in the lower line of impacts. This photo was taken 3:55 EDT on July 21, about 2.5 hours after R's impact. A large, dark patch from the impact of fragment H is visible just rising on the morning (left) side. Proceeding to the right, other dark spots were caused by impacts of fragments Q1, R, D and G (now one large spot), and L, which covers the largest area of any seen thus far. Small, dark spots indicated by B, N, and Q2 are present, but are not very visible in the image. The spots are very dark in the ultraviolet because a large quantity of dust is being deposited high in Jupiter's stratosphere, and the dust absorbs sunlight. The black dot on the upper half is the moon Io in front of Jupiter.

The cometary fragments smashed into Jupiter with a force of 300 million atomic bombs. Huge plumes 2,000 to 3,000 kilometers (1,200 to 1,900 miles) high heated Jupiter's atmosphere to 30,000° to 40,000° C (54,000° to 72,000° F). Dark, ringed marks in the atmosphere eventually were erased by Jupiter's winds.[15] It would be all over for the human race, not to mention most life on Earth, if something similar happened to the Earth. Is there anyone out there who still thinks that money is wasted on planetary science?

7. A recent event, and a near miss coming in 2029.

(a) Chelyabinsk. In the morning of February 15, 2013, a 20-meter-wide near-Earth asteroid streaked across the sky and exploded about 14 miles above the ground near Chelyabinsk, a Russian city with a population of about one million. This was the most significant asteroid collision with Earth since the 1908 Tunguska event. Thousands of buildings were damaged and more than 1,600 people were injured (most from shattering glass) with some having to be hospitalized. The shock wave blew out windows over a 200-square-mile area. The asteroid entered Earth's atmosphere at about 65,000 kilometers per hour (40,000 mph). The energy released in the blast was the equivalent of approximately 440,000 tons of TNT—30 to 40 times the energy of the Hiroshima atom bomb. Though this constituted a small event compared to Tunguska, it is a reminder that in the blink of an eye everything can change.[16,17,18]

(b) Apophis—the most dangerous asteroid known to date. This asteroid is named after the Egyptian serpent god Apophis, the enemy of the Sun god Ra. Apophis is associated with chaos, storms, thunder, darkness, death, and destruction. On April 13, 2029, (which happens to be Friday the 13th) the asteroid will pass as close as 31,900 kilometers (19,800 miles) to the Earth's surface, closer than the geostationary communication satellites at an elevation of 35,800 kilometers.[22] For observers in the Eastern Hemisphere, Apophis, about 340 meters in diameter,[23] will be visible to the naked eye as a fourth-magnitude star as it traverses the night sky.[22,23] MIT planetary scientist Richard Binzel characterized Apophis as "...300 times more massive than Tunguska, 5,000 times more massive than Chelyabinsk, so this is an object that certainly gets your attention."[22]

When it was discovered in 2004 it was briefly thought to have a three percent chance of colliding with the Earth in 2029.[26] A revised orbit including pre-discovery images quickly ruled out such an impact. But astronomers realized a danger still existed. If, in its 2029 passage, Apophis passes through a narrow, 600-meter-wide *keyhole window*, it would collide with the Earth in 2036.

What is a *keyhole window*?[19] It is a narrow window of space which, if the asteroid passes through it near closest approach to Earth, its orbit will be altered so that the object is now in resonance with the Earth, and, in the future, it will again pass nearby. The asteroid's passing through such a resonance keyhole would potentially imperil life on Earth.

That Apophis actually would pass through this window in 2029 was given a low probability. But there was concern that the Yarkovsky effect (see Topic 41) might play a role in permitting the asteroid to pass through the keyhole. But by 2013 it was known that Apophis would certainly not hit Earth in 2036.[25] It was determined that the asteroid is elongated and tumbling, two properties that diminish the Yarkovsky effect. But again, there was left a small, but non-zero, probability that there could be a collision in 2068. It's an asteroid that keeps coming back as a potential problem. However, on March 5, 2021, when Apophis passed by Earth at a distance of 17 million kilometers, using ground-based radar astronomers ruled out a collision in 2068.[24]

References

1. Cowen, R. (March 6, 1999). Asteroids get solar push toward Earth. *Science News,* p 151.
2. Chapman, Clark R. (October 5, 2000). The long trip to Earth. *Nature, 407*, p 573-576.
3. Cowen, R. (October 21, 2000). Nudging asteroid fragments toward Earth. *Science News,* p 264.
4. NASA. (Updated December 19, 2019). *The Oort cloud: what is it?* https://solarsystem.nasa.gov/solar-system/oort-cloud/in-depth/
5. NASA. (Updated March 28, 2019). *Near-Earth object observations program.* https://www.nasa.gov/planetarydefense/neoo
6. Scoles, Sarah (June 2021). The broken shield. *Scientific American,* p 37-43.
7. Eicher, David J. (January 2019). Meteor crater inside and out. *Astronomy Magazine,* https://astronomy.com/bonus/crater
8. Poelchau, Michael, Kenkmann, Thomas, & Kring, David (January 2009). Rim uplift and crater shape in Meteor Crater: Effects of target heterogeneities and trajectory obliquity. *Journal of Geophysical Research, 114*(E1), AGU.org https://agupubs.onlinelibrary.wiley.com/doi/full/10.1029/2008JE003235
9. Yarris, Lynn (March 9, 2010). Alvarez theory on dinosaur die-out upheld: Experts find asteroid guilty of killing the dinosaurs. Lawrence Berkeley Laboratory News Center. https://newscenter.lbl.gov/2010/03/09/alvarez-theory-on-dinosaur/

10. Alvarez, Luis, Alvarez, Walter, Asaro, Frank, & Michel, Helen (June 6, 1980). Extraterrestrial cause for the Cretaceous-Tertiary extinction. *Science, 208*(4448), p 1095-1108. https://science.sciencemag.org/content/208/4448/1095.abstract
11. Desch, Steve, Jackson, Alan, Noviello, Jessica, & Anbar, Ariel (June 1, 2021). The Chicxulub impactor: Comet or asteroid? *Astronomy & Geophysics, 62*(3), p 3.34–3.37. https://doi.org/10.1093/astrogeo/atab069
12. NASA/JPL. (March 6, 2003). PIA03379: Shaded relief with height as color, Yucatan Peninsula, Mexico. *Photojournal.* https://photojournal.jpl.nasa.gov/catalog/PIA03379
13. Monroy-Ríos, Emiliano (n.d.). Chicxulub crater and ring of cenotes. Karst geochemistry and hydrogeology. Department of Earth and Planetary Sciences, Northwestern University. https://sites.northwestern.edu/monroyrios/some-maps/chicxulub-and-ring-of-cenotes/
14. O'Dale, Charles P. T. (n.d.). Pingualuit impact crater. *Crater Explorer: My exploration and documentation of craters (presently only on this planet).* http://craterexplorer.ca/pingualuit-impact-crater/
15. NASA Science Solar System Exploration. (July 27, 2021). *P/Shoemake*r-*Levy 9.* https://solarsystem.nasa.gov/asteroids-comets-and-meteors/comets/p-shoemaker-levy-9/in-depth/
16. NASA. (February 15, 2018). *Five years after the Chelyabinsk meteor: NASA leads efforts in planetary defense.* nasa.gov. https://www.nasa.gov/feature/five-years-after-the-chelyabinsk-meteor-nasa-leads-efforts-in-planetary-defense
17. Howell, Elizabeth (January 9, 2019). *Chelyabinsk meteor: A wake-up call for Earth.* space.com. https://www.space.com/33623-chelyabinsk-meteor-wake-up-call-for-earth.html
18. Atkinson, Nancy (February15, 2013). *Airburst explained: NASA addresses the Russian meteor explosion.* universetoday.com. https://www.universetoday.com/100025/airburst-explained-nasa-addresses-the-russian-meteor-explosion/
19. Durda, Dan (December 2010). How to deflect a hazardous asteroid. *Sky & Telescope,* p 22-28.
20. Temming, Maria (August 15, 2020). Newly named asteroid is about to get schooled in hard knocks. *Science News,* p 5.
21. Peplow, Mark (November 9, 2005). *Gravity tractors beat bombs.* nature.com. https://doi.org/10.1038/news051107-7
22. Bartels, Meghan (February 11, 2021). *Scientists prepare for their last good look at asteroid Apophis before 2029 flyby.* space.com. https://www.space.com/asteroid-apophis-march-2021-flyby-science-preparations
23. ONeill, Ian J., & Handal, Joshua (March, 26, 2021). *NASA analysis: Earth is safe from asteroid Apophis for 100-plus years.* nasa.gov. https://www.nasa.gov/feature/jpl/nasa-analysis-earth-is-safe-from-asteroid-apophis-for-100-plus-years
24. Beatty, J. Kelly (July 2021). Asteroid Apophis will miss Earth in 2068. *Sky & Telescope,* p 11.
25. Beatty, J. Kelly (July 2013). News Notes, *Sky & Telescope,* p 14.
26. Beatty, J. Kelly (April 2013). Asteroids—easy miss by Apophis in 2036. *Sky & Telescope,* p 10.
27. Longo, Giuseppe (2007). The Tunguska event. Chapter 18 in: P.T. Bobrowsky & H. Rickman (Eds.), *Comet/Asteroid Impacts and Human Society: An interdisciplinary approach.* (p 303-330). Springer. https://www.researchgate.net/publication/282256596_The_Tunguska_Event The chapter is an especially thorough, scholarly, and authoritative source.
28. Phillips, Tony (Ed.). (June 30, 2008). *The Tunguska impact—100 years later.* https://science.nasa.gov/science-news/science-at-nasa/2008/30jun_tunguska
29. Marie, Ana (October 6, 2016). Tunguska event: The truth is out there. wnyc (New York public radio); Annotations: The NEH Preservation Project. https://www.wnyc.org/story/tunguska-event-truth-out-there/
30. Moore, R. Christopher, et al. (September 21, 2021). *The conversation.* https://theconversation.com/a-giant-space-rock-demolished-an-ancient-middle-eastern-city-and-everyone-in-it-possibly-inspiring-the-biblical-story-of-sodom-167678
31. O'Meara, Stephen (February 2022). A Martian comet crash? *Astronomy Magazine*, p 15.

The elusive and hard-to-study nature of TNOs makes them some of the most intriguing places in the solar system.

—Carly Howett, planetary scientist, February 2019

TOPIC 43. A rogues' gallery of the larger trans-Neptunian objects (TNOs).

Trans-Neptunian objects are solar system bodies farther from the Sun than Neptune. Being smaller bodies and far from Earth, little is known about most of them. Here we list the larger known ones and say a little about each.

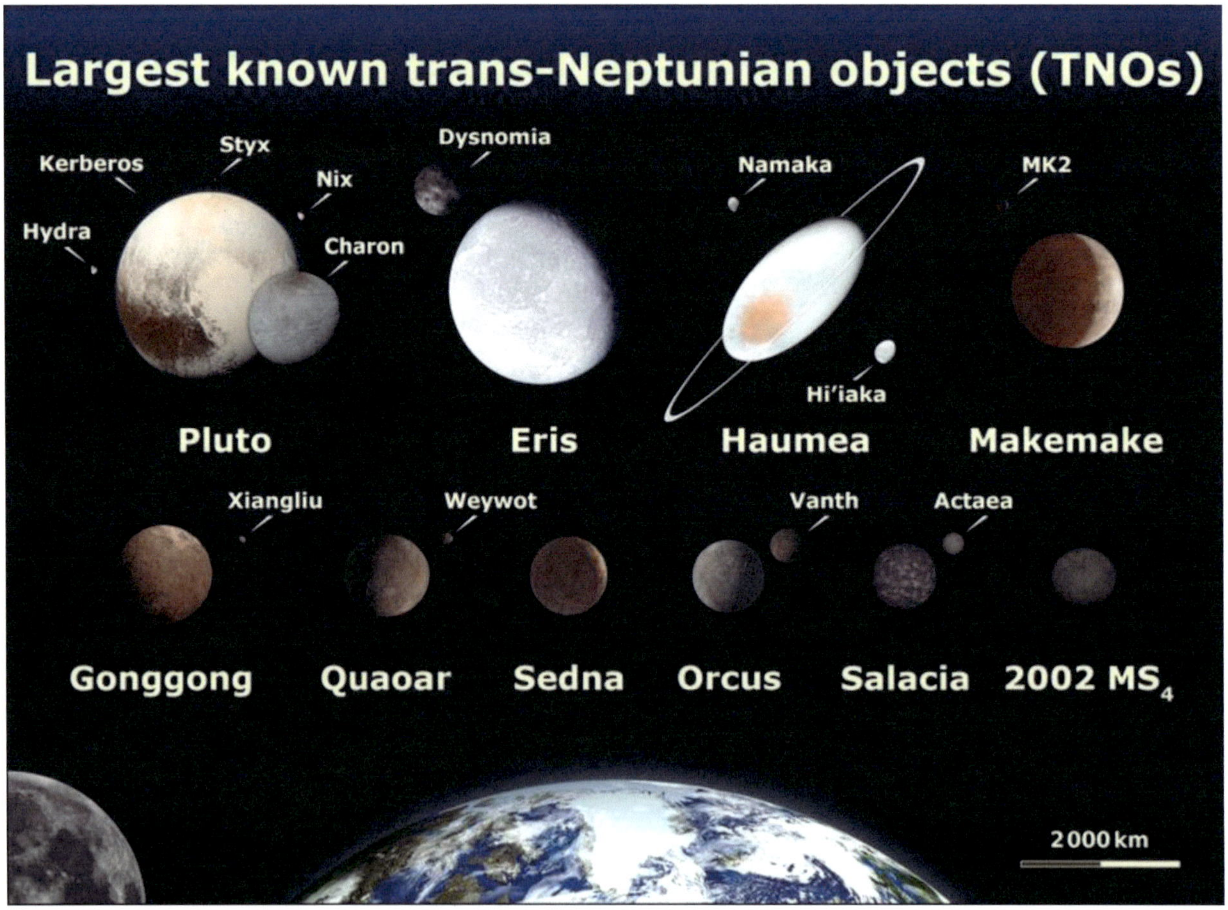

Figure 1. Largest known Trans-Neptunian Objects (TNOs). More recent data have revised upward the diameter of Gonggong; the illustration should depict Gonggong larger than Makemake. All the TNOs are artistic illustrations except for the photo of the Pluto system. Credit: Lexicon, CC BY-SA 3.0 <http://creativecommons.org/licenses/by-sa/3.0/>, via Wikimedia Commons

Below is information on each of the largest known TNOs: diameter, perihelion and aphelion, year of discovery, moons, and any special information. Objects are listed from the largest to the smallest in diameter. The list includes several objects in addition to those in Figure 1, and also lists Ceres (not a TNO) for comparison purposes.

1. Pluto
Has the largest diameter of all known TNOs.
The diameter of its major moon, Charon, is about half that of Pluto.
Diameter: 2,377 kilometers (1,477 miles).
Perihelion-Aphelion: 30-49 AU.
Discovered: February 18, 1930.
Moons: Charon, Hydra, Styx, Kerberos, Nix.

2. Eris
Slightly smaller in diameter than Pluto, but about 28 percent more massive.
Its moon's diameter, 700 ± 115 kilometers, is 25 to 35 percent that of Eris' diameter.
Diameter: 2,326 kilometers (1,445 miles).
Perihelion-Aphelion: 38-97 AU.
Discovered: January 5, 2005.
Moon: Dysnomia.

3. Haumea
Different sources give slightly different dimensions, but it is ellipsoidal—an oddity.
Haumea has rings, which were discovered in 2017 after it passed in front of a star.
Dimensions: approximately 2,100 x 1,680 x 1,074 kilometers (1,300 x 1,040 x 667 miles).
Perihelion-Aphelion: 35-52 AU.
Discovered: 2004.
Moons: Hi'iaka, Namaka.

4. Gonggong
Gonggong is the first major solar system body to have a Chinese name.
Diameter: 1,535 kilometers (954 miles).
Perihelion-Aphelion: 34-101 AU.
Discovered: July 17, 2007.
Moon: Xiangliu.

5. Makemake
Diameter: 1,430 kilometers (889 miles).
Perihelion-Aphelion: 38-53 AU.
Discovered: March 31, 2005.
Moon: S/2015 (136472) 1 (nicknamed MK2).

6. Quaoar
Diameter: 1,110 kilometers (690 miles).
Perihelion-Aphelion: 42-45 AU.
Discovered: June 4, 2002.
Moon: Weywot.

7. Sedna
Sedna is one of the reddest objects in the solar system—almost as red as Mars.
Diameter: approximately 1,000 kilometers (621 miles).
Perihelion-Aphelion: 76-937 AU. A very elongated orbit!
Discovered: November 14, 2003.
No moons.

8. Ceres
Ceres is not a TNO, but a dwarf planet and the largest object in the asteroid belt between Mars and Jupiter. It is inserted here only for reference. It is believed Ceres originated in the Kuiper belt, but moved inward to its current orbit.
Diameter: 952 kilometers (592 miles).
Perihelion-Aphelion: 2.56-2.98 AU.
Discovered: January 1, 1801.
No moons.

9. Orcus
Diameter: approximately 915 kilometers (569 miles).
Perihelion-Aphelion: 30-48 AU.
Discovered: February 17, 2004.
Moon: Vanth.

10. Salacia
Diameter: approximately 850 kilometers (528 miles).
Perihelion-Aphelion: 38-47 AU.
Discovered: September 22, 2004.
Moon: Actaea.

11. 2002 MS_4
Diameter: 778 kilometers (483 miles).
Perihelion-Aphelion: 36-47 AU.
Discovered: June 18, 2002.
No moons.

12. 2002 AW_{197}
Mean diameter: approximately 770 kilometers (478 miles).
Perihelion-Aphelion: 41-53 AU.
Discovered: January 10, 2002.
No known moons.

13. Varuna
Varuna is reddish in color and ellipsoidal in shape due to rapid rotation.
Mean diameter: approximately 670 kilometers (416 miles).
Perihelion-Aphelion: 40-45 AU.
Discovered: November 28, 2000.
Moons: Possible close-in satellite. Unconfirmed.

14. Ixion
Diameter: 710 kilometers (441 miles).
Perihelion-Aphelion: 30-50 AU.
Discovered: May 22, 2001.
No known moons.

15. 2002 UX_{25}
Diameter: 663 kilometers (412 miles).
Perihelion-Aphelion: 36-48 AU.
Discovered: October 30, 2002.
Moon: Has a moon with a diameter of about 210 kilometers.

16. Chaos
Diameter: approximately 600 kilometers (373 miles).
Perihelion-Aphelion: 41-51 AU.
Discovered: November 19, 1998.
No known moons.

17. 2002 TX_{300}
Diameter: approximately 286 kilometers (178 miles).
Perihelion-Aphelion: 38-48 AU.
Discovered: October 15, 2002.
No known moons.

Various references were used. The diameters of the TNOs given are the best estimates at the time of this printing, but have undergone past revisions and will probably be revised in the future as new measurements are taken.

The search for truth
takes you where the evidence leads you,
even if, at first, you don't want to go there.

—Bart D. Ehrman—Professor, UNC at Chapel Hill

TOPIC 44. The fate of the Sun.

Here we discuss the evolution of the Sun from the present time until it becomes a white dwarf.

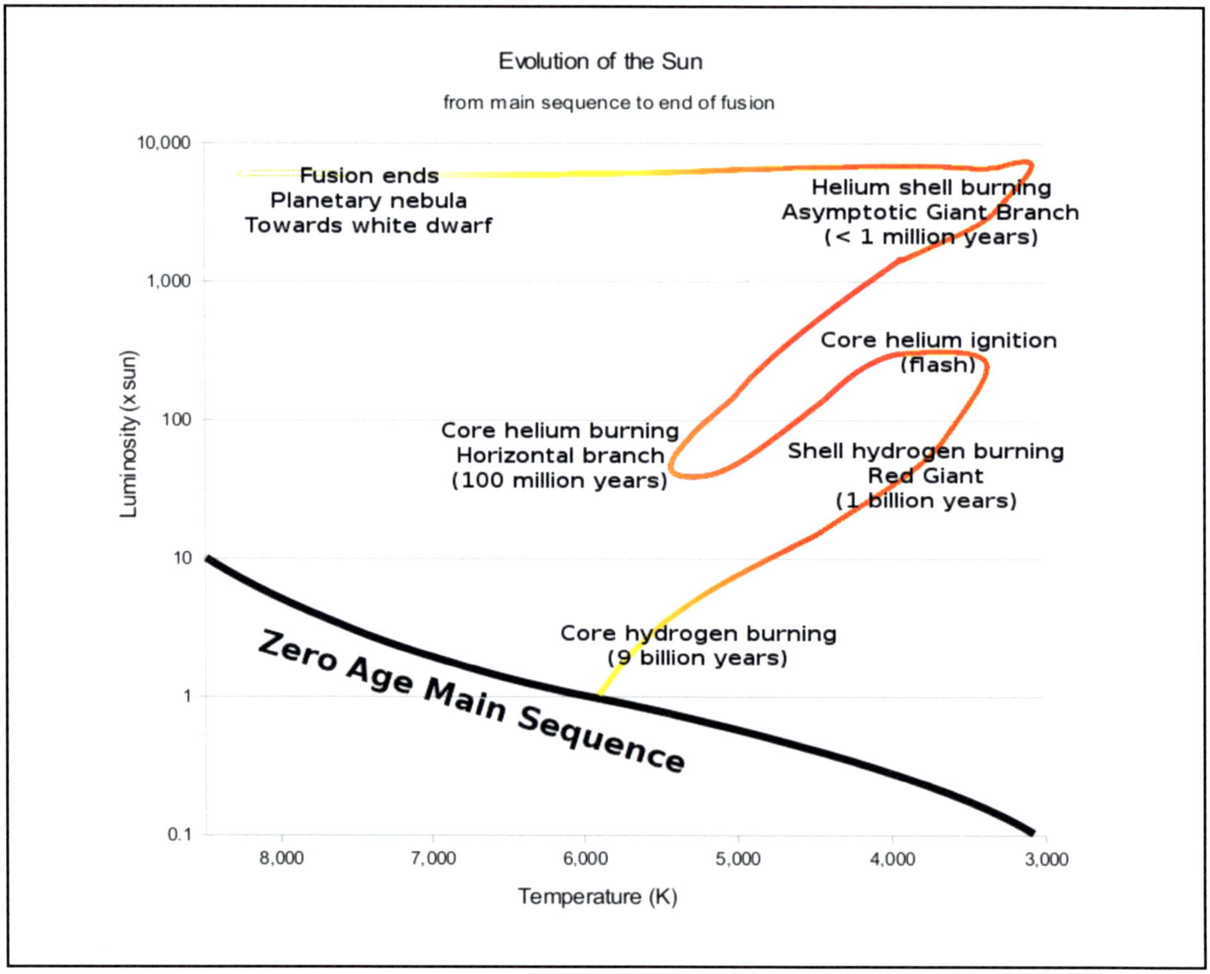

Figure 1. Evolution of the Sun: Luminosity vs Temperature. The diagram starts with the Sun's current position on the Hertzsprung-Russell diagram's main sequence (see Topic 49) which is where the bold, black curve crosses the horizontal luminosity line 1 (a unit of one indicating the current luminosity of the Sun). The Sun has been in this same position for about 4.5 billion years and leaves this position in about 4.5 billion more years, traversing the yellow/red curve upward until all fusion ends and we have a planetary nebula. The Sun will become a white dwarf. If we were to continue the curve from the upper left (not shown in the diagram) the curve would circle downward and to the right until it crosses the bottom horizontal line (indicating a luminosity of one-tenth of the current luminosity), its temperature continuously diminishing. Credit: Lithopsian, CC BY-SA 3.0 https://en.wikipedia.org/wiki/File:Evolution_of_a_sun-like_star.png

The Sun will live in a fixed position on the *main sequence*, where its energy comes from the fusion of hydrogen nuclei (or protons) into helium nuclei (each consisting of two protons and two neutrons), for about nine billion years. We are about halfway through this period. The core of the Sun, about 20-25 percent of the solar radius, is where almost all fusion takes place.

The main fusion process (called a proton-proton process) involves hydrogen nuclei under tremendous pressure and temperature (and requires a phenomenon called *tunneling* in quantum mechanics) so that two protons can get close enough to each other for the strong force between the protons to overcome the electric repulsion between like charges (each proton has a positive charge), resulting in a fusion of the two. The main proton-proton process (there actually are three) is given in Figure 2.

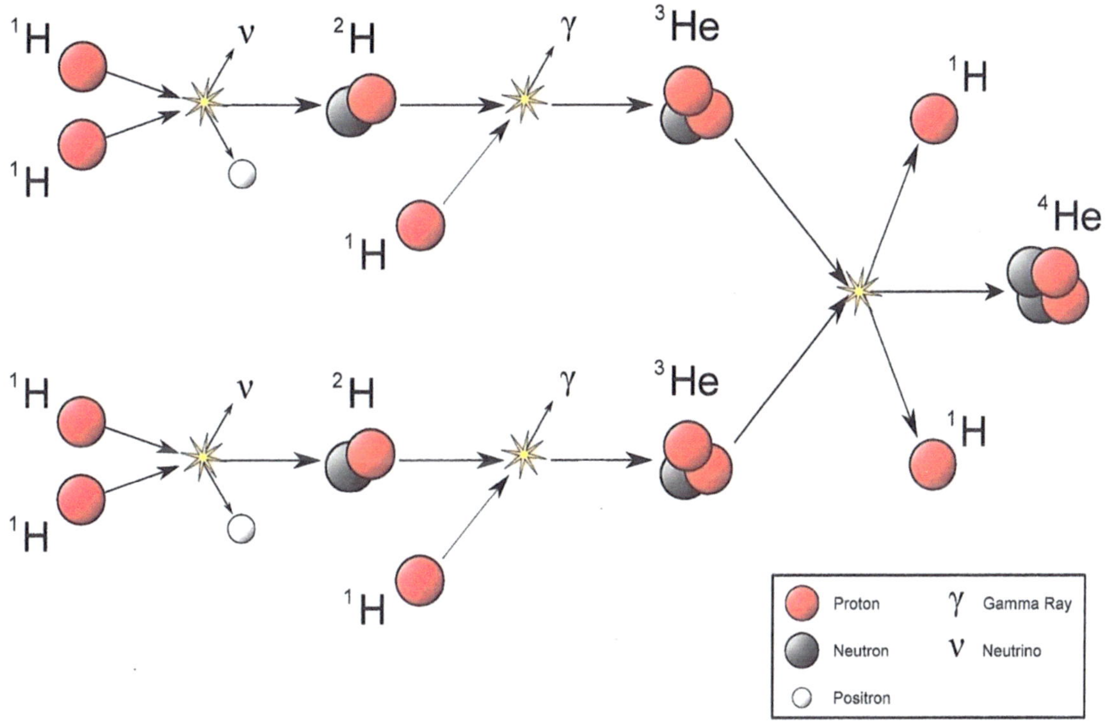

Figure 2. The main proton-proton fusion process producing a helium nucleus from hydrogen nuclei. Credit: Doctor C - Own work, CC BY-SA 4.0, https://commons.wikimedia.org/w/index.php?curid=48960438

A positron is an election in every respect, except it has a positive charge. Neutrinos are almost (but not quite) massless particles that very weakly interact with other matter. Almost all neutrinos striking the Earth pass completely through the planet. A deuterium nucleus consists of a proton and a neutron. A helium-3 nucleus consists of two protons and one neutron.

Where does all the energy come from? If one subtracts the mass of four hydrogen nuclei from the mass of a helium nucleus, one discovers the answer is not zero. Mass is lost in the process. (The mass of the two positrons and two neutrinos is negligible compared to the mass of a proton or neutron.) From Einstein's equation $E = mc^2$, this lost mass m is equivalent to energy. This energy is in the form of light (at various wavelengths).

A helium nucleus occupies less space than the four hydrogen nuclei. So as helium is created the core shrinks and the material surrounding the core moves farther inward. This increases the pressure inside the core, which increases the rate of fusion. Consequently the luminosity of the Sun increases. The increased pressure causes the Sun to expand. The Sun leaves its main sequence phase, entering its *red giant phase* where it remains for about one billion years. During this period, the core temperature and pressure is not high enough to initiate fusion of the helium in the core into carbon. With little hydrogen in the core, hydrogen fusion *in the core* diminishes, occurring instead in a shell surrounding the core.

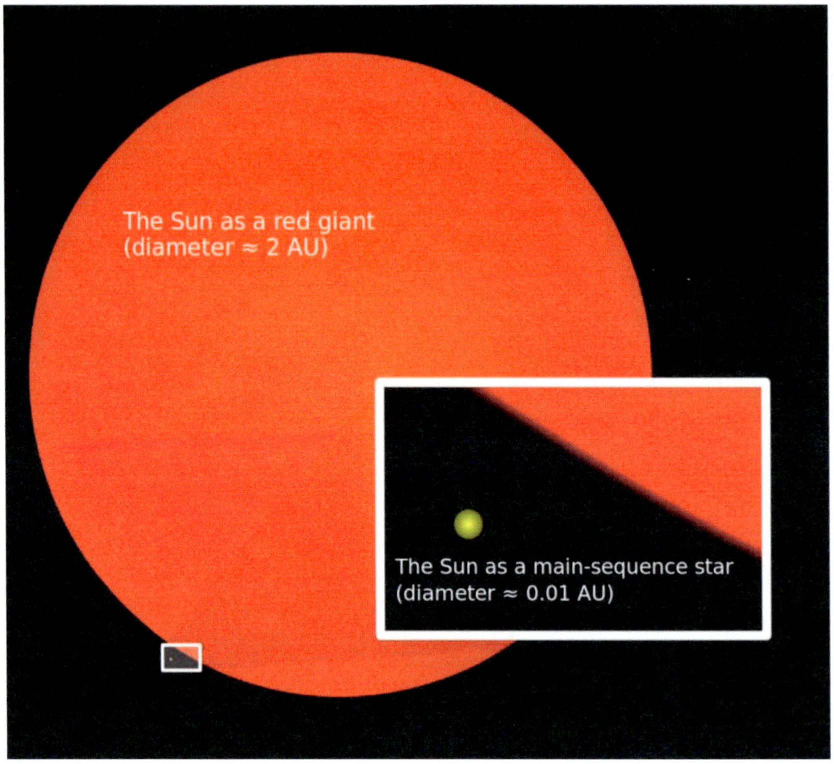

Figure 3. Comparison of the size of the Sun now and when it is a red giant. Credit: CC BY- SA 3.0

At the end of this approximately one-billion-year period something very dramatic occurs: the *helium flash*. Fusion of helium into carbon begins and within minutes an estimated six percent of the electron-degenerate helium core (about 40 percent of the Sun's mass) is converted into carbon.[1] (See the end of this topic for a discussion of electron-degeneracy.) Some of the carbon together with helium produces oxygen. The Sun undergoes a metamorphosis.

It will then enter the *horizontal branch (or subgiant stage)*, a brief period (astronomically speaking) of about 100 million years, where core helium burning occurs.

Afterward comes *the asymptotic giant branch period* (AGBP). The early part of this period lasts less than one million years. Hydrogen burns in a shell with helium burning in a deeper shell all surrounding a central core of carbon and oxygen. After a period of time the Sun becomes unstable and pulsates, shedding gas, and creating a strong, solar wind. This later period of the AGBP lasts about 20 million years.

Finally, with perhaps half of its mass ejected as ionized gas, the Sun is a *planetary nebula*. When all fusion ceases, what is left is a *white dwarf* whose size is about that of the Earth. Its energy then comes not from fusion but from stored thermal energy. The white dwarf will slowly lose luminosity over many billions of years—virtually the life of the universe.

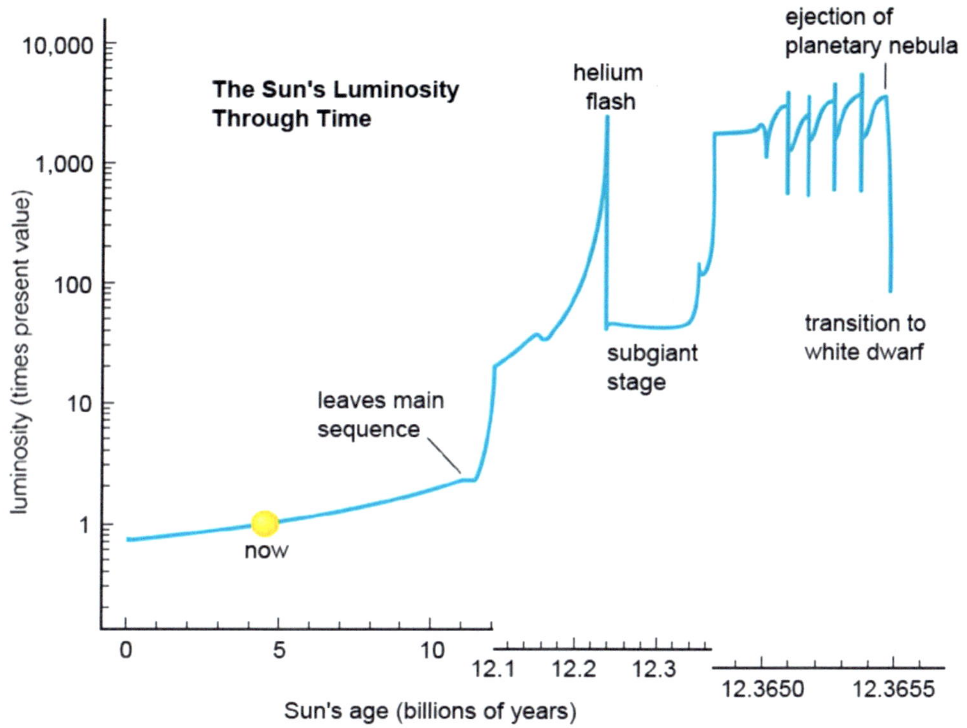

Figure 4. Luminosity of the Sun vs the age of the Sun. The subgiant stage in this diagram is identical to the horizontal branch in Figure 1.
Credit: https://faculty.wcas.northwestern.edu/~infocom/The%20Website/end.html

Prior to becoming a white dwarf, the Sun will expand to engulf Mercury, Venus, and probably Earth. In some locations in the solar system, e.g., the Galilean moon Europa, the ice cover may sublimate, creating an atmosphere, and maybe this moon will become a liquid-water world. Long after all life on Earth has perished, if there is life in Europa's ocean under the current ice covering, it might be that this (and perhaps life in water under ice coverings of some of the other outer moons in the solar system) will be the last life left in the solar system. And when the Sun is a white dwarf and the outer solar system is again in deep freeze, life in the water of Europa could continue to exist, since the energy source generating heat under the water comes not from the Sun but from the tidal stretching of Europa by the gravitational fields of Jupiter and its moons Io and Ganymede.

Planetary nebulae and white dwarfs.
Why the name planetary nebula? When planetary nebulae were first observed with telescopes, their disk-like shapes and substantial diameters resembled planets, though they clearly were not. At the time it was unknown that they were a dying phase in the life of a star. Figure 5 is an example.

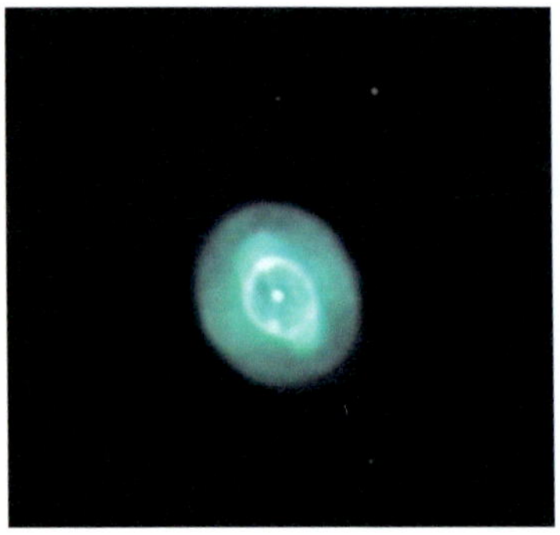

Figure 5 (left). The Ghost of Jupiter Planetary Nebula NGC 3242. Credit: Asam Block/NOAO/AURA/NSF

Figure 6 (right). Sirius A with Sirius B at 7 o'clock. Credit: NASA, ESA, H. Bond (STScI), and M. Barstow (University of Leicester)

The closest white dwarf to Earth is Sirius B. See Figure 6. It is a companion to Sirius A, about 8.6 light years distant and the brightest star in apparent magnitude seen from Earth. Sirius B's diameter is slightly less than that of Earth, yet it has the mass of our Sun. Could you pick up a cubic inch of this star's matter? It weighs about 41.5 tons. Sirius B orbits Sirius A in 50 years and is only about 220 million years old. It is visible through amateur telescopes, especially if the brilliant Sirius A is covered by an occulting bar and it is observed when at its maximum distance from Sirius A, about 11.3 arc seconds.

Further discussion—degenerate electron matter.
For those with physics backgrounds, here is one more point—a technical one, but quite interesting. Just previously it was mentioned that the *tunneling* effect in quantum mechanics is vital in getting hydrogen nuclei in the core of a star close enough to cause fusion. Without this effect, high temperatures and pressures would not suffice, and we would have no hydrogen fusion and thus no stars. Yet another idea from quantum mechanics is relevant here: *degenerate electron matter.*

In the current discussion this pertains to the helium nuclei in the core of the Sun along with many "free" electrons before the helium flash occurs. A class of particles called fermions, which includes protons, neutrons, and electrons, obeys a quantum mechanical rule called *Pauli's Exclusion Principle*. This says that no two fermions can simultaneously have identical values for all their quantum states. For electrons this means no two electrons can simultaneously be in the same location, have the same momentum and have the same energy. It is best to think of an electron as a cloud. Yes, that's right. By Heisenberg's Uncertainty Principle an electron can never have its location and momentum simultaneously precisely determined. This has nothing to do with how clever the experiment is to determine the location and momentum of an electron. It is impossible. Period. The location of an electron can only be given with a probability. If the location of an electron is known with a high probability, then the momentum becomes less certain, and vice-versa. Also, the energy of an electron, e.g., when in an atom, can only take on specific, discrete values. In the helium core of the Sun electrons are not in atoms. We have helium nuclei and electrons not bound to any of these nuclei. But they are all in close contact and with similar momenta. The mixture is more like a liquid, and Pauli's Exclusion Principle applies here. The higher the pressure and temperature, the higher the probability that two electrons can have the same location and momentum. In this case, two such electrons must have different energy states. Trillions and trillions of electrons are in the same tiny volume of space. By Pauli's Exclusion Principle, many electrons are forced to have different energy levels. But only one electron can be at the lowest energy level. And only one at the next highest level. And so on for trillions of values. This situation causes the energy level in the core to rocket upward, igniting the fusion of helium into carbon.

References

1. The end of the Sun. https://faculty.wcas.northwestern.edu/~infocom/The%20Website/end.html

All these worlds are yours, except Europa.
Attempt no landing there.
Use them together. Use them in peace.

—Arthur C. Clarke, 2010: Odyssey Two

TOPIC 45. Surface flexing and strange cycloidal loops on the surface of Europa.

Europa circles Jupiter every 3.55 Earth days or about 85 Earth hours, always keeping the same face toward its planet. The gravitational pull of Jupiter results in the surface ice shell flexing upward by perhaps 30 meters,[1,2,3,4] the height of a seven-story building, at points on Europa's surface directly facing Jupiter and at points on Europa 180 degrees from those points (i.e., on Europa's opposite side).

You might think that at any point on Europa's surface the bulging up doesn't change (so no tides) since the moon always has one face locked to Jupiter. But the situation is not stable. While there are no tides in the sense of tides on Earth caused by our Moon and Sun, Europa's orbit around Jupiter is not a perfect circle, which means that Europa changes its speed in circling Jupiter. This causes the bulges to shift their positions slightly. It is thought that the shifting can result in tension cracks opening up and propagating along the surface creating cycloidal loops or scallops similar to those on Europa's surface.

Gregory Hoppa and Randall Tufts at the University of Arizona used a computer model of tidal stresses to explain how these loops could form.[1,2,3,4] It required assuming there was a water ocean under the ice as otherwise the bulging would only be at about one meter. Their work supported the idea (now widely believed) that there is a liquid-water ocean not far below the ice.

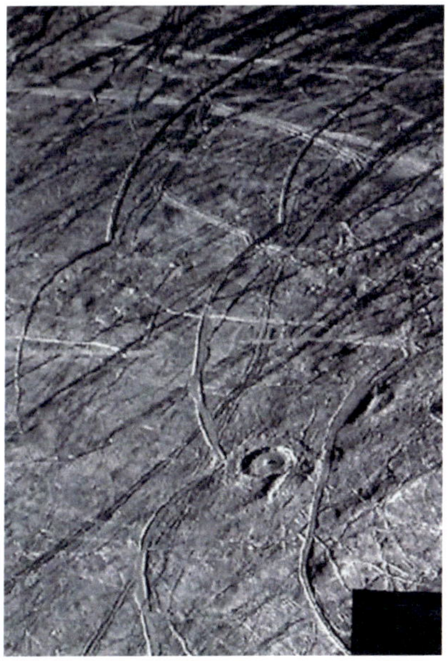

Figure 1. Scallops on the surface of Europa. Credit: Moses Milazzo (PIRL/University of Arizona)

References

1. (December 1999). Europan cracks tied to tides. *Sky & Telescope,* p 17.
2. Anderson, Charlene (January/February 2000). Searching for life on Europa. *The Planetary Report,* p 12.
3. Pappalardo, Robert (July 2004). Jupiter's water world. *Astronomy Magazine,* p 34.
4. Mackenzie, Dana (August 2001). Is there life under the ice? *Astronomy Magazine,* p 33.

In a building with apartments...you want to make connections. Life is easier that way. There's salt if you don't have salt; you can knock at someone's door....

—French actress Juliette Binoche

TOPIC 46. Chaotic terrain and table salt on the surface of Europa.

When planetary scientists at Caltech and NASA's Jet Propulsion Laboratory in Pasadena, California, simulated conditions on Jupiter's moon Europa in the laboratory, they found that white table salt turned yellow when irradiated, allowing them to deduce that the yellow color visible on portions of the surface of Europa is actually sodium chloride.[1] The JPL experiments matched temperature, pressure, and electron radiation conditions at Europa's surface. Tara Regio is the yellow area in Figure 1, a region of chaos that is geologically younger than the surrounding area. The images in Figure 1 were taken in June 1997. The JPL study, "Sodium chloride on the surface of Europa," was published 22 years later in *Science Advances* on June 12, 2019.

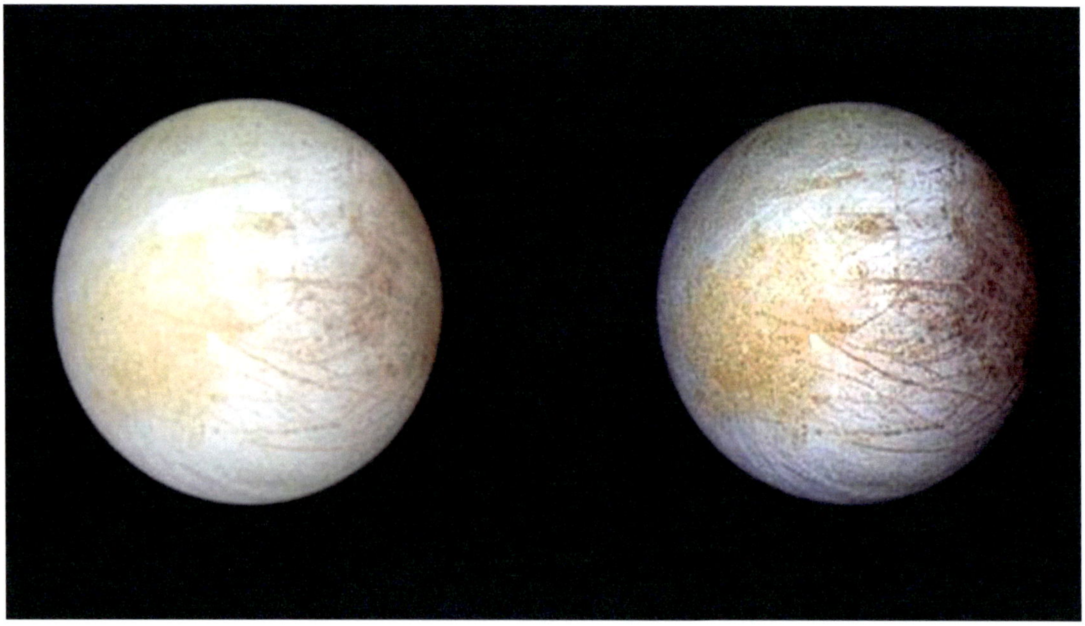

Figure 1. Tara Regio is the yellowish area to left of center in this NASA Galileo image of Europa's surface. An abundance of sodium chloride was discovered in this region of geologic chaos. The view on the left is in natural color and on the right is in enhanced color designed to bring out subtle color differences in the surface.
Credit: NASA/JPL/University of Arizona

Figure 2. A mosaic of Galileo images shows the region Tara Regio containing the mitten-shaped Murias Chaos on Europa. Image data is from Galileo's E15 flyby on May 31, 1998. Credit:NASA/JPL/ Processed by Kevin M. Gill

It is known that Europa has a global ocean under its icy, billiard-ball surface. While it has not been established that the salt comes from water below the ice, this is a strong possibility. It is believed that the bottom of the ocean consists of rocky material (not another layer of ice) and this material is heated due to the tidal stretching of Europa by Jupiter's gravity and by the gravity of the moons Io and Ganymede. This action could produce fumaroles similar to those found at Earth's mid-ocean ridges. We know many life forms are found at such fumaroles on Earth, deriving their energy from minerals ejected into the water. And once you have microbial life you have initiated a vertical food chain leading to larger creatures. Could this be happening now on Europa?

In addition to Tara Regio there are several other areas of chaotic terrain on Europa. A famous early example is that of the Conamara Chaos area, a 28 x19 mile portion of which is shown below. It is named after the rugged part of western Ireland named for Conmac, son of the Queen of Connacht.[2]

Figure 3. Conamara Chaos *"iceberg" region. Credit: NASA*

The image above strikes one as that of individual sheets of ice having once been floating on liquid water, moving and rotating in various directions, and then refreezing in place. Perhaps upwelling liquid water or warmer ice breached the surface and created this pattern.

References

1. NASA. (June 18, 2019). *Table salt compound spotted on Europa.* nasa.gov. https://www.nasa.gov/feature/jpl/table-salt-compound-spotted-on-europa
2. Gazetteer of Planetary Nomenclature. (1997). *Conamara Chaos.* USGS Astrogeology Research Program. https://planetarynames.wr.usgs.gov/Feature/1282

TOPIC 47. Carbon stars (which are giants) and dwarf carbon stars (the first of which was found in the 1970s).

Ordinary carbon stars (giants).
Carbon stars are beautiful. They can be a deep red and are the outstanding examples of night sky objects capable of being seen as having color through amateur-sized telescopes. There exist bright red, giant stars such as Antares, Betelgeuse, and the Garnet Star (Mu Cephei), but these are not carbon stars, our eyes seeing them as being closer to orange. Ordinary carbon stars are red giant, variable stars which have an excess of carbon in their atmospheres. They constitute a relatively rare class of stars, but being red giants—and thus bright—many of them are easily visible in small, amateur telescopes.

Figure 1. The carbon star V Crb (= HD 141826) in the constellation Corona Borealis. Credit: Posted on Stargazers Lounge by Martin Meredith on November 11, 2016. https://stargazerslounge.com/topic/281150-herschels-garnet-star/

Figure 2. A contrast in colors—the carbon star DY Cru at 1 o'clock from the first-magnitude star Beta Cru in the southern constellation Crux. The size of Beta Cru is exaggerated by an over-exposure in order to image DY Cru. Credit: Mick Laws, Astronomical Society of New South Wales

Why are carbon stars so much redder than ordinary red giant stars?
Ordinary red giants get their color from their cooler surface temperatures. In Topic 44 (The fate of the Sun) we discussed the fusion of hydrogen into helium and then later the fusion of helium into carbon plus some oxygen. Convection inside the star transports the carbon and oxygen from the core to the star's surface where carbon monoxide gas (CO) is formed. CO is a colorless gas. But if the star has considerably more carbon than oxygen, the excess carbon surrounds the star with a shell of carbon

soot/dust and smoke, consisting of particles of amorphous carbon (free, reactive carbon that does not have any crystalline structure) and silicon carbide or carborundum (SiC). These particles scatter the shorter wavelengths of light at the blue end of the spectrum. Only the red/orange light that doesn't get scattered can escape, giving the star a ruby-red color. Red sunsets and sunrises on the Earth are similar phenomena. Here *Rayleigh scattering* (the instantaneous absorption and emission of blue wavelength photons by atmospheric nitrogen molecules) lets only reddish wavelengths pass straight through to our eyes. (The scattered, blue photons make our sky blue.)

In addition to the shell of carbon particles scattering blue light in carbon stars, various carbon molecules in the atmosphere of a carbon star absorb light in the blue portion of the spectrum (their *Fraunhofer lines* have wavelengths in this part of the spectrum). C_2, CN, and CH are among the more common of such molecules.[2,7] (A Fraunhofer line in the continuous spectrum of a star is a dark absorption line due to an element in the star's photosphere absorbing a photon of a particular wavelength and then re-emitting it in a random direction. So photons of this wavelength that would have reached our eyes are absorbed and then scattered in all directions, with very few of them reaching our eyes. This causes the spectral line to be dark. Every element has its own unique set of Fraunhofer lines, which allows astronomers to discover what elements and molecules are present in stars. In this way the element helium was first discovered in the Sun and then on Earth.)

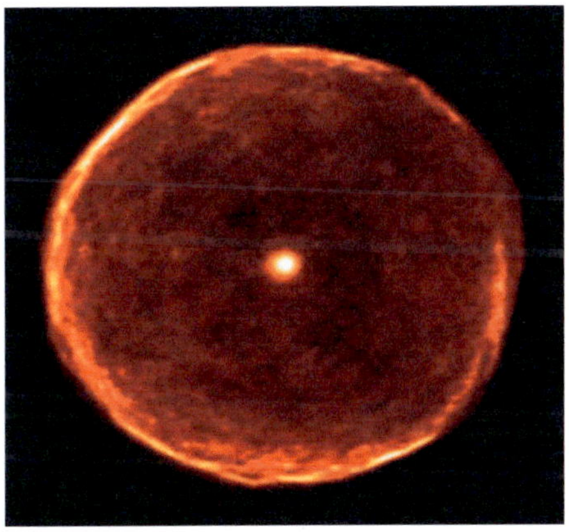

Figure 3. This ALMA (Atacama Large Millimeter/ submillimeter Array) image shows a very thin, spherical shell around the carbon star U Antliae in the southern constellation Antlia (The Air Pump). About 2,700 years ago the star ejected the material in this shell at high speed over a period of only a few hundred years. Credit: ALMA/ESO/NAOJ/ NRAO/F. Kerschbaum

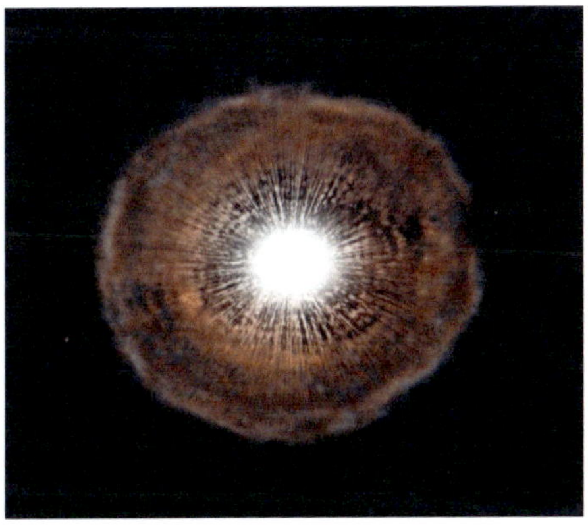

Figure 4. Hubble image of the carbon star U Cam and its bubble located in the northern constellation Camelopardalis (the Giraffe). Image taken in 2012. Credit: ESA/Hubble, NASA, and H. Olofsson (Onsala Space Observatory)

Note: U Cam itself is actually much smaller than it appears in Figure 4. In fact, the star would easily fit within a single pixel at the center of the image. Its brightness, however, is enough to saturate the camera's receptors, making the star look much bigger than it really is. The star is running out of fuel and is becoming unstable. Every few thousand years, it coughs out a nearly spherical shell of gas. The gas ejected in the star's latest burp is clearly visible in this picture as a faint bubble surrounding the star.

Color.

Most objects in the night sky have little or no color to them. In contrast, during the day we are surrounded by color. That is because so few photons strike our retinas from deep-sky objects. The larger the diameter of the mirror or the lens on our telescope, the more photons are directed to our retina, hence the brighter we perceive the object and the more likely we are to detect color. The retina consists of two general types of light-sensitive receptors—rods and cones. The cones surround the center of our visual field (called the fovea) and are capable of detecting wavelengths of light that we perceive as color. The rods cover the rest of the retina and are far less sensitive to color, perceiving basically black and white. However rods are more sensitive to faint light than cones. That increased sensitivity of the rods to faint light is why using *averted vision* (not looking directly at an object, but looking at it from the side) is a technique for making a faint celestial object brighten. Still, for detecting color one should use the cones by looking directly at the object.

Dwarf carbon stars.[1]

The first dwarf carbon star, G77-61 in the constellation Taurus, was discovered in the 1970s. Though only 255 light years distant, it is quite dim. Being red and being dim at such a relatively short distance, G77-61 was at first thought to be a red dwarf, the most common type of star in our galaxy. As the name implies, red dwarfs are small, slowly fusing hydrogen into helium over a life span of trillions of years, many times the age of the universe. They are almost immortal. But G77-61 is not a red dwarf. In January 1976 the spectrum of G77-61 was taken at the Lick Observatory and, surprise, surprise, it was revealed to be a carbon star. Red dwarfs (unlike red giants) don't produce carbon. What could possibly be going on? Astronomers now believe they have the answer. Appearing dim from a relatively short distance, the star must be small. And the considerable carbon in its atmosphere must have been acquired from an external source. An intricate, convoluted process is believed to have led to its current state:

1. A red dwarf and a blue star (a few times as massive as the Sun) form as a binary.
2. After using up the hydrogen in its core, the blue star expands into a red giant. Helium in its core ignites to make carbon and oxygen.
3. Some of the carbon rises to enter the giant's atmosphere (perhaps turning the star into an ordinary carbon star).
4. The giant sheds mass, some of which falls onto the red dwarf, transforming it into a carbon dwarf.
5. The giant then casts off its atmosphere, exposing its hot core as a white dwarf.
6. Billions of years later the white dwarf fades so that observers can no longer see it.

And indeed, the recent discovery of a white dwarf companion to G77-61 confirms the scenario suggested above. The two stars orbit each other in 245 days.

Since then, many dwarf carbon stars have been discovered, and it is now believed that they vastly outnumber the giant carbon stars. Observational results show that the number of dwarf carbon stars with white dwarf companions is consistent with the theory that all of them have such a companion. The stellar halo of our galaxy contains as many as 30 percent to 60 percent of the dwarf carbon stars; most of the remainder lie in the thick, inner part of the galactic disk. All are older stars. It is interesting that halo stars possess little oxygen as they were formed before supernova explosions ejected much oxygen into interstellar space. So it is easier for a blue halo star to have much more carbon than oxygen. When it becomes a red giant it doesn't have to transfer as much gas (including oxygen and carbon) to its future dwarf carbon star companion. Remember that any carbon atoms that combine with oxygen atoms (forming carbon monoxide) are not carbon atoms that are available for making the star a carbon star.

A few more facts about the dwarf carbon star G77-61.
It is the brightest (13.9 in apparent magnitude) and the closest of the known dwarf carbon stars. It is in the galaxy's halo.

Its motion against the background stars is easily detected in amateur telescopes on a *monthly* basis. Why should this be? It is not because the halo star is moving so fast with respect to the center of the galaxy. The Sun and with it the Earth are revolving in the disk of the galaxy, as are most of the stars that we see in the sky. Since all the disk stars are revolving in the same direction about the center of the disk, our motion relative to any other nearby disk star is small. But a halo star is above or below the center of the galactic disk and roughly stationary with respect to the galaxy's center. Hence we and the nearby background stars are moving fast with respect to halo stars. In fact, detecting a star that moves rapidly with respect to the background stars is a clue that it might be a halo star.

Viewing carbon stars.
With most variable stars when the star is brighter the color pales. So carbon stars are a darker red near their minima and less so near their maxima.[6]

The Purkinje effect or shift. [3,4,5,8]
This reaction is a shift in maximum sensitivity of the eye to the wavelength of light from 550 nanometers during the day to 500 nanometers at night. The shift in sensitivity is toward the blue direction of the spectrum and is due to the combined action of the cones and rods.

An example: a red rose dims as evening approaches, going from red to a darker red to black. But as evening comes on, a blue flower will brighten. The combined action of rods and cones changes as light dims.

The Purkinje effect can also be noticed if one stares at a red star at night. The star will become a brighter red, as in staring we are mainly using the cones (so we are changing the combined action of the rods and cones). Suppose one wishes to estimate the brightness of a carbon star (it being a variable) to that of a nearby star of a different color, e.g., white. The Purkinje effect makes it more difficult to get an accurate estimate. More generally, suppose two different colored stars are perceived to be equally bright. If the background sky changes in illumination, they will no longer seem to be equally bright. Likewise, if both stars increased in brightness, say by 20 percent, they would no longer be perceived as equally bright. In the case of a red star and a white one, this effect is magnified.

B-V color values.
A value can be given to a star that indicates its color. One measures a star's brightness by passing its light through a blue ("B") filter, filtering out all wavelengths not close to blue. Then one does the same using a visual ("V") filter. This filters out all colors except for wavelengths close to the green-yellow band. This band is where the human eye has its greatest sensitivity. It turns out that the B-V value correlates well with the temperature at the star's photosphere (surface). The higher the B-V value, the redder the star. The hottest blue stars have a value of about -.25; white stars have a value of about 0.2

to 0.3; the reddish M type supergiants have a value of about 1.85. Carbon stars have an approximate range from 2 to 5.8.

Here are some sample B-V values of familiar bright stars: The Sun 0.65, Vega and Sirius 0.0, Polaris 0.60, Antares and Betelgeuse about 1.85, and bluish Rigel is at -0.03. Delta Orionis (Mintaka) at the top of Orion's belt is even bluer at -0.22.

A few of the brighter and redder carbon stars.

1. R Leporis (Hind's Crimson Star) in Lepus (The Hair)

In 1845 John Hind described this red star as "like a drop of blood on a black field." Its magnitude varies every 432 days from 5.5 to 11.7. Its maximum also varies over a long, 40-year secondary cycle from 5.5 to 6.5 and back. Like other red variables in general and carbon stars in particular, the star's color at maximum is not as intense as at minimum: at maximum it is only copper-hued, but near minimum it approaches true crimson. B-V = 5.74.

2. V Hydrae in Hydra (The Water Snake)

V Hydrae fluctuates in a period of roughly 17 months superimposed on a longer cycle spanning 17 years. The effect of the two periodicities yields an overall magnitude range of 6.0 to 12.3. B-V = 5.5.

In 2016 the Hubble Space Telescope detected giant blobs of gas, each twice the mass of Mars, being ejected near V Hydrae.[9] It is estimated such blobs are fired every 8.5 years and have been ejected for at least 400 years. These could not emanate from V Hydrae itself. The leading theory is that they are fired from a companion star in an elliptical orbit around V Hydrae, coming close to V Hydrae every 8.5 years. In fact, its closest approach would occur inside V Hydrae's bloated atmosphere. See the depiction below. If true it is another illustration that truth can be stranger than fiction.

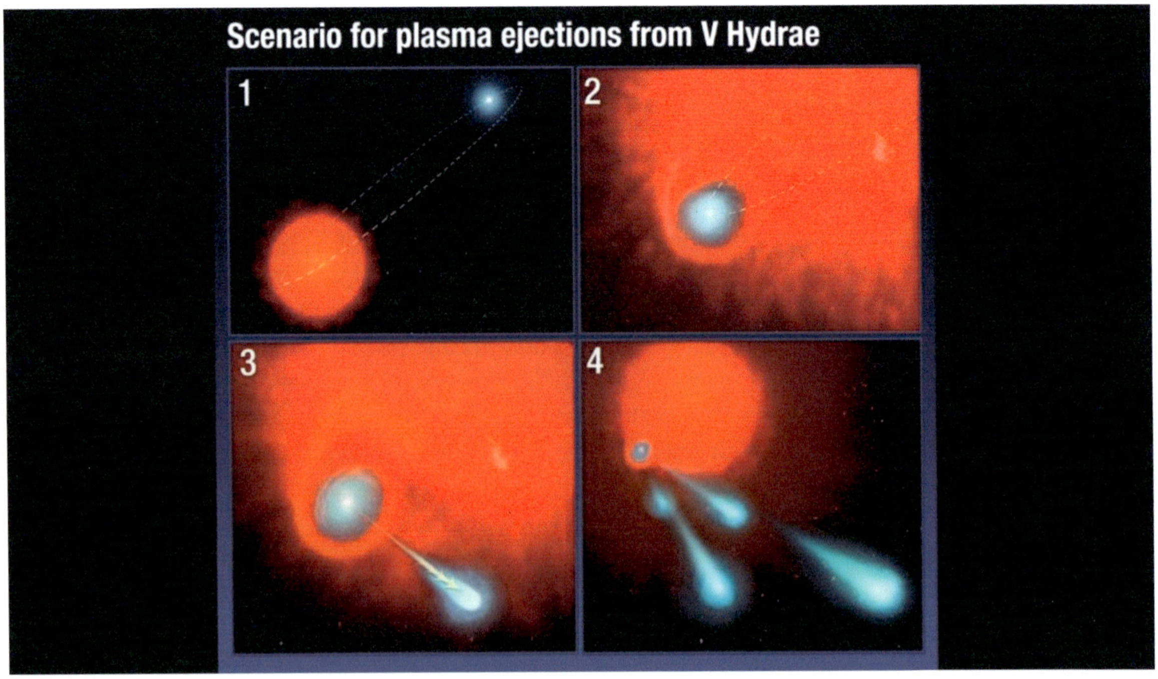

Figure 5. Scenario for plasma ejections from V Hydrae. Credit: NASA, ESA, and A. Feild (STScI)

3. T Lyrae in Lyra (The Lyre)
Magnitude varies from 7.5 to 9.2 with an irregular period. B-V = 5.5.

4. TX Piscium = 19 Piscium in Pisces (The Fish)
An irregular variable star which fluctuates between magnitudes 4.8 and 5.9 every 220 days or so. While its B-V is only 2.6, it is the brightest carbon star in the sky.

5. DY Cru = Espin-Birmingham 365 (Ruby Cru) in Crux (The Cross)
DY Cru has an irregular or unknown period, and its magnitude varies from 8.4 to 9.9. B-V = 5.8. It is only 2.4 arc minutes west of Beta Crucis. See Figure 2 above.

Since images of V Crb, U Antliae, and U Cam appear above, we list their respective B-V values: 4.4, 2.9, and about 4.2.

A word of caution regarding B-V values. Different sources quite often list different values for the same star due to variations in the filters and the photomultipliers that are used. And of course if a star is a variable, as is the case for all carbon stars, there is the issue of at what point in the period of variability the measurements were taken.

References

1. Croswell, Ken (February 2020). Dwarf carbon stars. *Sky & Telescope,* p 58-61.
2. Allen, Jesse S. *The classification of stellar spectra.* http://www.star.ucl.ac.uk/~pac/spectral_classification.html
3. Levy, David H. (December 2002). Red lights in the sky. *Sky & Telescope*, p 81-83.
4. King, Bob (December 3, 2014). Carbon stars will make you see red. *Sky & Telescope.* https://skyandtelescope.org/astronomy-blogs/carbon-stars-will-make-see-red1203201401/
5. Sidgwick, J. B., & Gamble, R. (1980). *Amateur Astronomer's Handbook*, Dover Publications. p 429.
6. MacRobert, Alan (January 2010). The long rise of R Leporis. *Sky & Telescope*, p 61-62.
7. Skiff, Brian A. (May 1998). Carbon stars: Reddest of the red. *Sky & Telescope*, p 90-95.
8. Davson, Hugh, & Perkins, Edward S. (August 7, 2020). Human eye—anatomy. *Encyclopedia Britannica.* https://www.britannica.com/science/human-eye
The Purkinje shift has an interesting psychophysical correlate. It may be observed, as evening draws on, that the luminosities of different colors of flowers in a garden change. The reds become much darker or black, while the blues become much brighter. What is happening is that, in this range of luminosities, called mesopic, both rods and cones are responding, and, as the rod responses become more pronounced—i.e., as darkness increases—the rod luminosity scale prevails over that of the cones.
9. NASA/JPL. (Oct 6, 2016). *Hubble detects giant 'cannonballs' shooting from star.* jpl.nasa.gov. https://www.jpl.nasa.gov/news/news.php?feature=6639

We must heap up a great pile of doing for a small diameter of being.

—Henry David Thoreau

TOPIC 48. How do you determine the diameter of a star?

A clever way to determine a star's diameter uses the luminosity and temperature of a star together with the star's distance from Earth. In fact, this method is the main way stars' diameters are determined.

Alternatively, one might think it possible to determine a star's diameter by directly measuring the angle that the star's disk subtends at the Earth (in which case the diameter of the star is simply the angle in radians times the star's distance). But this is not practical as, even with a space telescope, the angle would be tiny and diffraction would limit accuracy. However, this direct method has given an angle of 0.042 arc seconds for the supergiant star Betelgeuse.[1] Because of uncertainties in the star's distance, its diameter can only be said to be roughly 700 times the diameter of our Sun.

The equation for the diameter D of a star in terms of its luminosity L and temperature T is

$$D = (\sigma\pi)^{-1/2} L^{1/2} / T^2 . \qquad (1)$$

The terms and symbols in the equation need some explaining.

L is the luminosity which is the total energy per second emitted by the star over its entire surface. L is also called the total energy flow of the star. The luminosity of the star does *not* depend on its distance from the Earth. Luminosity is a constant for a star (ignoring the aging of the star). The brightness b of the star as seen from Earth *does* depend on the star's distance from us. b is directly proportional to the luminosity of the star and inversely proportional to the square of the star's distance d.

$$b = L/4\pi d^2 . \qquad (2)$$

Since $4\pi d^2$ is the surface area of a sphere of radius d centered on the star, b is the luminosity of the star per unit area of this sphere.

T is the star's temperature using the Kelvin scale; σ is the Stefan-(Ludwig) Boltzmann constant, whose value doesn't concern us now as we aren't going to do any calculations.

Equation (1) is an immediate consequence of the *Stefan-Boltzmann Law*

$$L = \pi D^2 \sigma T^4 . \qquad (3)$$

Notice that this equation indicates the luminosity is very sensitive to a temperature change as the right side of the equation involves the fourth power of the temperature.

T can be directly determined spectroscopically, and b can be determined photometrically. Knowing T, b, and d, we then obtain D from equations (1) and (2).

References

1. Young, Monica (November 6, 2020). *How big is Betelgeuse, really?* https://skyandtelescope.org/astronomy-news/how-big-is-betelgeuse-really/

TOPIC 49. The Hertzsprung-Russell Diagram (HR diagram).

The HR diagram plots a star's absolute magnitude on the vertical axis against its temperature on the horizontal axis. In Figure 1 below, the green band is the main sequence and applies to most stars including our Sun. Regions of white dwarfs and giant stars appear elsewhere in the diagram.

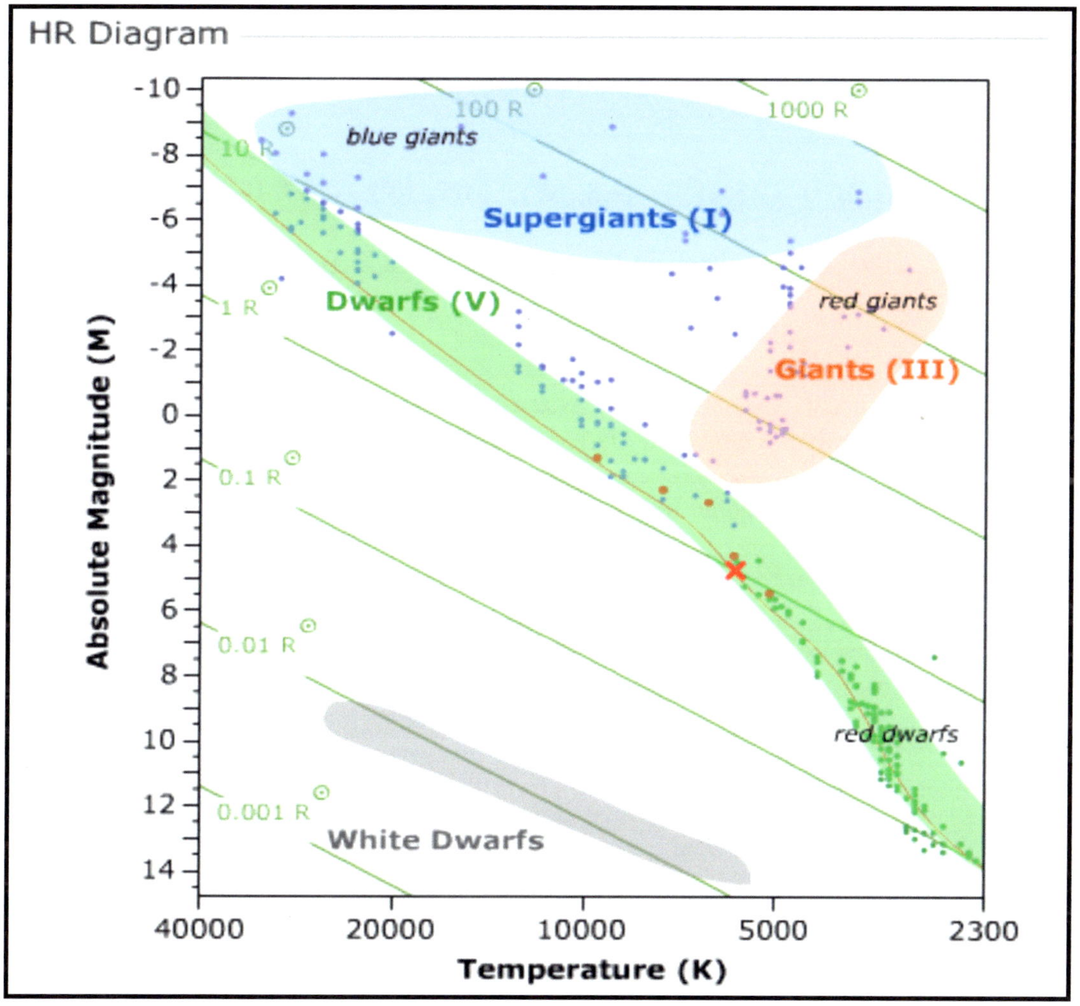

Figure 1. Hertzsprung-Russell Diagram. X marks the place of the Sun (where it has been for about 4.5 billion years and where it will be for about another 4.5 billion years, after which it leaves the main sequence). (See Topic 44, The fate of the Sun.) The parallel diagonal lines are isoradial lines—any two stars on the same isoradial line have the same radius. The radii are given in terms of the radius of the Sun, e.g., .01R^S (in green) means the star has a radius only 1/100th that of the Sun. The vertical scale gives the absolute magnitude of a star. Credit: University of Nebraska-Lincoln Astronomy education. https://www.physics.rutgers.edu/analyze/hr-diagram.html

The vertical axis in Figure 1 gives the absolute magnitude M of a star. Instead, this axis could just as well be replaced by the luminosity L of a star. This is because of the direct relationship between M and L, which is

$$L = 400\pi 10^{-M/2.5} . \qquad (1)$$

And since the Sun's absolute magnitude is +4.8,

$$L' = L / L_S = 10^{-(M - 4.8)/2.5} \qquad (2)$$

gives the relationship between a star's absolute magnitude M and its luminosity L' in terms of how many times its luminosity is greater or less than the Sun's luminosity L_S, which is $L_S = 400\pi 10^{-4.8/2.5}$. So $L' = 10^0 = 1$ if the star has the same absolute magnitude (4.8) as the Sun. Here are some values for (M, L'), going from the bottom to the top of the vertical axis of Figure 1 (where the values of L' are powers of 10):

$$(12.3,\ 10^{-3}) \quad (7.3,\ 10^{-1}) \quad (4.8,\ 10^{0}) \quad (-0.2,\ 10^{2}) \quad (-7.7,\ 10^{5}) .$$

Replacing D by 2R in the Stefan-Boltzmann Law (see equation 3 in Topic 48) we obtain

$$L = 4\pi R^2 \sigma T^4 , \qquad (3)$$

where R is the radius of a star. By holding R constant and in the HR diagram plotting the absolute magnitude M, or equivalently the luminosity L, as the temperature T decreases, we generate a set of parallel, diagonal lines, called *isoradial lines*. These constant radius lines have negative slope, i.e., in the HR diagram they go from the upper left to the lower right.

A caveat: You might think that by using the HR diagram we can get a star's temperature from its luminosity and vice-versa. This is not actually true. It is true if and only if we know the star's radius. Note that the bands in the HR diagram have a thickness. So, given the temperature, we can only determine a range for the luminosity (unless we also know the radius) and vice-versa. To show how absurd it is to say that from the temperature alone we could determine the luminosity, consider the following argument: We take the spectrum of a star and this gives the star's temperature T. Say from this (alone) we could obtain the luminosity L from the HR diagram. Then from

$$R = (4\sigma\pi)^{-1/2} L^{1/2} / T^2 \qquad (4)$$

we have R and from

$$b = L/4\pi d^2 \qquad (5)$$

we obtain d, since the star's apparent brightness b is easily determined. So by just taking the spectrum of a star (which gives us its temperature) we would be able to determine the star's luminosity L, its distance d from the Earth, and its radius R. Not really!

But it's all right now, in fact, it's a gas
But it's all right. I'm Jumpin' Jack Flash.

—*The Rolling Stones 1968*

TOPIC 50. How high could an astronaut jump straight up (spring up) from the surface of various solar system bodies?
Could an astronaut reach escape velocity?
Could an astronaut jump from Dactyl, the small moon of asteroid 243 Ida, to Ida itself?
What would an astronaut weigh on various solar system bodies?

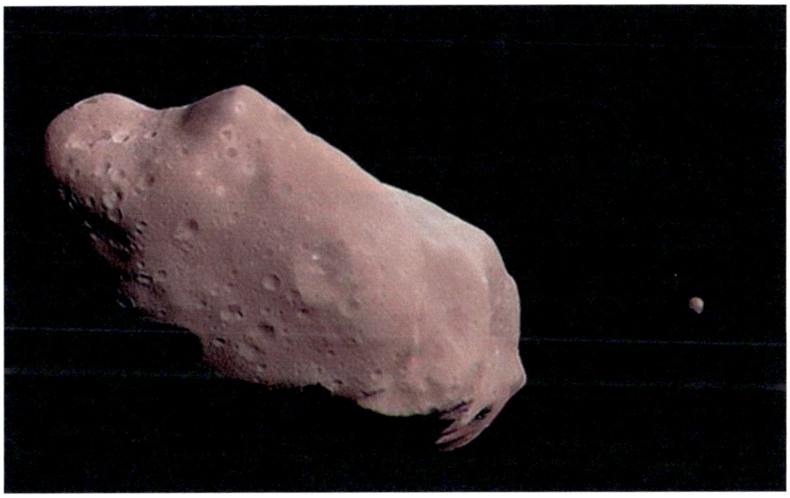

Figure 1. The asteroid 243 Ida and its small moon Dacytl as seen by the Galileo probe (headed for Jupiter) on August 28, 1993. Credit: NASA/JPL/ Processed by Kevin M. Gill—CC BY-SA 2.0

I. Let us first discuss how high an astronaut could jump up off of a solar system body.

Of course this depends on how skilled the person, call her Alice, is at doing this on the Earth. Let's say that Alice can jump h_e = two feet straight up from the surface of the Earth. Perhaps if she were a basketball star this would be higher, but two feet is pretty good. The height of her jump also depends on the acceleration of gravity g_e on the surface of the Earth and g on the surface of the body. The equation giving the height h that Alice can jump up from the body is

$$h = (g_e/g)h_e \ . \qquad (1)$$

For a proof that this is true, see *Appendix 50.1*.

A word of caution. This equation holds if we can assume that the acceleration of gravity g is essentially the same at height h as it is at the surface of the body. If the body is very small one might be able to jump a long way off the surface (perhaps Alice could even reach escape velocity), and the force of gravity, which weakens with distance, could be noticeably weaker at height h. In this case the equation does not hold. (To derive an equation that gives h where this decrease in gravity is taken into consideration can be done with an argument using integral calculus.)

Let us use this equation to obtain values for h when the solar system body is the Moon, Mars, the dwarf planet Ceres, and Saturn's largest moon, Titan, respectively.

$g_e = 9.807$ meters/second^2. In meters/second^2, the values of g for the four bodies are 1.62 for the Moon, 3.720 for Mars, 0.27 for Ceres, and 1.352 for Titan.

Using equation (1), we thus obtain the following distances that Alice could jump up from the four bodies, remembering that we assumed she could jump up two feet on the Earth:

Object	Height in feet
Moon	12.1
Mars	5.3
Ceres	72.6
Titan	14.5

In 2020 the world record for pole vaulting (not just springing up off the ground) was 20.18 feet. Clearly, having the Olympic Games held on Ceres would cause many world records to be broken.

II. Now we discuss escape velocity.

The escape velocity of a body is the minimum *initial* velocity required to cause an object to leave the surface of the body and (with no acceleration) travel to infinity. As soon as the object leaves, say the Earth, the Earth's gravity starts slowing the object down. The initial velocity must be enough to never allow the speed of the object to drop to zero. Of course gravity weakens as the object gets farther away.

The escape velocity v_{esc} of an object from a body is given by the equation

$$v_{esc} = \sqrt{2GM / R} \qquad (2)$$

where M is the mass of the body, R is the radius of the body, and G is Newton's gravitational constant. Notice that the mass of the object leaving the body is not relevant. Centrifugal force on the object due to the rotation of the body is ignored. A proof of this equation (which requires integral calculus) is given in *Appendix 50.1*.

Could Alice jump up from either of the two moons of Mars—Phobos and Deimos—at escape velocity? How about for the small moon Dactyl of the asteroid 243 Ida?

To answer the questions, we need to know what Alice's initial velocity is when jumping up from the surface of any of these moons. For any object given an upward velocity v_o *from the Earth* we have that

$$v_o^2 = 2g_e h_e \qquad (3)$$

where h_e is the maximum height the object rises before it stops and then drops back down. (See *Appendix 50.1* for a proof.)

This is the *same initial* velocity she has when she jumps off the surface of *any object.* Why?
If Alice's physiology allows her to apply a downward force F(t) over time interval t_0 to t_1, then her *initial* velocity, *regardless of the body she pushes off from*, is the same, namely

$v_o = \Delta v = (1/m)\int_{t_o}^{t_1} F(t)dt$. m is Alice's mass. If you are not familiar with integration in calculus, consider the case where F(t) is a constant F. Then $F = m\Delta v/\Delta t$, so $v_o = \Delta v = (1/m)F\Delta t$, with $\Delta t = (t_1 - t_o)$. Of course the rate at which she slows down depends on which body she jumps up from.

For Alice, we assumed $h_e = 2$ feet, which in meters is 2(.3048) = .6096 meters.
From (3) we obtain $v_o^2 = 2(9.807)(.6096) = 11.957$, whence

$$v_o = 3.46 \text{ meters/second.}$$

From equation (2), plugging in the known values of the radius and mass of each moon, we obtain the following escape velocities:

11.39 meters/second for Phobos, 5.56 meters/second for Deimos.

So Alice could not jump up at escape velocity from either of the Martian moons.

Now let us consider the moon Dactyl orbiting the much larger asteroid 243 Ida in the inner asteroid belt. In Greek mythology Dactyls are creatures which inhabit Mount Ida in Crete and guard the infant god Zeus in a cave on that mountain. Dactyl is 1.6 x 1.4 x 1.2 kilometers in size.[4] So we could say its radius R is about 0.7 kilometers. That is such a small radius that you might believe that surely Alice can attain escape velocity from it. And you would be right. But let us continue the calculation anyway. We could guess at the density of Dactyl, and knowing R, get its mass M, and then use equation (2) to get the escape velocity.

But we will do something slightly different to reach our conclusion, and at the same time come up with an interesting equation. We ask the question

What is the smallest radius R for a moon (or any small body) of density ρ, such that a person who can jump up and reach a height h_e (= 2 feet for Alice) on Earth, can attain escape velocity when jumping off the moon? (Here Alice's initial velocity would equal the escape velocity for the moon of radius R and density ρ.)

The equation for R is given by

$$R^2 = R_e h_e(\rho_e/\rho) \tag{4}$$

where R_e is the radius of the Earth and ρ_e is the density of the Earth. This equation is valid if the escape velocity of the moon in question is much less than the escape velocity of the Earth, which we are assuming is true. A derivation of (4) is given in *Appendix 50.1* and also in reference [1].

While it is not known if Dactyl was captured by 243 Ida or if it was created from debris when a strike on Ida created a crater, if we assume the latter, then its density is most likely about the same as that of Ida. The density of Ida is approximately known: Ida's volume is known from data taken in the Galileo flyby; Ida's mass is known, as the mass of any body can be ascertained if it has a moon orbiting it. And of course density is mass divided by volume.

Ida's density is 2.6 ± 0.5 grams/centimeter^3.[2] Therefore, by our assumption, we set $\rho = 2.6$ grams/centimeter^3. Using $R_e = 6{,}378$ kilometers, $\rho_e = 5.51$ grams/centimeter^3, and $h_e = 2/3281$ kilometers (which is two feet), we obtain from (4) that

$$R = 2.87 \text{ kilometers.}$$

So Alice could reach escape velocity for a moon with the same density as that of Dactyl and with a radius no larger than 2.87 kilometers. Since Dactyl's radius is ~0.7 kilometers, Alice can easily reach escape velocity.

This means that Alice could jump up from Dactyl and fly the approximately 85 kilometers[2] that separates Ida from Dactyl and that she could land on Ida.

Figure 2. Alice jumps off Dactyl.

It is to be noted that here we have not even taken into consideration the gravitational attraction of Ida on Alice. Alice would be advised to carry a small thruster in case she misses Ida, going into orbit around it instead, or, alternatively, heads into deep space. And if she has no desire to leave Dactyl, she should be careful not to exert too much downward force in walking around its surface.

Finally, a few more comments. We have only considered springing straight up off a body. A greater launch speed could be obtained with a running start. Also we have ignored the centrifugal force on an astronaut due to the spinning of the body. By adding this upward force to the upward force from Alice's pushing down on the surface, she can increase her initial upward velocity. In thinking about this, remember that centrifugal force only acts on Alice when she is standing on the surface; once she is off the surface—even by a millimeter—it is non-existent. The centrifugal force only affects her *initial* velocity.

Consider 243 Ida itself. It has an ellipsoid shape with best-fit principal dimensions 59.8 x 25.4 x 18.6 kilometers.[2] Its axis spin period is a quite fast 4.63 hours.[2] The spin axis is about halfway between the two far ends, with the maximum centrifugal force at these two ends (and outward along the long axis of Ida). Because of Ida's irregular shape and its rapid spin, the surface accelerations vary greatly (0.3 to 1.1 centimeters/second^2).[3] The downward force (gravitational force minus centrifugal force) is lowest at the far ends and about halfway between the ends, at Ida's thin, lower-mass "waist." Note: Ida's waist is not adequately revealed in Figure 1 above. As for Alice's weight on 243 Ida, if she wished to reduce her weight she would only have to walk to a different part of the asteroid.

III. Finally we discuss how much a person would weigh on the surface of various solar system objects. We give the weight as a percentage of what that person weighs on Earth.

The equation giving the ratio of a person's weight on the surface of the solar system body to that person's weight on the Earth is

$$\text{Ratio} = g/g_e = (R/R_e)(\rho/\rho_e) \quad (5)$$

where (as before) R_e, ρ_e, and g_e are respectively the radius, (average) density, and acceleration of gravity at Earth's surface, and R, ρ, and g are the same quantities, but for the other body. It is assumed all bodies are at least approximately spherical and the density is approximately uniform at any given fixed depth below the surface. See *Appendix 50.1* for a derivation of this equation.

For convenience we again call the astronaut Alice. Using equation (5), with appropriate values for the various bodies, we obtain the following table.

Solar system object	Alice's weight as a percentage of her weight on Earth
Earth	100.0
Moon	16.5
Mercury	37.7
Venus	90.2
Mars	37.8
Ceres	2.8
Io	18.3
Europa	13.4
Ganymede	14.5
Callisto	12.6
Titan	13.8
Pluto	6.3
Charon (Pluto's innermost and largest moon)	2.9

Alice's weight on Mercury is over twice that on any of the Galilean moons or on Titan. That is due to Mercury's high density—it has a large iron core.

References

1. Sinnott, Roger W. (July 1984). Jumping off of an "asteroid" and other puzzles. *Sky & Telescope,* p 62.

2. Belton, M. J. S., Chapman, Clark R., Klaasen, Kenneth P., Harch, Ann P., Thomas, Peter C., Veverka, Joseph, McEwen, Alfred S., & Pappalardo, Robert T. (March 1996). Galileo's encounter with 243 Ida: An overview of the imaging experiment. *Icarus, 120*(1), p 1–19. http://www.planetary.brown.edu/pdfs/1764.pdf

3. Thomas, P. C., et al. (March 1996). The shape of Ida. *Icarus, 120*(1), p 20-32. https://www.sciencedirect.com/science/article/abs/pii/S0019103596900330?via%3Dihub

4. Geissler, Paul E., Petit, Jean-Marc, Durda, Daniel D., Greenberg, Richard, Bottke, William F., Nolan, Michael, & Moore, Jeffrey (March 1996). Erosion and ejecta reaccretion on 243 Ida and its moon. *Icarus, 120*(1), p 140-157. https://www.boulder.swri.edu/~bottke/Reprints/Geissler_Icarus_1996_120_140_Erosion_Ejecta_Ida.pdf

...a mathematician among geographers,
and yet a geographer among mathematicians;
...consequently on both sides he offers
his opponents occasions for contradiction.

—Eratosthenes, describing himself

TOPIC 51. In the Third Century BCE Eratosthenes of Cyrene was the first to estimate the circumference of the Earth. His figure was fairly accurate. How did he do it?

Figure 1. Eratosthenes of Cyrene (276 - 194 BCE).

Eratosthenes, born in Cyrene (now in Libya), was a scholar in a range of subjects including astronomy, mathematics, geography, poetry, and music theory.

He traveled to Athens to further his studies.

In 245 BCE the Egyptian pharaoh Ptolemy III Euergetes (the benefactor) invited Eratosthenes to serve as librarian at the Library of Alexandria, the greatest library of antiquity. Shortly thereafter he became its chief librarian. He spent the remainder of his life in Alexandria and was a friend of Archimedes.

In his treatise *On the Measurement of the Earth* he determined the circumference of the Earth. While this treatise has been lost, some details of his calculations appear in other works.

In particular, Cleomedes in his book *On the Orbits of the Celestial Bodies* credited Eratosthenes with the calculation and gave a simplified example of how Eratosthenes did it.[3] Cleomedes is our primary source for Eratosthenes on this subject and wrote his book sometime after 50 BCE. In his calculation, Eratosthenes correctly assumed the Sun was so far away that all its rays reaching Earth were parallel.

In his argument Cleomedes used the two cities Alexandria and Syene (near modern-day Aswan). They are almost (but not exactly) on the same meridian of longitude, Syene being south of Alexandria. And Syene is almost on (it's slightly north of) the Tropic of Cancer. So, on summer solstice the Sun at Syene would be almost directly overhead. (Incidentally, Eratosthenes also invented the system of latitude and longitude.) At noon on summer solstice the Sun would shine straight down a well in Syene. On that same day in Alexandria the length of the shadow cast by an upright pole of known length was measured. From this information, and the approximate distance between the two cities, the circumference of the Earth can be determined. The details are given below.

I. The determination of the circumference of the Earth using two cities (such as Alexandria and Syene) on the same meridian with the Sun at the zenith at one of them.

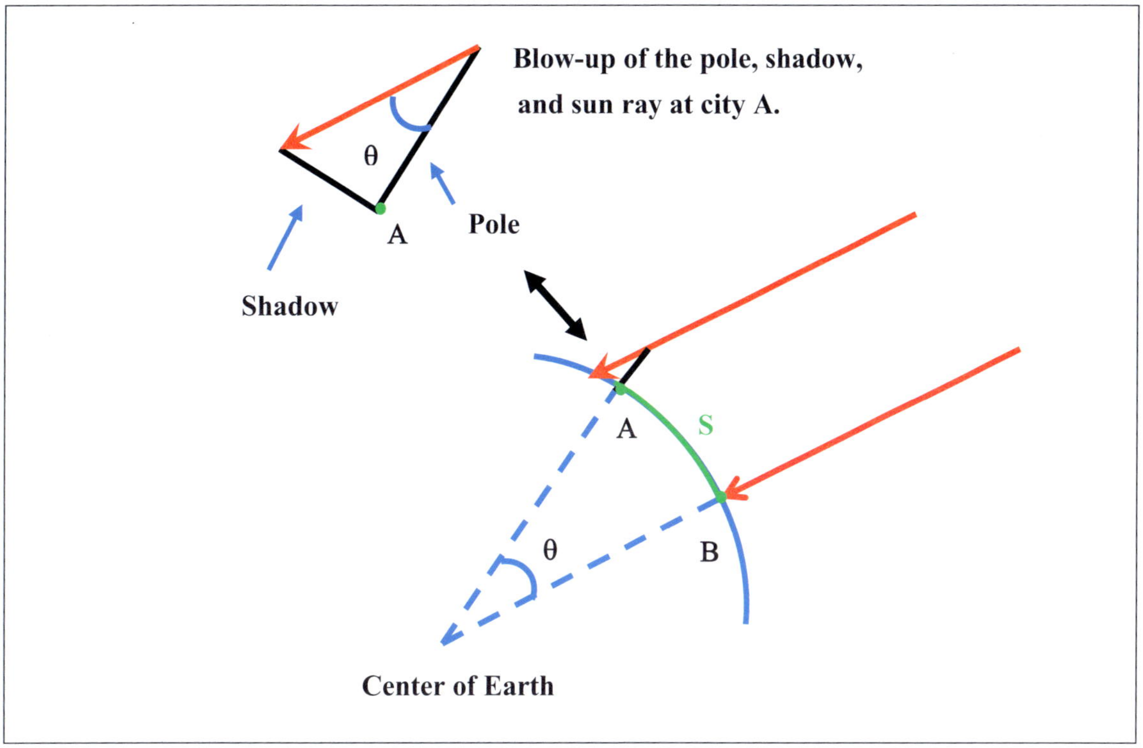

Figure 2. A and B are the cities of Alexandria and Syene respectively. Parallel Sun rays (in red) strike the two cities. At B the Sun is directly overhead at noon. At noon, at A, an upright pole casts a shadow whose length can be measured. The angle θ subtended at the center of the Earth by the arc of length S (in green) between the two cities is the same angle as between the upright pole and the ray of the Sun passing over the top of the pole and striking the ground at the end of the shadow. This is because if two parallel lines (the red ones) are crossed by a transverse line (the upper blue, dashed line), the interior, acute angles formed by the transverse line with the parallel lines are equal.

S, the distance between the cities, was known. If C is the circumference of the Earth, then the ratio of S to C is the same as the ratio of θ to 360 degrees, i.e.,

$$S/C = \theta/360^{\circ} .$$

Knowing the length of the pole and its shadow, the angle θ is determined. For example, one could quite simply draw (on papyrus or on the ground) the right triangle using these lengths and then directly measure the angle. (For those living in the 21st century using a calculator, θ is the arctangent of the ratio of the shadow's length to the pole's length.) We have

$$C = 360 \times S/\theta .$$

The value of the angle θ was found to be about 7.2 degrees, and the distance between the cities was about 5,000 stadia.[1] From this (using the last equation) the circumference would be about 250,000 stadia. A stadion (plural: stadia), a unit of length, is thought to be somewhere between 500 and 600 feet[1] making Earth's circumference somewhere between 23,700 and 28,400 miles. The actual circumference of the equator is about 24,900 miles. The circumference of the great circle passing through the two poles is somewhat less as the Earth is not a perfect sphere. Because the Earth rotates on its axis, bulging occurs at the equator resulting in a slightly flattened sphere.

The above discussion assumes the two cities are on the same meridian and that the southern city B is at a latitude such that at some time during the year the Sun would be directly overhead. (So city B would have to be between the Tropic of Cancer and the Tropic of Capricorn.) The shadow measurement at city A would be made at noon on the same day as the noon zenith appearance of the Sun at city B. It is not necessary to assume that city B is on the Tropic of Cancer.

At this point the reader might very well ask the following two questions:

II. How can the Earth's circumference be determined if the two cities are on the same meridian, but it is *not* assumed that the Sun is necessarily directly overhead at one of them?

III. Same question, but where it is *not* assumed that the cities are necessarily on the same meridian.

In *Appendix 51.1* the answers are given.

See references [2, 4] for more information on Eratosthenes and Cleomedes.

References

1. American Physical Society News. (June 2006). June, c. 240 B.C. Eratosthenes measures the Earth. *APS Physics, 15*(6) aps.org. https://www.aps.org/publications/apsnews/200606/history.cfm
2. O'Connor, J. J., & Robertson, E. F. (January 1999). Eratosthenes of Cyrene. *MacTutor.* https://mathshistory.st-andrews.ac.uk/Biographies/Eratosthenes/
3. Pearse, Roger (April 15, 2017). *Cleomedes: How big is the earth?* https://www.roger-pearse.com/weblog/2017/04/15/cleomedes-how-big-is-the-earth/
Pearse's post includes Heath's English translation of the section on Eratosthenes and the size of the Earth taken from Cleomedes' *On the circular movement of the heavenly bodies.*
4. Wikipedia has interesting information on Eratosthenes and Cleomedes. (https://en.wikipedia.org/wiki/Eratosthenes) and (https://en.wikipedia.org/wiki/Cleomedes).

...in some respects this work of Aristarchus is the most impressive example of the application of mathematics to natural science in the ancient world.

—Steven Weinberg, 1979 Nobel laureate in physics, in his book "To Explain the World"

Topic 52. Aristarchus of Samos, c. 310-230 BCE, was able to estimate the sizes of the Sun and Moon and their distances from Earth using geometry and a few measurements. How could this be done in the third century BCE?

Aristarchus was an ancient Greek mathematician and astronomer—born on the Greek island of Samos. While not a great amount is known about his life, he did spend at least a few years at the Library of Alexandria. In his only surviving work, *On the Sizes and Distances of the Sun and Moon,* he cleverly showed how to derive the Moon's and Sun's diameters and their distances from the Earth *in terms of the Earth's diameter*. And, given that Eratosthenes (see Topic 51) gave a value for the Earth's diameter, this meant actual values could be given for all these quantities. (Eratosthenes was born after Aristarchus, but their lives overlapped a few decades.)

In the interest of showing how such seemingly difficult results could have been ascertained as early as the third century BCE, geometrically based equations are derived that give the diameters and distances of the Moon and the Sun in terms of the Earth's diameter. These equations contain four constants that had to be experimentally determined. Measurements to discover the value of these constants are discussed. Although the exact formulae that we obtain are not likely those of Aristarchus, they probably approximate what he did and expose the geometric relations he had to consider.

The values he obtained are at considerable variance with current values, but the method was sound and quite an accomplishment for the third century BCE. The discrepancies with modern values are due to the difficulty of obtaining accurate measurements for the values of the four constants (especially one of them) required by the equations. But Aristarchus' results did show that the Sun was far larger than the Earth and the Moon.

Before proceeding, it should be mentioned that Aristarchus also was the first to propose a heliocentric theory for the solar system. Alan Batten speculates that Aristarchus' deduction that the Sun is much larger than the Earth might have suggested to him that the Sun was at the center of the solar system.[4] Aristarchus had the planets revolving around the Sun in circular orbits, with the Sun and the stars fixed in their positions. He assumed the Earth rotated on its axis and suspected the stars were distant suns, but much farther away from Earth than the Sun. The original text has been lost, but in Archimedes' book *The Sand Reckoner* (*Archimedis Syracusani Arenarius & Dimensio Circuli*) Aristarchus' work is discussed.

Now we expose the method by which Aristarchus obtained his results.

Let m = the radius of the Moon, M = the distance of the Moon from the Earth, e = the radius of the Earth, s = the radius of the Sun, and S = the distance of the Sun from Earth. (So lower case letters for radii; uppercase for distances.)

The geometrically based equations in the unknowns m, M, s, and S will contain four constants and e itself. The constants are

(a) the angle α subtended by the Moon's *radius* at the Earth. (So 2α is the angular size of the Moon as seen from Earth.)

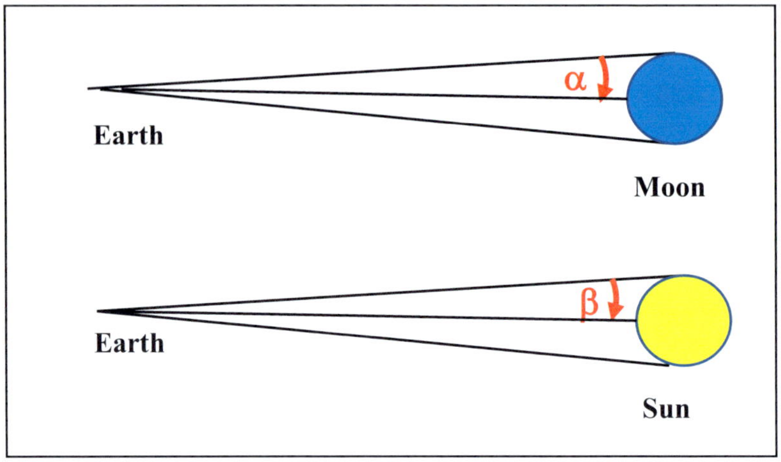

Figure 1. The angles α and β are illustrated. In reality, α and β are almost equal (about 15 arc minutes).

(b) the angle β subtended by the Sun's *radius* at the Earth,
(c) the ratio u/m, where u is the radius of the circular cross-section of Earth's umbral cone at the Moon's location during a total lunar eclipse (see Figures 2, 3, and 4 below),
(d) an angle θ when the Moon is at first quarter phase (and thus half lit) that will be discussed below (see Figure 6). This angle determines the ratio S/M.

Concerning (a), α and β are virtually the same, namely about 15 arc minutes. That of course is why at a total solar eclipse the Moon just covers the face of the Sun.

We next discuss (c).

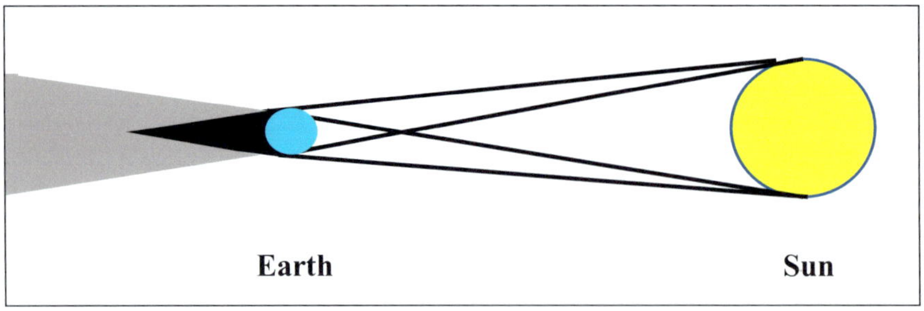

Figure 2. The Earth's shadow consists of two regions: The black region is the umbra (Latin: shadow); the grey region is the penumbra.

The umbra is a cone-shaped region into which all sunlight is blocked by the intervening Earth. Some sunlight penetrates the penumbra. For a total lunar eclipse the Moon must be in the umbra.

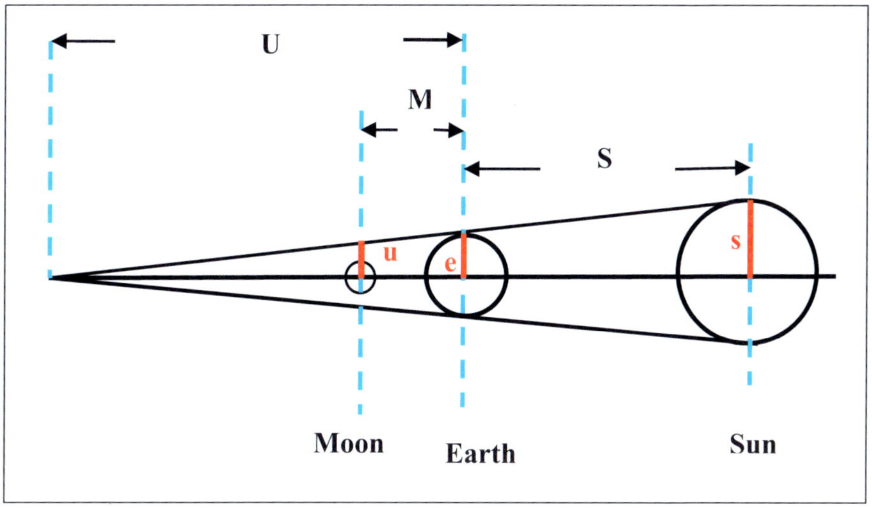

Figure 3. Situation at a total lunar eclipse where the Moon is inside the umbra of the Earth's shadow. Sizes and distances are not to scale (to say the least). The radii sizes u, e, and s are illustrated and so are distances M and S as well as the distance U from the Earth to the tip of the conical cone of the umbra.

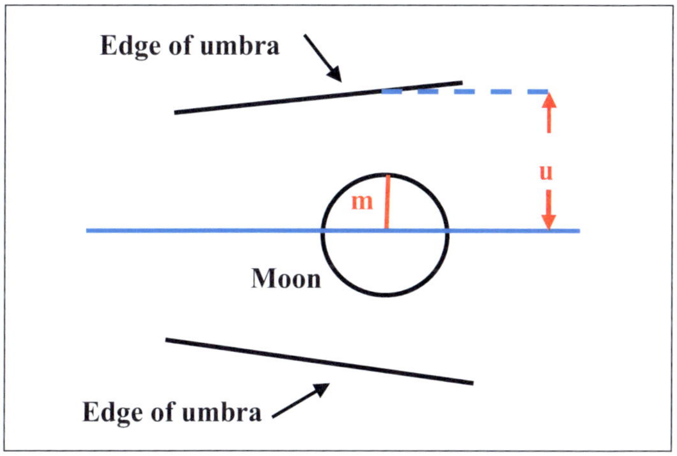

Figure 4. An enlargement of the area around the Moon in Figure 3. The radius u of the cross-section of the umbral cone at the Moon's position is indicated, as is the Moon's radius m.

How can u be approximated in terms of m, i.e., how can u/m be approximated?

We assume a total lunar eclipse in which, ideally, the (center of the) Moon passes across the umbra's center. Thus the Moon doesn't pass mainly through just the northern or southern parts of the umbra. Let T be the time between when the Moon first enters the umbra to the moment when the Moon begins to exit the umbra. Put another way, T is the time between when the spot on the Moon's surface that first touches the edge of the umbra begins to exit the umbra. T can be measured.

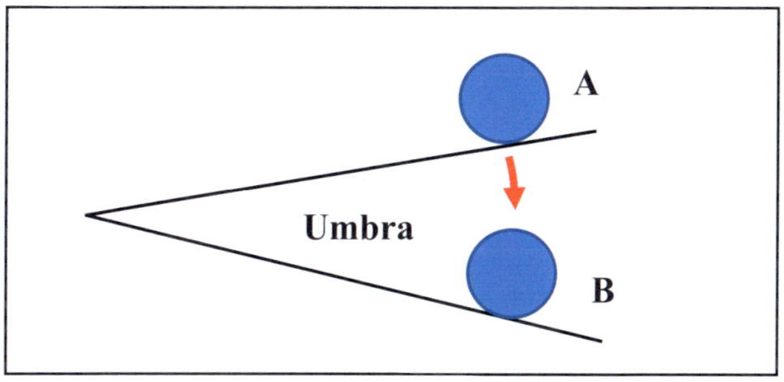

Figure 5. T is the time for the Moon to move from position A to position B.

Let T' be the time for the Moon to make one revolution (with respect to the fixed stars) about the Earth. Then the Moon will move through the angle α (which is defined above and is about 15 arc minutes) in time $T'' = T'\alpha/(360 \times 60)$ where α is in arc minutes. Since $2u/m = T/T''$, we obtain

$$u/m = (T/2) / [T'\alpha/(360 \times 60)] . \quad (1)$$

Another method of determining u/m is as follows. With some device that measures angles, determine the angle γ between the point in the night sky where the Moon first touches the umbra and the point in the night sky at which the Moon begins to exit the umbra. Then

$$2u/m = \gamma/\alpha . \quad (2)$$

Thus we can determine u/m.

At this point, from similar triangles in Figure 3, we also can obtain two equations:

$$e/U = s/(S+U) \quad (3)$$

and

$$u/(U-M) = e/U . \quad (4)$$

We now discuss (d).

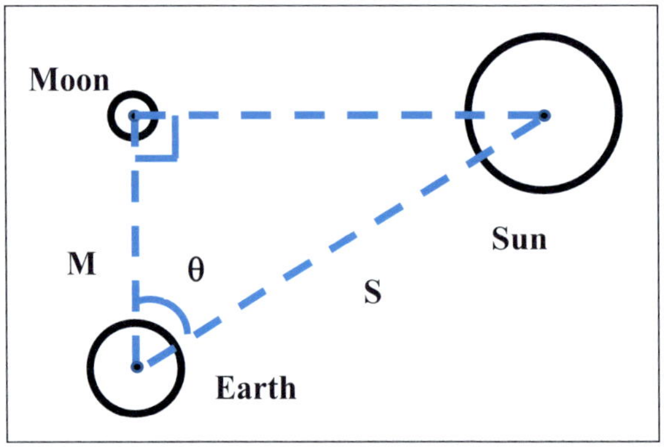

Figure 6. The relationship of the Earth, Moon, and Sun when the Moon is at first quarter.

The angle θ between the Sun and Moon at first quarter is almost 90 degrees. From Figure 6 this doesn't seem to be true, but that is because the illustration is not to scale. Actually, the rays from the Sun to the Earth and Moon are almost parallel because of the great distance of the Sun from the Earth-Moon system. We have

$$S/M = \sec\theta = 1/\cos\theta . \quad (5)$$

So it seems easy to determine the ratio S/M. Just measure the angle θ and take its secant.

But determining an accurate value for θ is the biggest difficulty in Aristarchus' scheme. Obviously the measurement has to be done in daylight, and neither the Sun nor the Moon should be close to the horizon as diffraction would change the true angle. In addition, the measurement must be done when the Moon is *exactly* at first quarter, which is not easy to determine. Aristarchus effectively used the value $\theta = 87$ degrees, which gives $S/M \approx 19$.[4] (And of course Aristarchus did his work before the invention of trigonometry—no trig tables were available to him.) The correct value for θ is about 89 degrees 52 minutes which gives S/M ≈ 400.[4] Assuming no error at all was made in measuring the angle, this huge discrepancy for S/M would occur if Aristarchus had made a six-hour error in timing when the Moon was exactly half dark.[4] Alan Batten tried several times to obtain a value for θ, but his best result was about that of Aristarchus.[4] Using a telescope would have been very useful, but of course it was not invented until some 18 centuries later. Even if the error in θ were reduced to only one arc minute, the error in S/M would still be about 10 percent.

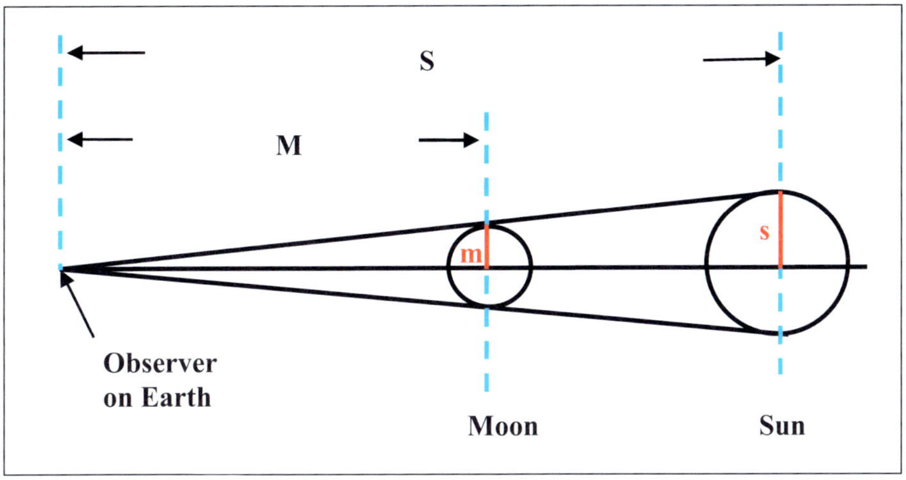

Figure 7. Total solar eclipse. In this diagram, the observer's location on Earth's surface, the spot where the umbral cone comes to a point, and the center of the Earth are all considered essentially the same. This approximation is valid as M and S are far greater than e (the radius of the Earth).

Similar triangles in Figure 7 give us the equation

$$m/M = s/S = \alpha \,, \qquad (6)$$

if α is in radians. Thus

$$s/m = S/M \,. \qquad (7)$$

We are now finished with geometric considerations. Let us summarize:

By taking measurements we have values for S/M and u/m. Let

$$A = S/M \text{ and } B = u/m \,. \qquad (8)$$

We know that the angles α and β, subtended respectively by the *radius* of the Moon and the Sun, are the same, namely about 15 arc minutes.

We also have equations (3), (4), (6), and (7). Using a little algebra (see *Appendix 52.1* for the calculations) we obtain the following equations that give S/e, s/e, M/e, and m/e in terms of A, B, and α:

$$s/e = (A+1)/(B+1), \qquad m/e = (A+1)/A(B+1) \,, \qquad (9)$$

and

$$S/e = (s/e)/(\pi\alpha/180), \quad M/e = (m/e)/(\pi\alpha/180), \tag{10}$$

where α is in degrees.

Aristarchus obtained the values $S \approx 380e$ and $M \approx 20e$.[4] Thus $S/M \approx 19$. The true value of S/M is 390. The true value of s/e is 109. He didn't obtain this value, and there is some issue with what value he used for α.[4,5] But, as was pointed out above, even though Aristarchus' values were at variance from the true values, his method revealed that the Sun was much larger than the Earth (i.e., s/e is large), perhaps the reason he proposed a heliocentric theory. And it is believed that his effort was the first attempt to use mathematics and experiment to ascertain the relationship between the sizes and distances of different solar system bodies. I think Galileo would have liked to have met Aristarchus. But that would have required his traveling 18 centuries back in time. Galileo had access to telescopes, but not to a time machine.

See [1, 2, 3, 4, 5] for further material on this topic.

In antiquity, natural phenomena on the Earth and celestial objects and events in the night sky were commonly explained as being caused by "the gods." The transition to giving explanations based on conducting experiments and using mathematics was met with resistance and persecution. This was true even in ancient Greece. Thales of Miletus (c. 624/625—548/545 BCE) is often cited as the first to free explanation of phenomena from godly intervention, paving the way toward scientific endeavor. Anaxagoras (c. 500—428 BCE) was accused of impiety and put on trial c. 450 BCE. Pericles spoke in his defense. But Anaxagoras was forced to leave Athens. Aristotle (384—322 BCE) was denounced in 322 BCE for impiety, prompting him to flee to his mother's family estate, where he died (of natural causes) the same year. It was perilous not to believe in the Greek gods and instead seek natural explanations.

References

1. Heath, Thomas (1913). *Aristarchus of Samos, the ancient Copernicus.* Clarendon Press. https://archive.org/details/aristarchusofsam00heat/mode/2up This volume was later reprinted by Oxford University Press (1997) ISBN-13: 978-0-19-858111-6 and ISBN: 0-19-858111-4 and by Dover Publications (2004) ISBN-13: 978-0-486-43886 and ISBN: 0-486-43886-4.
2. Aristarchus. *On the Sizes and Distances*. Wikipedia. https://en.wikipedia.org/wiki/On_the_Sizes_and_Distances_(Aristarchus)
3. Evans, James. *Aristarchus of Samos, Greek astronomer*. britannica.com https://www.britannica.com/biography/Aristarchus-of-Samos
4. Batten, Alan H. (1981). Aristarchos of Samos. *Journal of the Royal Astronomical Society of Canada, 75,* p 29. http://articles.adsabs.harvard.edu/cgi-bin/nph-iarticle_query?1981JRASC..75...29B&defaultprint=YES&filetype=.pdf Also, http://articles.adsabs.harvard.edu/full/1981JRASC..75...29B
5. Weinberg, Steven (2015). *To explain the world: The discovery of modern science.* Harper Perennial, p 300-301.

...when we're in resonance singing together,
we become one for that 3 1/2 or four minutes the song lasts.

—Mary Gauthier, American folk singer-songwriter

TOPIC 53. Orbital resonance of the Galilean moons Io, Europa, and Ganymede. Orbital stability of planets.

In celestial mechanics, *orbital resonance* occurs when orbiting bodies exert regular, periodic gravitational influence on each other, usually because their orbital periods are related by a ratio of small integers.[3] Examples of orbital resonances are 2:3 for Pluto and Neptune (in that order), and 1:2:4 for Ganymede, Europa, and Io (in that order). This means Pluto goes twice around its orbit for every three round trips for Neptune, and Ganymede goes around its orbit once for every two for Europa and every four for Io. (Another way to express the 1:2:4 resonance is to say that Europa's *period* is twice that of Io's, and Ganymede's *period* is four times that of Io's.) There are many Kuiper Belt objects that share a 2:3 resonance with Neptune; others are in different resonances such as 4:7, 1:2, and 2:5 with Neptune.

Each term in the 1:2:4 ratio is accurate to nine significant digits. No other three bodies in the solar system participate in a resonance. Titan and Hyperion have a resonance ratio of 3:4 around Saturn. This resonance and the 2:3 resonance of Pluto and Neptune are ratios which are accurate to three digits.[1]

Periodically the three inner Galilean moons are in a straight line with Jupiter, with two of the moons on one side of Jupiter and the third moon on the opposite side. This is known as the great conjunction. Ganymede is never the lone moon in this alignment. If this could occur, then in half a revolution of Ganymede all three moons would be in alignment on the same side of Jupiter, which never happens. By googling "orbital resonance of Jupiter's moons" one can find an animation of the 1:2:4 Laplace resonance between Ganymede, Europa, and Io.

Pierre-Simon de Laplace, an 18th century French mathematician who did much work in celestial mechanics, discovered the resonance of the three moons. In his five-volume treatise on celestial mechanics, Laplace proposed the nebular hypothesis of how planets form around a star, and this is still the accepted theory. He also solved a perplexing problem in orbital stability that Newton had unsuccessfully worked on. This was the problem that the orbits of Jupiter and Saturn showed a long-term shift that seemed to violate the Newtonian laws of gravity. Jupiter's orbit appeared to be shrinking, while Saturn's was expanding. In fact, Newton suggested that periodic divine intervention might be required in order to have Jupiter and Saturn return to their proper orbits. In the 1780s Laplace successfully showed that Newton's laws indeed can explain this orbital deviation and that after a 929-year cycle the orbits of the two planets return to where it was thought they should be. The cyclical deviation is due to the gravitational interaction between Jupiter and Saturn. Napoleon is reputed to have asked Laplace why he did not mention God in explaining the creation of the solar system and the stabilizing of the orbits of the planets. Laplace's reply: "I had no need for that hypothesis."[2]

References

1. Resonant Explanations. (July 1981). Science and the citizen. *Scientific American*, p 76.
2. Gribbin, John (2002). *Science: A history.* Penguin Books, p 295-296. Highly recommended.
3. https://en.wikipedia.org/wiki/Orbital_resonance

A playground swing was one of Tesla's favorite examples of a resonant system. It's easy to measure its natural frequency. Each time the swing moves forward and then returns to its starting position counts as one cycle.

—*intuitor.com*

TOPIC 54. How can orbital resonance between two planets or between two moons either promote or prevent the more massive body from changing the orbit of the less massive body? A case in point: Resonance between Pluto and Neptune prevents Pluto from coming too close to Neptune. In fact, Pluto comes closer to Uranus than to Neptune.

In the case of Neptune and Pluto, Pluto is sometimes closer to the Sun than Neptune. It is sometimes said that the orbits of Neptune and Pluto cross. But that doesn't mean that there is point of intersection of the orbits themselves. There is none. This is because the orbits of the two bodies are not in the same plane. Neptune's orbital plane is close to the ecliptic plane (Earth's orbital plane), while Pluto's is inclined about 17 degrees to the ecliptic.

Let us illustrate the situation using Neptune and Pluto, where the resonance is 2:3. See Figure 1 below. It is only at points in Neptune's orbit that are close to Pluto's orbit where the gravitational pull of Neptune might disrupt Pluto's motion. So we focus our attention on those points. Looking down on the solar system we see both Neptune and Pluto revolving in a CCW direction around the Sun.

Let C_N and C_P be points respectively on Neptune's and Pluto's orbits such that the distance between C_N and C_P is at least as small as the distance between any point on one of the orbits and any point on the other. Here "C" stands for "closest" point. So C_N is a closest point in Neptune's orbit to Pluto's orbit and C_P is a closest point in Pluto's orbit to Neptune's orbit. And the distance between C_N and C_P is the closest distance between the two orbits.

We need a coordinate system in the orbital plane of Neptune to describe points in Neptune's orbit. Let L be the line segment from the center K of Neptune's elliptical orbit to the point C_N. This will be the zero-degree direction. For a point N on Neptune's orbit let θ be the CCW angle from L to a line from K to N.

Now consider a point P on Pluto's orbit that is *close* to C_P. When Pluto is at point P at some specific time t, say $t = 0$, assume Neptune is at point N on Neptune's orbit.

In Figure 1 below, Neptune's orbit is the small ellipse and the lower curve is a portion of Pluto's orbit. Consider Neptune's orbital plane to be the book's flat paper page. Pluto's orbit lies partly above and partly below the book's page. It makes no difference to the proof.

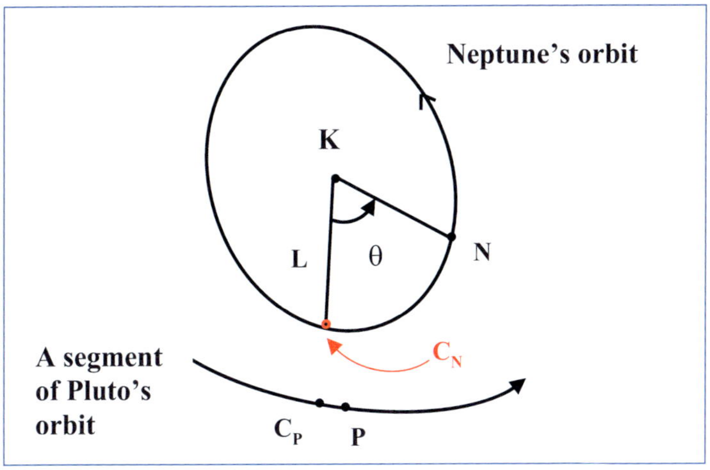

Figure 1. Neptune's orbit and a portion of Pluto's orbit.

Case 1. At time t = 0, θ is close to being either 0 degrees or 180 degrees. At t = 0, the planets are very close, or in one more orbit of Pluto they will be close, and the resonance could be destroyed. (At θ = 180 degrees, Neptune makes 1.5 revolutions for one revolution of Pluto and thus would be close to Pluto.)

Case 2. At time t = 0, θ is not close to being either 0 degrees or 180 degrees. We wish to show that the 2:3 resonance will guarantee that this condition will hold into the indefinite future (excluding exogenous factors), and thus Pluto's orbit will not be altered by Neptune's gravity (which is actually the case).

For illustrative purposes choose θ = 50 degrees. Now, since Neptune makes 1.5 orbits for one of Pluto, after one orbit of Pluto, Pluto will be back at point P and Neptune will be at a point given by θ = 50 +180 = 230 degrees. After another complete orbit by Pluto, Neptune's position will be given by θ = 230 + 180 = 410 degrees which is equivalent to 410 - 360 = 50 degrees. So we have Neptune being at 50 degrees, then 230 degrees, then back to 50 degrees CCW for every return of Pluto to point P. *This pattern will repeat indefinitely*. Thus when Pluto is at point P, Neptune is always either 50 degrees or 360 - 230 = 130 degrees away from its closest point, C_N, to C_P. So when Pluto is at P, which can be *any* point in its orbit that is close to C_P, Neptune is always at a discrete distance away, preventing massive Neptune from altering Pluto's orbit.

In the above argument, it should be noted that the point N is determined by the *choice* of the point P *close* to C_P, resulting in the angle θ. But choosing a different point P' *close* to C_P on Pluto's orbit would determine a point N' on Neptune's orbit that is very close to N, so that the resulting angle θ' would be very close to θ. Therefore θ satisfying the above restriction would mean that so does θ'.

See *Appendix 54.1* for a more general proof (where we don't use the 50-degree angle).

The actual distance between the orbits of Neptune and Pluto (which in the above argument is the distance between C_N and C_P) is 2.4 AU.[1] But resonance prevents the planets (dwarf planet in the case of Pluto) from getting closer than about 16 AU.[1] In fact, Pluto actually comes closer to Uranus than to Neptune.

References

1. Wasserman, Larry (August 2005). How close does Pluto's orbit come to Neptune? Ask Astro, *Astronomy Magazine.*

The aurora borealis is a fickle phenomenon.
A week can pass without a flicker
then bang! The Northern Lights come on
like a celestial lava lamp.

—Nigell Tisdall, British travel writer

TOPIC 55. The auroras.

An explanation of an aurora (in the northern or southern hemisphere) involves three actors: the Sun, the Earth's magnetic field, and a little quantum mechanics.

Figure 1.
Northern lights.

The Sun and an aurora.
The Sun is constantly streaming charged particles in all directions from the corona, its upper atmosphere. This *solar wind* consists mainly of electrons, protons, and alpha particles (helium nuclei). Sometimes there is a large increase of particles emitted in a particular direction, called a *coronal mass ejection* (CME). This large particle "bubble" collectively contains an enormous amount of energy and may head in Earth's direction.

The Sun has its own magnetic field called the *Interplanetary Magnetic Field* (IMF).[3] The solar wind causes the IMF to extend well beyond the solar system, and the magnetic orientation of the field often changes. An aurora can occur when this magnetic field is oppositely aligned with the Earth's magnetic field (so the south pole of the IMF is aligned with the Earth's north pole). This creates shimmering and moving lights that circle Earth's north or south geomagnetic pole. At some time in their life everyone should travel to witness an aurora (and the same can be said of a total solar eclipse).

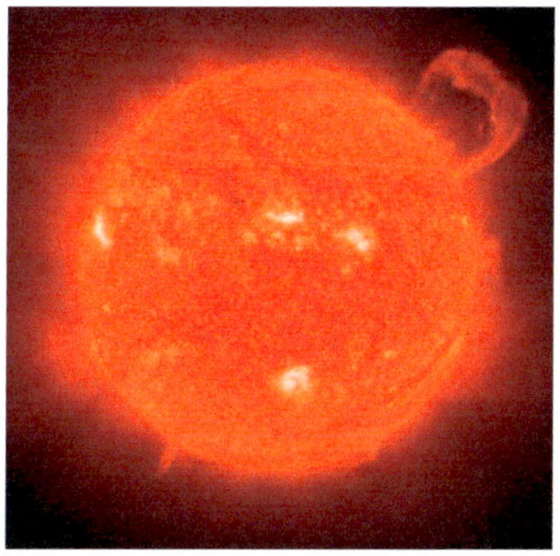

Figure 2. Coronal Mass Ejection from the Sun. Credit: Jean Beaufort

The THEMIS project (Time History of Events and Macroscale Interactions during Substorms). Launched in 2007, THEMIS' goal was to determine what physical process in near-Earth space initiates the violent eruptions of the auroras that occur during substorms in Earth's magnetosphere. Five spacecraft probes were used.[5]

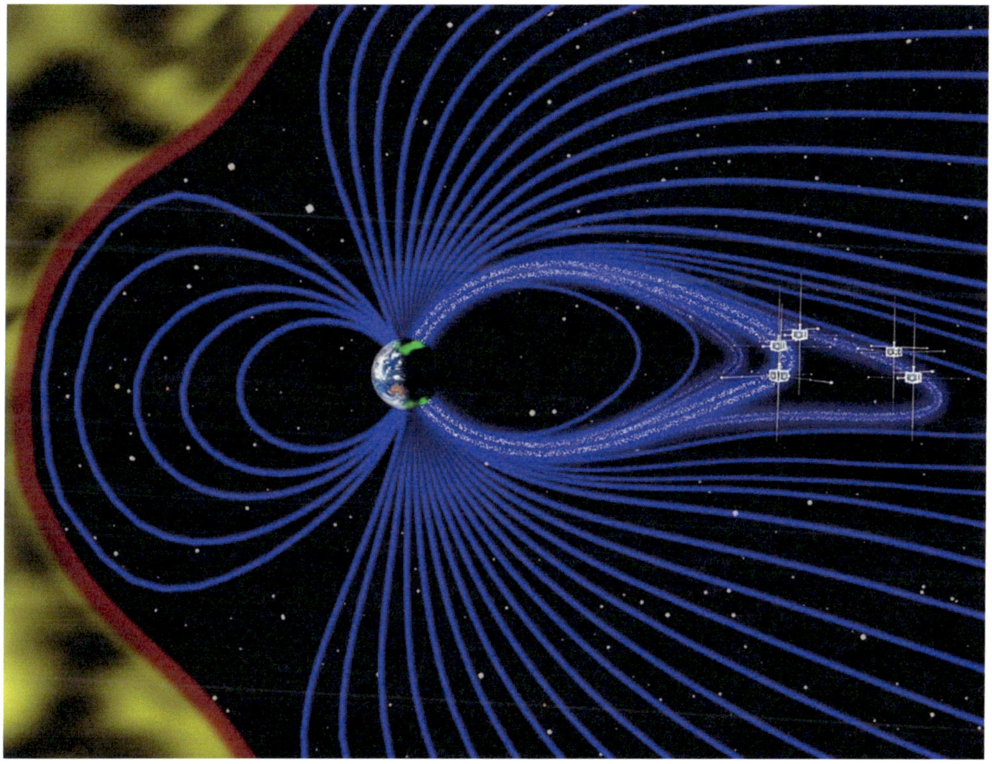

Figure 3. The Earth's magnetosphere and the five THEMIS probes. An artist's rendering (not to scale) of a cross-section of the magnetosphere, with the solar wind on the left in yellow and magnetic field lines emanating from the Earth in blue. The five THEMIS probes were well positioned to directly observe one particular magnetic field line as it oscillated back and forth roughly every six minutes. In this unstable environment, electrons in near-Earth space, depicted as white dots, stream rapidly down magnetic field lines toward Earth's poles. There they interact with oxygen and nitrogen particles in the upper atmosphere, releasing photons and brightening a specific region of the aurora.[1] Credit: Emmanuel Masongsong/UCLA EPSS/NASA

Encounter with the Earth.[1,3]

The volume of space containing Earth's magnetic field is called the *magnetosphere* and is in the shape of a teardrop with the tail trailing behind the Earth, i.e., in the direction away from the Sun. This shape is created by the solar wind. The boundary of the magnetosphere is called the *magnetopause*. The tail of the teardrop is called, not surprisingly, the *magnetotail*.

When CME electrons and ions approach Earth they encounter Earth's magnetic field (EMF). When the EMF is oppositely aligned with the IMF, the two three-dimensional magnetic fields intertwine (i.e., the field lines close in on each other). A similar two-dimensional effect is produced when the north pole of one bar magnet is placed on a flat surface containing iron filings near the south pole of a second bar magnet. If the north poles of two bar magnets are placed near each other their fields (visualized by the iron filings) will not intertwine (the field lines diverge from each other). With the EMF and IMF fields oppositely aligned, some of the CME electrons are diverted around Earth and into the magnetosphere.

The magnetosphere on the daylight side is compressed (creating dayside auroras invisible to us) and the magnetotail is stretched with the field lines getting squeezed closer together. Then the stretched magnetotail snaps back toward the Earth and begins to vibrate back and forth, creating a *substorm*. The snapback starts when Earth's magnetotail field lines touch each other (referred to as *magnetic reconnection*), releasing heat and energy. The start of the snapback takes place at about 20 Earth radii behind the Earth,[3] i.e., on the opposite side of the Earth from the Sun.

NASA scientists working on the THEMIS project positioned five probes at strategic locations along the magnetotail enabling them to observe the aurora wax and wane in step with the vibrations mentioned above. The field lines oscillated in approximately a six-minute cycle in sync with the brightening and fading of the aurora.

Near-Earth electrons are accelerated Earthward, forming a *plasma sheet*. The magnetic field lines bounce backward (away from Earth) at 10–14 Earth radii,[2] which is a little farther from Earth's center than the location of geosynchronous orbits. (A satellite has a geosynchronous orbit if its distance from the Earth's center—about 6.6 Earth radii[3]—is such that it revolves in sync with the Earth's rotation. It may drift a little north or south, but it maintains a constant longitude.) The plasma sheet continues on toward Earth, hitting the upper atmosphere, and electrons travel down the Earth's magnetic field lines toward one of the Earth's geomagnetic poles. There they interact with nitrogen and oxygen molecules, resulting in auroral lights of different colors. Auroras can be produced by a strong solar wind, but the more spectacular ones are generated by CMEs.

How are the colors produced?

The colors (wavelengths) produced depend on two factors—the type of molecule the electron interacts with and the altitude of that molecule. Quantum mechanics enters the scene when one examines the interaction between an atom and an approaching electron.

Most weather phenomena take place at an altitude below 20 kilometers (12.4 miles). In contrast, auroras occur at an altitude between 80 and 400 kilometers (50 and 250 miles). Green is the most common auroral color; blood red is rare.

Just as it is in neon lights, auroral light is produced when electrons excite gas molecules. In the case of auroras, the atmospheric gases are neutral oxygen atoms and nitrogen molecules. The electrons are free electrons driven toward the geomagnetic poles along Earth's magnetic field lines. A *neutral* atom is one that is electrically neutral, i.e., the number of protons (each with a positive charge of +1) in its nucleus equals the number of electrons (each having a negative charge of -1) surrounding the nucleus.

We know from quantum mechanics that the electrons surrounding the nucleus of an atom are in *shells*. (Warning: here we are making a simplification as shells themselves have subshells.) Think of an onion, but with its various shells having space between them. Each shell can only have a fixed maximum number of electrons in it, and that number increases as the radii of the shells increase. The first shell can only hold one or two electrons. A hydrogen atom's one electron is in its first shell and a helium atom's two electrons are in its first shell. All other atoms have two electrons in their first shell, with the rest in higher shells.

Each shell represents an energy level, and the energy levels increase as the radii of the shells increase. Every atomic electron is in one of the shells. The change in the energy of an atomic electron *cannot* be by any amount: its energy levels are discrete and can only change to the energy represented by another shell. This situation is a big break from classical physics where continuous change in energy is allowed. The highest-energy electrons of an (unexcited) atom occupy what is called the *outer shell* and are called *outer electrons*; the other shells/electrons are called *inner* shells/electrons. In general (but not always) each inner shell contains its predetermined maximum number of electrons. The outer shell usually is only partially filled.

If external energy is input to the atom (e.g., from interaction with a free electron approaching the atom), an outer electron can be boosted to a higher energy level. When this occurs, this electron frequently drops back to a lower energy level and emits a photon—i.e., light. This process obeys the law of conservation of energy. If the atom's total energy is increased due an outer electron being raised to a higher energy level, then when this electron drops back to a lower energy level, the atom's total energy is decreased. The difference in energy levels is the energy of the emitted photon. Of course it is possible that an outer electron, instead of just moving to a higher shell, can completely escape from the atom. In this case the atom is now called an *ion*: it has an electrical charge of +1, having one more proton in its nucleus than it has electrons in all its shells.

As an interesting aside, all chemical reactions involve only the outer electrons of atoms. An element is determined by the number of protons in an atom's nucleus; chemical reactions do not affect it. So the alchemists' goal to convert lead into gold was doomed to fail. It is well known that Isaac Newton spent considerable time trying to convert one element into another using chemistry. But he can be forgiven as he lived over two centuries before the atomic theory of matter was fully accepted. It took Einstein's 1905 paper on Brownian Motion to remove the last doubters. And the discovery of the existence of the nucleus, much less protons and neutrons, occurred later in the 20th century.

Most oxygen molecules, O_2, at high elevations are dissociated into atoms by solar extreme ultraviolet light. Molecular nitrogen, N_2, is exceptionally stable.

Oxygen atoms produce green light (557.7 nanometers) at altitudes between 100 and 150 kilometers (62 and 93 miles). At altitudes above 150 kilometers they can produce bright red light. Nitrogen molecules can produce blue light at very high altitudes and deep red (purple-red) light at lower altitudes.

Protons, electrons, and other ions approaching an atom of oxygen or nitrogen in the atmosphere can excite the atom. By "excite" we mean boost the energy level of an outer electron. This, as we have seen from above, can result in the emission of a photon (with the electron dropping to a lower level). This process is called *radiative decay* (which produces light).

In addition, ions can *ionize* a nitrogen atom. Here a nitrogen atom doesn't just have an outer electron get boosted to a higher energy shell. It *loses* an outer electron. When this happens, the atom has a net positive charge and can then capture a free electron while radiating a photon.

A nitrogen *molecule*, N_2, that is ionized and then captures a free electron emits a *blue* photon. Ionization occurs at very high altitudes. Radiative decay occurs at lower altitudes. With radiative decay a nitrogen molecule emits a *red* photon. There are few nitrogen *atoms*, N, below 400 kilometers. Though they are capable of emitting a green light, it is masked by the green of oxygen.[4]

How about oxygen atoms? Radiative decay will produce either *green* or *red* light. Which color depends on the amount of energy absorbed by the oxygen atom.[4] It is green if the atom goes from a very exited energy level to a lower, but still excited energy level (i.e., not to the ground state). It is red light if the atom goes from a less excited state to the ground state. Above, it was stated that the green light is from oxygen atoms at an altitude of 100 to 150 kilometers and that red light is from altitudes above 150 kilometers. Why the difference in altitudes given that both colors are produced by the same process, namely radiative decay?

The answer is very interesting.[4] It involves *collisionally induced quenching,* which produces *no* light. The radiative lifetime (or decay time) of a green photon emission is about three quarters of a second—a long time by ordinary electronic transition standards. But for a red photon it is about 110 seconds—a lifetime. If, during this time, the oxygen atom "collides" with another *neutral* atom or molecule no photon is emitted. Instead of the oxygen atom losing energy by emitting a photon, it loses energy by increasing the kinetic energy or internal energy of the particle that collided with it. Here "internal" can refer to angular momentum or to vibrational energy. Green light is possible only in the near vacuum of space, at least 100 kilometers high, where collisions are less likely. For red light the altitude must be at least 150 kilometers.

A word on the use of the term "collides" when spoken of as action between a neutral atom/molecule with an oxygen atom: this event is not a hard-shell collision. It is a Coulomb-induced (charge-induced) perturbation of the outer electrons of the oxygen atom (with no photons being emitted). The perturbation causes outer electrons in the oxygen atom to drop to a lower energy level but, as stated, instead of photons being emitted, the energy of the colliding particle is increased.

Prime time for aurora viewing.
The best time to see an aurora is when the Sun is near an equinox: March/April or September/October.

References

1. Tran, Lina (September 12, 2016). *NASA's THEMIS sees auroras move to the rhythm of Earth's magnetic field.* nasa.gov https://www.nasa.gov/feature/goddard/2016/nasa-s-themis-sees-auroras-move-to-the-rhythm-of-earth-s-magnetic-field This web article is based on [2].
2. Panov, E. V., et al. (September 12, 2016). Magnetotail energy dissipation during an auroral substorm. *Nature Physics, 12*, p 1158-1163. https://www.nature.com/articles/nphys3879 is the URL for the abstract.
3. Brekke, Pal (February 2013). Secrets of the northern lights. *Sky & Telescope,* p 18-25. A very informative article.
4. Atmospheric Optics. Glowing gases—aurorae. atoptics.co.uk _https://www.atoptics.co.uk/highsky/auror3.htm
5. NASA. *THEMIS and ARTEMIS*. nasa.gov https://www.nasa.gov/themis-and-artemis

There's no twilight in the tropics.
Night falls like a curtain.

—*Waldemar Young, American writer*

TOPIC 56. Twilight in the tropics and why twilights are shorter there than at other locations. If you are located on the equator, then every night of the year the Sun rises and sets vertically with respect to the horizon. At any other location on the Earth the Sun never sets or rises vertically. As your latitude increases, the angle at which the Sun sets with respect to the horizon gets smaller.

And, of course, if the angle is vertical, i.e., 90 degrees, then the Sun will drop below the horizon by any fixed *vertical* distance faster than if the angle is less than 90 degrees. So at any day of the year at sunset it gets dark quicker if one is on the equator than at any other location on the globe.

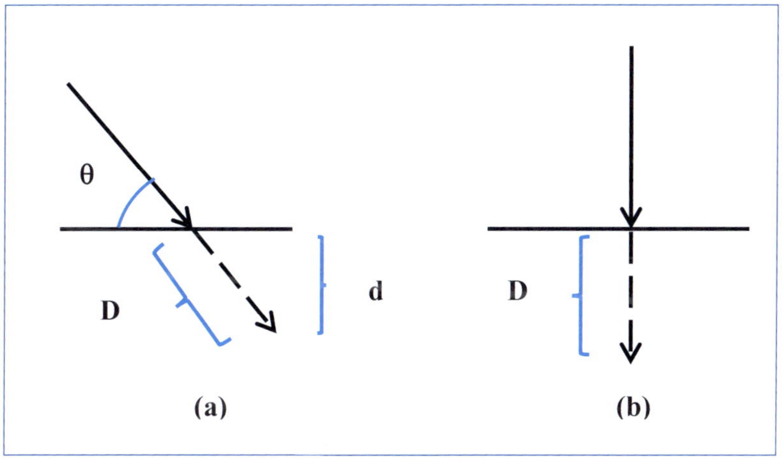

Figure 1. The Sun setting at an angle vs vertically.

In (a) the Sun sets at angle θ, and in (b) the Sun sets vertically. The Sun has moved a distance D (in the same amount of time) in both cases, but in (a) the Sun is only distance $d = D\sin\theta < D$ below the horizon and so the sky is not as dark as in (b).

At the north pole the Sun can be seen to set/rise only when the Sun is at an equinox and the angle of the Sun with respect to the horizon is zero. When this is the case, the Sun circles the horizon during one 24-hour period. At all other days at the north pole the Sun is either always above or below the horizon and thus doesn't rise or set.

Let J be the angle that the Sun makes with the horizon at sunset/sunrise, i.e., J is the angle between the diurnal circle of the Sun (the daily path of the Sun) and the horizon.

Let L be the latitude of an observer on the Earth, and D be the declination of the Sun, i.e., the Sun's celestial latitude. The equation

$$\cos(J) = \sin(L)/\cos(D) \qquad (1)$$

relates the three variables.[1,2]

The horizon curve on the celestial sphere is the curve where the tangent plane to the Earth at the position of an observer intersects the celestial sphere. Neither it nor the diurnal circle are great circles. A great circle is one whose center is at the center of the Earth. So J is defined as the angle between two *non*-great circles. It is interesting and useful to realize that

The angle J at which the Sun sets can be expressed as the angle between two great circles on the celestial sphere.

One can see this with the following argument (see Figure 2 below):

Let the great circle passing through the zenith (for position O on the Earth) and thc Sun be denoted by ZS. Let the hour circle of the Sun (which is the great circle passing through the North Pole and the Sun) be denoted by NS. The angle q between ZS and NS (called the *parallactic angle*) is well defined whether or not the Sun is on the horizon (rising or setting). (This angle is useful in rotating a camera attached to an altazimuth-mounted telescope to keep the sunspots on the Sun crossing the face of the Sun in a straight line, or the bands of Jupiter from rotating, or any astronomical field of view from getting rotated during a time exposure.)

The angle J is only defined for when the Sun is rising or setting. Suppose we only consider the case of the Sun rising or setting. *Then J is the same as q.*

It is easy to see this with the following geometric argument. See Figure 2 below. S is the setting Sun for an observer O. N is the North Pole. NS is the great circle (in red) passing through N and S. ZS is the great circle (in blue) passing from O's zenith Z through S. The diurnal circle is the green dashed arc. The equator is the black dashed arc. q is the parallactic angle.

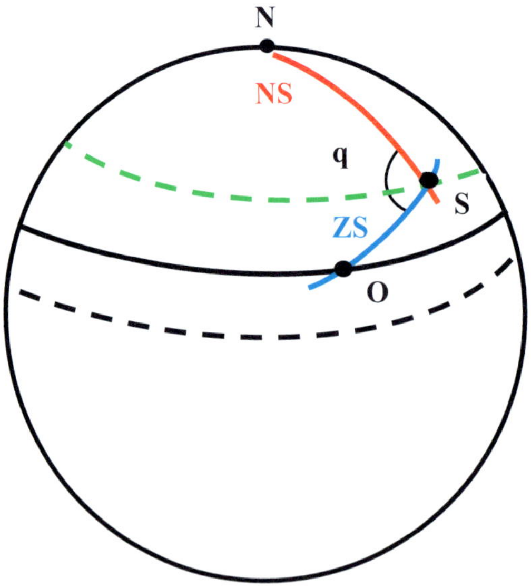

Figure 2. Angle q is the *parallactic angle.*

The great circle ZS is perpendicular to the horizon for O, and the great circle NS is perpendicular to the diurnal circle of the Sun. If two lines L_1 and L_2 (here tangent to circles) are respectively perpendicular to lines L_1' and L_2', then the angle between L_1 and L_2 is the same as the angle between L_1' and L_2'. This means that the angle q between ZS and NS is the same as the angle between the diurnal circle of the Sun and the horizon, which is the angle J at which the Sun sets.

In Figure 2 the observer O is at latitude L and the Sun is setting at point S with declination (i.e., celestial latitude) D where D > L. To repeat our conclusion: The angle J of the setting Sun, namely the angle between the horizon and the diurnal circle, equals the angle q between NS and ZS.

The setting angle of the Sun is always 90 degrees when the observer is on the equator.
This of course follows from the equation as then L = 0: cos(J) = 0 and thus J = 90 degrees. But it is also easy to see this from Figure 2: If O is on the equator, then N is on the horizon. But S is on the horizon as the Sun is setting. Thus NS is the horizon. Since ZS is always perpendicular to the horizon we have that q (the angle between NS and ZS) is 90 degrees, and q = J.

The setting angle of the Sun is never 90 degrees for an observer not on the equator.
The equation shows that J = 90 degrees (i.e., the Sun sets/rises vertically) implies that sin(L) = 0, hence L = 0 (i.e., O is on the equator). But again, we can see that the statement is true from Figure 2: Say O is north of the equator. With S on the horizon and ZS perpendicular to the horizon, the angle q between ZS and NS must be less than 90 degrees unless N is on the horizon, which it is not.

See *Appendix 56.1* for a further discussion.

References

1. Meeus, Jean (1998). *Astronomical Algorithms.* Willmann-Bell, p 98-100.
q is the parallactic angle; J is the angle of the diurnal path of a celestial body (e.g., the Sun) relative to the horizon at the time of its rising/setting. On page 99, Meeus gives the formula cos(q) = sin(L)/cos(D) for the parallactic angle *when* the Sun is on the horizon. (He uses symbol ϕ for L and δ for D.) But above we showed that, when the Sun is on the horizon, q and J are the same (and of course, J is only defined for the case of the Sun being on the horizon). We thus obtain our equation (1) by substituting J for q into Meeus' equation cos(q) = sin(L)/cos(D). Meeus' equation (14.1) on page 98 gives the general formula for q. It is interesting to note that at the bottom of page 100 he gives a formula for the angle J that is seemingly different from equation (1). One can show using trigonometry that his equation and equation (1) are in fact equivalent.
2. Smart, W. M. (1977). *Textbook on spherical astronomy.* (6th ed. revised by R.M. Green). Cambridge University Press. p 50. The second equation involves the parallactic angle q (Smart uses the symbol η in lieu of q). For the case when the Sun is on the horizon we know that q = J. In his equation, just set z = 90 degrees, which is the case if the Sun is on the horizon, to obtain our equation (1).

To collapse or not to collapse—that is the question.

—Shakespeare-(not really)

TOPIC 57. Why does a rotating gas cloud around a star become a disk when it collapses?

In asking this question we make two assumptions:

We assume the gas cloud is in thermal equilibrium (temperature is the same everywhere in the cloud).

We assume the cloud (which we can think of as being spherical, but that is not necessary) is rotating about some axis through the star's center. If this is happening, it usually is the case that the cloud's rotational axis is the same as that of the star. While we assume this in Figure 1 below, the argument that follows is perfectly valid if the star's axis of rotation is tilted with respect to that of the cloud. However, if that is the case, then the resulting gas disk will not be in the planet's equatorial plane.

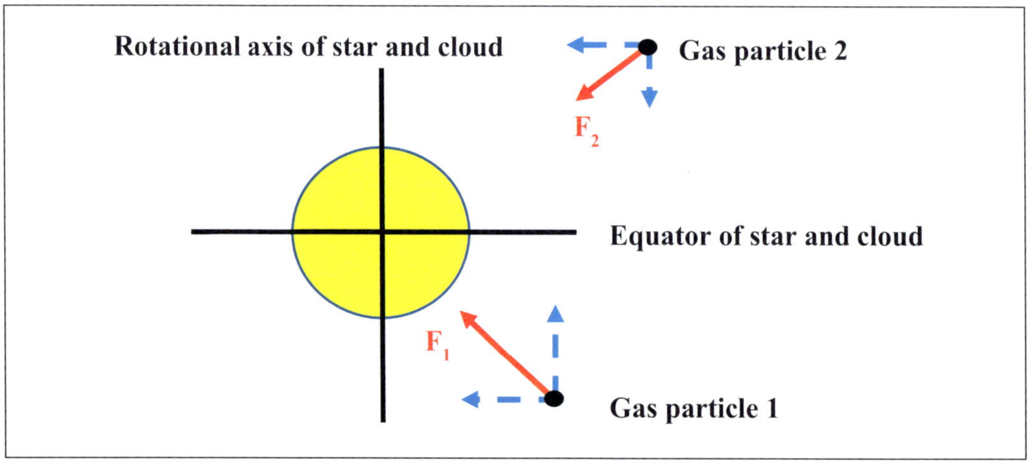

Figure 1. Particles in a gas cloud are attracted toward the center of the star with forces F_1 and F_2. The blue arrows are the parallel and perpendicular components of the forces. Both particles move toward the star's equatorial plane because the perpendicular components point in that direction no matter where a gas particle is located. So a disk is formed in the equatorial plane.

Look at the force acting on a particle near the star (the force vector, not the current velocity vector of the particle). It is a gravitational force vector pointing from the particle toward the center of the star. This force has a component parallel to the cloud's axis of rotation and a component perpendicular to this axis. The component parallel to the axis always points *toward* the cloud's equatorial plane (the plane perpendicular to the cloud's axis of rotation that passes through the center of the star). Thus the particle will move toward the cloud's equatorial plane. The component perpendicular to the axis will cause the particle to move toward the axis of rotation, and conservation of angular momentum will cause the particle to speed up its rotation about the star—remember the twirling skater pulling her arms inward to increase her rotational speed. When the perpendicular component balances the centrifugal force, the particle will stop moving closer to the axis of rotation. Thus we have a disk of gas swirling around the star.

From this distant vantage point,
the Earth might not seem of particular interest.
But for us, it's different. Consider again that dot.
That's here, that's home, that's us.

—Carl Sagan Pale Blue Dot

TOPIC 58. Views of the Earth (and Moon) from afar.

The photograph below, more than any image in history up to the time of its release, jolted people into realizing that we are all members of one race, the human race, and that we all share this sphere with our fellow creatures. Before this iconic image, the only wide views of the Earth were those taken in low Earth orbit where only sections of the Earth were visible.

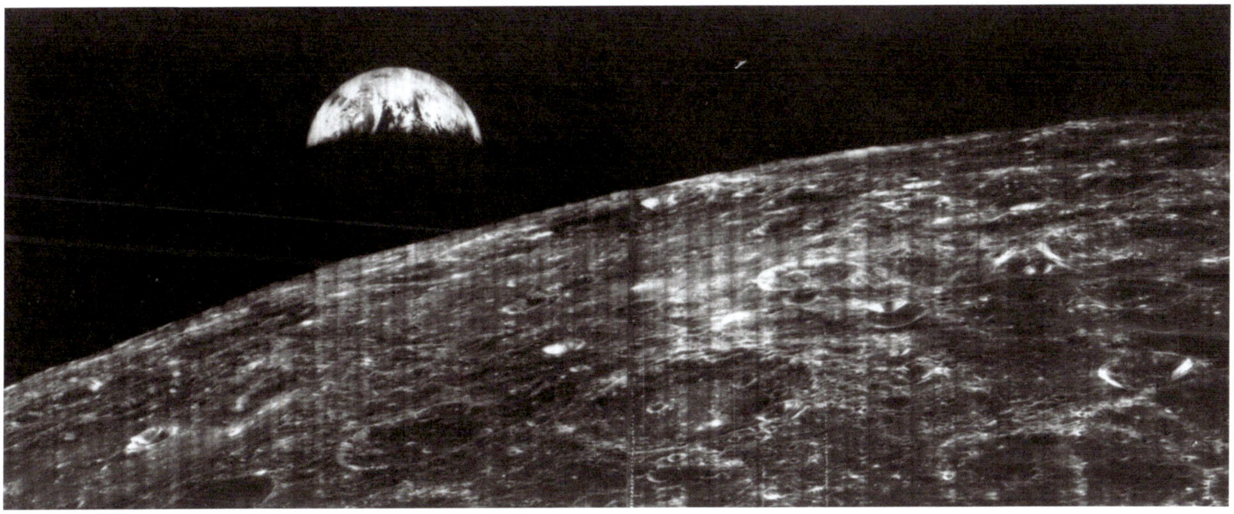

Figure 1. Lunar Orbiter 1's original iconic photograph—the first image of the Earth taken from the Moon. Imaged on August 23, 1966. Obtained from [1]. Credit: NASA

In 2008, using refurbished machinery and modern digital technology, NASA released a restored version of Figure 1. The new image, Figure 2 below, has a much higher resolution than was possible originally. Note: The field of view in Figure 1 is wider than in Figure 2 (which is a section from the restored image).

This image was taken from the Lunar Orbiter, not from the lunar surface. Interestingly, for an observer on the lunar surface facing Earth, the Earth would always remain stuck in virtually the same position in the sky, although it would wax and wane through phases.

But something more is going on. The Moon's orbit around Earth is not a perfect circle, so the Moon's orbital speed varies. This situation enables *Earth-bound* observers at certain times to see a little more "around" either side of the Moon than is usual—a phenomenon known as lunar libration (or wobble). Now consider a *lunar* observer on the Moon's far side, but very close to the Moon's left or right limb—at a point just "beyond" the Moon's edge (as seen from Earth). From this spot on the Moon, sometimes the Earth would be just below the lunar horizon and other times just above it. Now-you-see-it, now-you-don't.

Figure 2. A 2008 refurbished section of Figure 1. Credit: NASA, Lunar Orbiter Image Recovery Project

Figure 3. "Earthrise" 1968. Credit: Bill Anders/NASA, taken from Apollo 8

Figure 4. "Blue Marble," 1972. Credit: NASA/Apollo 17 crew

Figure 5. The Earth and its Moon as seen on May 6, 2010, from the MESSENGER spacecraft near Mercury at a distance of 183 million kilometers (114 million miles) from Earth. Credit: NASA image provided by the MESSENGER science team, NASA/Johns Hopkins University Applied Physics Laboratory/Carnegie Institution of Washington

Figure 6. A view of Earth through the rings of Saturn taken by Cassini in September 2006. Credit: NASA/JPL

Figure 7. The Moon transiting the Earth, seen by the Deep Space Climate Observatory satellite on August 4, 2015. The far side of the Moon is illuminated by the Sun as the Moon crosses between the satellite and the Earth. The satellite is one million miles from Earth. Credit: NASA/NOAA

For an animation showing the Moon passing across the face of the Earth, see https://www.nasa.gov/feature/goddard/from-a-million-miles-away-nasa-camera-shows-moon-crossing-face-of-earth.

In Figure 7 you might be struck by how dark the Moon looks. This is the side of the Moon we cannot see from Earth, and in this image it is being directly lit by sunlight. The far side of the Moon doesn't contain any of the dark maria dominating the side of the Moon facing the Earth, so it is the side with the higher albedo—the brighter side. Why is it so dark compared to the seemingly very bright full moon seen by humans on Earth? As explained by Bob Berman, "the mammalian eye resets brightness levels according to your surroundings." [2] Against a black night sky our brains overexpose the Moon. Berman states that the average lunar albedo is 11, that of charcoal is 5, and dark forest foliage has an albedo of 15.[2] So the Moon's surface is really quite dark. This is an example illustrating the fact that our brain assembles images that can be at variance with reality.

If we perceived the Moon as the dark body that it really is, most humans, and certainly all the werewolves, would not be pleased. But amateur astronomers, who prefer new moons to observe deep-sky objects because of the increased contrast of faint galaxies, etc., with the background night sky, would have more nights each month to venture out. Yet they too are in love with our bright, beautiful Moon.

References

1. Ferreira, Becky *(*August 23, 2016). The first photographs of the Earth from the Moon were taken 50 years ago today. *Motherboard* (Tech by VICE). https://www.vice.com/en/article/bmvyzw/the-first-photographs-of-the-earth-from-the-moon-were-taken-50-years-ago-today Motherboard is an online magazine and video channel dedicated to technology, science, and humans and launched by VICE Media Group in 2009.
2. Berman, Bob (October 2021). Facing reality—what's in a full moon. *Astronomy Magazine,* p 12.

A time will come when men will stretch out their eyes
They should see planets like our Earth.

—*Sir Christopher Wren, English architect and astronomer*

TOPIC 59. A discussion and some personal thoughts on the possibility of other life in the solar system and beyond, and space travel.

Perhaps the most important question in astronomy (or all of science) today, that is within our current reach of answering, is whether there was a second genesis in our solar system, i.e., whether life arose somewhere in the solar system independent of life on the Earth. If so, then life is probably common in other star systems.

We understand how life proceeds once it occurs—Darwinian evolution would be the process everywhere. But how did life first arise on Earth? We as yet don't know. If we had some idea of how to start life, perhaps in a laboratory, we could make a start at estimating the probability of the prevalence of life in the galaxy. But asking questions now about the probability of life arising on other worlds, or asking about the probabilities for the factors in the Drake equation,[9] is not time well spent. Currently we only have one example of life having arisen—namely life on Earth. One cannot generalize from an instance of one. We need more data. That can only come with exploration. Looking for a second genesis in our solar system is something humans can accomplish in the near term.

One can use Plato's cave analogy to make the point about exploration: People, deep in a cave, who only can see shadows on the cave walls due to a burning fire, speculate about the nature of the world. They mistake the shadows for reality. Without going out of the cave how could they ever learn about the outside world and the stars? In their speculations could they ever have hit upon the true nature of reality?

It is tragic that in the 2020s humans have not had a manned space mission since the last Moon landing in 1972. Here near-Earth missions to the space station are not considered as manned space missions. The United States needs to have a consistent, funded, planetary exploration policy that can span different administrations. NASA had the Constellation program to build a base on the Moon. Then, unfortunately, it was cancelled and there was a "we are going directly to Mars" policy—without funding any mission. Hundreds of engineers working on the Constellation program lost their jobs, left NASA, and scattered to various companies. Neil Armstrong, who by nature kept a low profile on public issues, spoke out against the cancellation. Funding of NASA at that time was only about one quarter of one percent of the national budget. Now it is a little more. To talk about "all that money spent on a space program" is a red herring. That's not where the money is going. Now we again have plans to build a base on the Moon, and hopefully there will not be another lengthy hiatus in manned missions.

Any time humans investigate the unknown, amazing discoveries are made. When Galileo turned his telescope toward the night sky he discovered among other things that Jupiter had moons. If the United States *simultaneously* funded a robust robotic mission to the planets and moons in the solar system, a human mission to Mars, and the establishment of a base on the Moon, with substantial and sustained funding, I suggest that the next generation would look back at the discoveries and have no doubt this was a wise policy.

Some always want to know the benefits (never mind satisfying human curiosity) of money spent on basic research *before* the research is done. But that is not how it works. Breakthrough discoveries are made *after* the research is done. There is a story (probably apocryphal) that the Lord Mayor of London, on hearing that Michael Faraday had invented a device (the magneto) that could generate electricity, asked Faraday of what use it could ever be, to which Faraday replied, "Perhaps you can tax it."

Besides satisfying human curiosity and the practical benefits that come from research, in the case of solar system exploration there is another benefit: survival of the human race. The solar system is a dangerous place. The dinosaurs and much of life on the planet were wiped out when an asteroid hit the Yucatan Peninsula 65 million years ago. A similar threat could—and eventually will—arise again. The dinosaurs *literally* became extinct because they didn't have a space program. Astronomers are searching for near-Earth asteroids, ones whose orbits cross the Earth's, and are considering which of those might pose a danger to us in the not-too-distant future. They also are devising strategies for intercepting and diverting these objects from a collision course with Earth. If the dinosaurs had diverted that asteroid 65 million years ago they might very well still be here, and we would not.

Some current solar system objects of special interest in the search for a second genesis: Enceladus, Europa, Titan, and Mars. The first three are briefly discussed below.

Enceladus. It is now known that the small (500 miles in diameter) Saturnian moon Enceladus has a global ocean whose salinity is similar to Earth's oceans[5] and contains the basic elements required for life as we know it. It is believed that below the ocean is a rocky layer most likely with geothermal vents releasing chemicals into the water similar to the vents on Earth's mid-ocean ridges. This is based on the Cassini spacecraft having passed through plumes of water vapor and icy grains ejected high above the surface of Enceladus from four "Tiger stripe" linear cracks in the icy surface near the south pole. The close encounters with Enceladus extended from 2005 to 2015. A close flyby in 2008 was the first of the passes in which samples of the plumes were collected and analyzed onboard the spacecraft—the results strongly supporting what is claimed above. This assumes the plumes' samples are from the global ocean. Recently it has been suggested that they may come from concentrated interstitial liquid—pockets of slushy brine in Enceladus' icy shell above the ocean itself.[10]

From analyzing the collected plume material[7,8] we can arrive at several conclusions. The Cassini spacecraft detected a relatively high concentration of certain molecules associated with hydrothermal vents on the bottom of Earth's oceans, specifically dihydrogen, methane, and carbon dioxide. The amount of methane found in the plumes was particularly unexpected. CHNOPS is an acronym for key elements (C, H, N, O, P, and S) believed needed for life. Data showed that Enceladus' ocean contains C, H, N, O, and possibly S. Phosphorous, P, is likely present in the rocky core, and probably partially dissolved into the ocean at abundances below the limit Cassini's instruments could detect. Simple and complex organic compounds were detected. The vapor contains carbon in the form of CO_2, CH_4, and perhaps CO. It contains nitrogen in the form of NH_3 and HCN, known precursors of amino acids and other more complex organic compounds of prebiotic relevance. Sulfur may be present as H_2S. N_2 may be present, but is indistinguishable from CO in Cassini data.

The interested reader might also wish to consult [6].

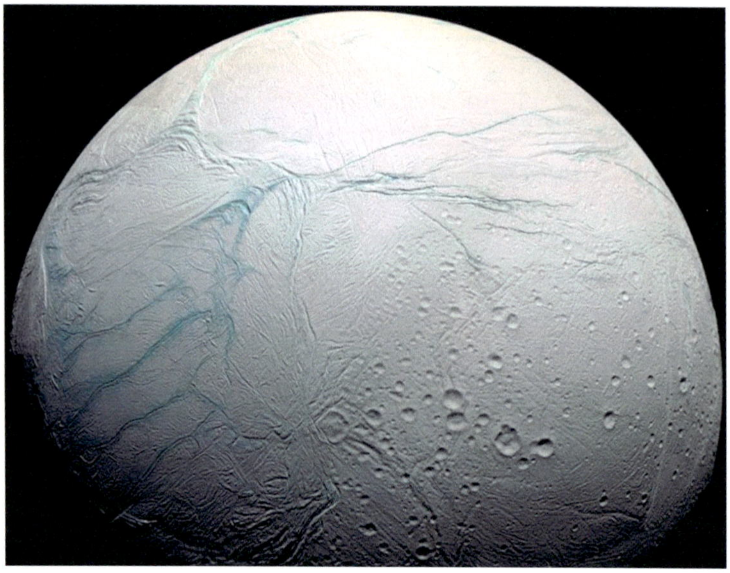

Figure 1. Four long, parallel scratches at the south pole of Enceladus spew icy material from the subsurface ocean. In this image captured by Cassini, the Tiger Stripes are visible in false-blue color. Credit: NASA, ESA, JPL, SSI, Cassini Imaging Team

How could such a small moon be hot enough in its interior to produce these geysers? One idea is that resonant perturbations with the nearby moon Dione could provide periodic eccentricity changes in the orbit of Enceladus around Saturn. This would induce tidal stretching providing the heat needed to maintain a global ocean.

Planetary scientist Chris McKay has stated that any life in the oceans of Enceladus would thrive in Earth's oceans and vice-versa.[5] This means that any spacecraft landing on Enceladus would present a serious contamination danger to Enceladus itself. Likewise, any samples brought back to Earth would have to be handled with great care.

Europa. (See Topics 45, 46, and 53.) Europa presents a situation similar to Enceladus: A global ocean over a rocky bottom (not ice) probably with geothermal vents. Geysers have been seen, but not as many as on Enceladus. No spacecraft have yet flown through a geyser on Europa to sample the contents. Europa, about the size of our moon, is much bigger than Enceladus. Perhaps life more complex than microorganisms exists in its ocean—possibly visibly large creatures as suggested in the science-fiction film *Europa Report*.

Titan. Much has already been said on Titan in Topic 10. Scientists, for good reason (to be discussed below), are biased toward carbon-based and water-based life. Titan certainly has many organics, but any surface life would most likely be based on liquid methane/ethane. And as has been noted in Topic 10, molecules of methane and ethane, unlike water molecules, are non-polar and thus cannot dissolve molecules such as amino acids, proteins, and DNA, allowing cells to use them. Hence any life on Titan would constitute a second genesis. Because life on Earth depends so much on water molecules being polar, we next discuss polar molecules.

A bit of Chemistry.

Water molecules and polar molecules in general.
A water molecule consists of one oxygen atom and two hydrogen atoms. The oxygen atom has two electrons in its inner-most shell and six in the next shell outward. A maximum of eight can populate this second shell. A hydrogen atom has its sole electron in its first shell, which can contain a maximum

of two electrons. All the atoms "want" to fill their outermost shell. Electrons in the outer-most shell are called *valence electrons*. Each hydrogen atom shares its one electron with the oxygen atom and the oxygen atom shares one of its valence electrons with each hydrogen atom. This gives the oxygen atom and each hydrogen atom its full complement of valence electrons. Such shared bonds are called *covalent bonds*.

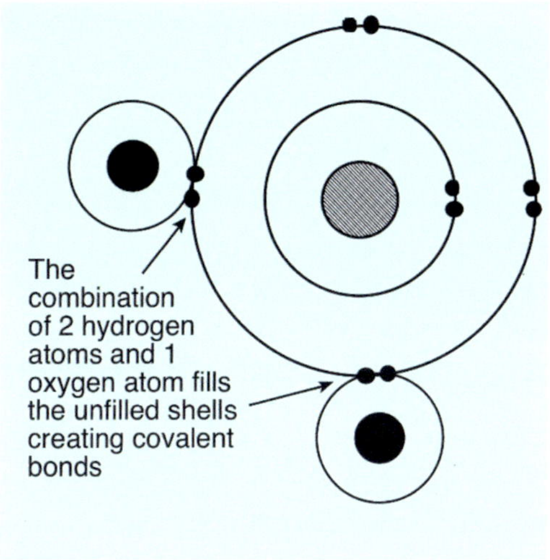

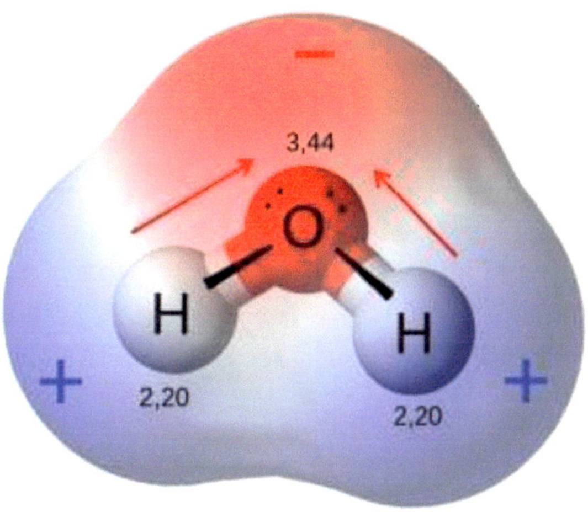

Figure 2. (left) Water molecule showing the covalent bonds; (right) Water molecule. Credit (left): http://www.earthsurfaceprocesses.com/3b3-E-WaterStructure.html (Cropped by author); Credit (right): https://biologydictionary.net/polar-molecule/

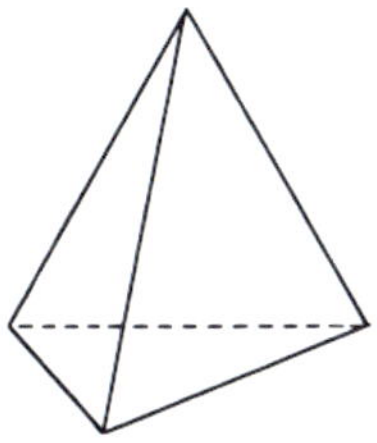

Figure 3. Tetrahedron.

The oxygen atom is at the center of a tetrahedron, with each hydrogen atom at one of its four corners. The four valence electrons of the oxygen atom that are not part of the covalent bonds with the hydrogen atoms form two pairs, each pair being at one of the other corners of the tetrahedron. In Figure 2 (right) each of these pairs is indicated with a pair of black dots.

Each atom has a property called *electronegativity*. This is a measure of how strongly the atom attracts electrons. Higher electronegativity means higher attraction. The electronegativity of oxygen and hydrogen respectively are 3.44 and 2.20. Each red arrow in Figure 2 (right) indicates that the two electrons (not explicitly depicted) forming a covalent bond between the oxygen atom and a hydrogen atom move closer to the oxygen atom because of its higher electronegativity. This results in the water molecule having two dipoles, one for each O-H bond, resulting in a net charge difference across the water molecule, as indicated in Figures 2 (right) and 4. Such a molecule is called a *polar molecule*.

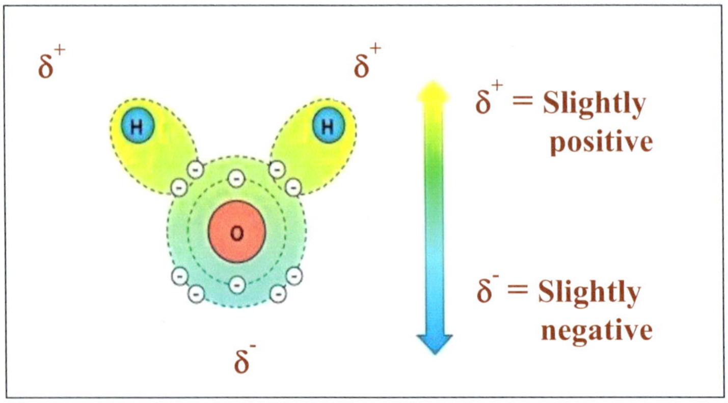

Figure 4. High oxygen electronegativity results in a polar water molecule. The covalent electrons are drawn closer to the oxygen atom resulting in a dipole for each O-H bond, and thus a charge difference across the water molecule. Credit: https://ib.bioninja.com.au/standard-level/topic-2-molecular-biology/22-water/water-structure.html (Cropped by the author)

In general, a *polar molecule* is a molecule in which one end of the molecule is slightly positive, while the other end is slightly negative, where this is due to electronegativity as in a water molecule. Ammonia, NH_3, and hydrogen fluoride, HF, are two other examples of polar molecules. Atoms in polar molecules that frequently play the same role as oxygen, O, in water are carbon, C, nitrogen, N, fluorine, F, and sulfur, S.

Polar molecules as solvents.

Substances that dissolve into a solvent are called *solutes*, and the resulting mixture is a *solution*. Water's polar molecules form hydrogen bonds with ions and polar molecules, allowing ionic and polar covalent compounds to disperse easily in water. What are ionic and hydrogen bonds and how is dispersal effected? In answering this, we first repeat the definition of a covalent bond.

Covalent bond: A sharing of electrons by two atoms. Examples: Cl_2, H_2, H_2O, and HCl. The reason that two Cl atoms stick to each other is that the shared electrons are attracted to both nuclei. Two atoms have a *single, double, or triple (covalent) bond* between them if *each* respectively shares one, two, or three electrons with the other. So a total of two, four, and six electrons are shared in the respective types of covalent bonds.

Ionic bond: A bond between two ions, e.g., Na^+ and Cl^-. Consider salt, i.e., NaCl. Na has one valence electron and Cl has seven. If atoms of Na and Cl come next to each other, Na transfers its lone valence electron to Cl, so that now each atom has a complete outer shell. The two atoms are now oppositely charged ions, Na^+ and Cl^-, and are held together because of this, forming a molecule of salt.

Here we took neutral atoms and ended with oppositely charged ions that "stick" to each other because they have opposite charges. But we could have started with *molecules* that *already have* opposite charges. If they are near each other they can form an ionic bond. Here is an example of ionic bonding between *molecular* ions. Consider ammonium sulfate $(NH_4)_2SO_4$. NH_4^+ is ammonium (an ion) where one of the four hydrogen atoms has lost its electron. SO_4^{-2} is the sulfate ion where each of two oxygen atoms has lost one electron. Hence the ammonium ion has a charge of +1 and the sulfate ion has a charge of -2. An ammonium sulfate molecule is formed when two ammonium ions are attracted to a sulfate ion.

Hydrogen bond: Let us first start with an example. Consider a water molecule. It is a polar molecule where a covalent bond joining the oxygen atom to one of its hydrogen atoms causes the two shared valence electrons to be tugged closer to the oxygen atom than to the hydrogen atom. This shifting of valence electrons produces weak positive charges at the two hydrogen ends of each water molecule and a weak negative charge at the oxygen end. Thus two water molecules can be weakly attracted to each other if one of the hydrogen ends of one water molecule gets close to the oxygen end of the other water molecule. This attraction is called a hydrogen bond. Hydrogen bonds are responsible for the bonding between water *molecules* in liquid and solid states, and are weaker than covalent and ionic bonds.

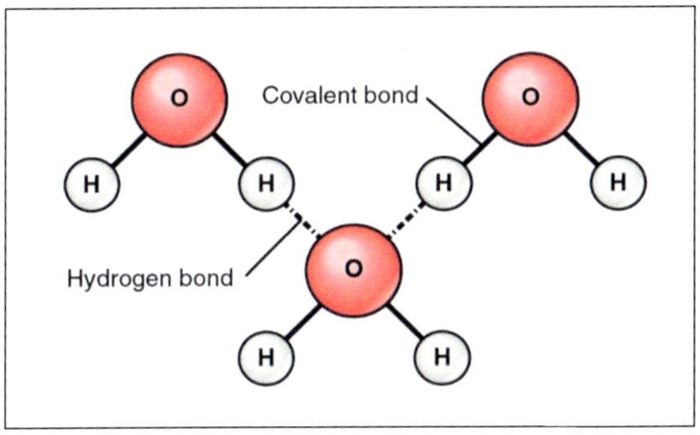

Figure 5. Hydrogen bonds (dashed lines) between water molecules. Credit: https://philschatz.com/anatomy-book/contents/m46000.html

More generally, a *hydrogen bond* is a bond that occurs when a hydrogen atom, call it H, that is covalently bonded to a strongly electronegative atom A_1 in a polar molecule M_1, forms a weak bond (the hydrogen bond) with another strongly electronegative atom A_2 in a polar molecule M_2. The valence electrons for the covalent bond between H and A_1 are tugged closer to A_1 than to H, producing a "polar" effect for molecule M_1. M_2 may be the same kind of molecule as M_1. Let us relate this to the specific example given in Figure 6 below: Here A_1 corresponds to the oxygen atom, M_1 to the water molecule, A_2 to the nitrogen atom, and M_2 to the ammonia molecule.

Other examples of a hydrogen bonding:

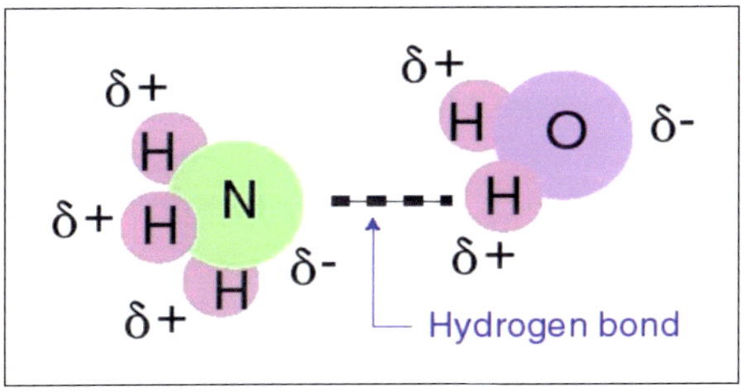

Figure 6. Hydrogen bond between an ammonia molecule NH_3 and a water molecule H_2O.
Credit: https://socratic.org/questions/what-are-some-examples-of-hydrogen-bonds

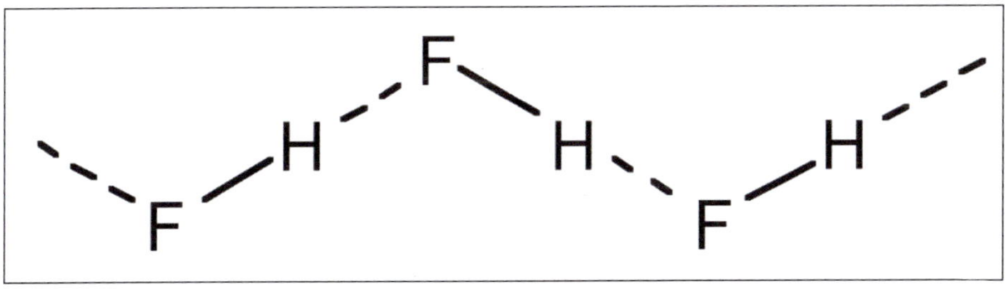

Figure 7. Hydrogen bonds (dashed lines) between hydrogen fluoride molecules, HF. Credit: chem.libretexts.org

What is the difference between a hydrogen bond and an ionic bond? An ionic bond occurs between two *ions*, e.g., Na^+ and Cl^-. In the case of a hydrogen bond we have bonding between (a hydrogen atom in) a polar molecule and (a strongly electronegative atom in) another polar molecule. Polar molecules are *not* ions, but there is a (weak) charge difference across a polar molecule. As was stated, hydrogen bonds are much weaker than ionic and covalent bonds.

Why does its polarity make water a good solvent?
As mentioned above, water can dissolve molecules that are essential to life on Earth. These include amino acids, proteins, and DNA. This allows cells to use them. Nonpolar molecules such as methane cannot do this.

First, when an ionic or polar compound (the solute) enters water, it is surrounded by water molecules. The relatively small size of water molecules typically allows many water molecules to surround one molecule of solute.

Interaction occurs between water molecules and salt molecules (or in general molecules held together by an ionic bond). The positively charged side of a water molecule is attracted to the negatively charged chloride ions and the negatively charged side of a water molecule is attracted to the positively charged sodium ions. Essentially, a tug-of-war ensues with the water molecules winning the match. The Na and Cl atoms are separated from each other. The dipole forces of water disrupt the attractive forces that hold the sodium and chloride in the salt molecule together, dissolving the salt.

What about interaction between water molecules and other polar molecules (the solutes)?
The partially negative dipoles of a water molecule are attracted to positively charged components of the solute, and vice versa for the positive dipoles of the water molecule. Thus the dipole forces of water disrupt the attractive forces that hold the solute molecules together and so dissolve the solute.

Other positive life-supporting features of water.
Liquid water is very prevalent in the universe. Water can serve as a medium for supporting complex molecules. The term *thalassogen* (Greek: sea-former) was coined by the science-fiction writer and biochemist Isaac Asimov to describe liquids capable of forming oceans.[1]

Water plays a role in the chemical processes of life. In the joining of two amino acids by a peptide bond (-CO-NH-), a molecule of water is created. Proteins are chains of amino acids held together by peptide bonds.

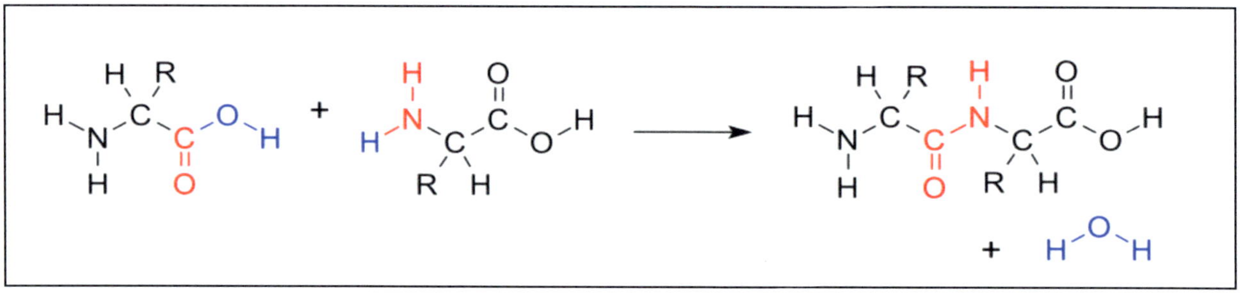

Figure 8. The dehydration condensation of two amino acids to form a peptide bond (red) with expulsion of water (blue). Credit: Wikipedia. https://en.wikipedia.org/wiki/Peptide_bond

The extra bonding between water molecules gives liquid water a large specific heat capacity. The *specific heat capacity* of a substance is the amount of heat that is needed to be applied to a unit mass of the substance to increase its temperature by one unit. Since water has a high specific heat capacity, a large body of water can more easily maintain a near-constant temperature, which is advantageous to anything living in the water.

The solid form of water, ice, has a lower density than liquid water. This is unusual and is advantageous for life living in water. As water freezes, ice forms at the surface of the liquid water and not below it. This helps to insulate the liquid water from further freezing. If the solid state of the substance is denser than its liquid state, the solid state would form at the surface and then sink to the bottom of the liquid, which would keep the surface at a lower temperature, allowing further freezing. This would not be beneficial for life living in the liquid.

Metabolic/chemical processes in life can proceed at a faster rate at higher temperatures. A general rule for chemists says reaction rates drop by half every time the temperature goes down by 18° F (10° C).[1] Thus any life in a substance that is a liquid only at very low temperature results in very slow reaction rates.

Temperatures at which certain substances are a liquid at one atmosphere of pressure:[1]
Water: 32° to 212° F (0° to 100° C)
Methane: -297° to -260° F (-183° to -162° C)
Ammonia: -108° to -28° F (-78° to -33° C).

Ammonia is a polar molecule and, for supporting life, is more favorably viewed as a water substitute than other liquids.

What is the advantage of basing life on carbon, as opposed to silicon and other elements?

Carbon atoms are capable of forming long chains, rings, loops, and other shapes. Two carbon atoms can get together by each sharing two of their four valence electrons—a *double covalent bond*. Double bonds are strong and are the reason carbon atoms are capable of forming stable and complex molecules involving many carbon atoms.

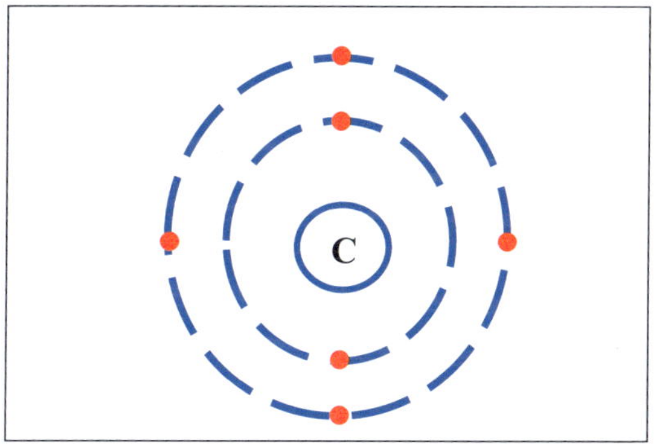

Figure 9. Electron configuration in the carbon atom. There are six electrons, four being valence electrons.

While on Earth there are about 30 silicon atoms for every carbon atom, in the solar system there are about 10 carbon atoms for every silicon atom.[1]

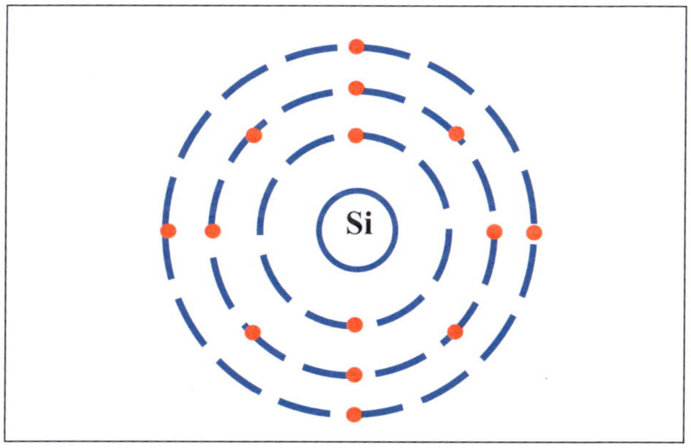

Figure 10. Electron configuration in the silicon atom. There are 14 electrons, four being valence electrons.

Silicon can also form double bonds, and thus chains, etc. But the silicon double bond is not as strong as the carbon double bond, and so its chains are not as long, and its molecules are less complex. Chemists have suggested the reason for the weaker double bond in silicon is that its valence electrons are farther apart than the valence electrons in carbon.[1] This farther separation is due to silicon's valence electrons being in its third shell as opposed to carbon's being it its second. The ability of carbon to form longer, stronger, and more complex molecules than silicon is its major advantage over silicon as a base element for life.

And carbon has a possible advantage over silicon when one considers metabolism of carbon-based life over silicon-based life. Consider the following:[1] Hydrocarbons are molecules of carbon and hydrogen. The metabolism of hydrocarbons involves burning them, which uses atmospheric oxygen. As an example, let us burn the hydrocarbon methane, CH_4. Here we have

$$2O_2 + CH_4 \Rightarrow CO_2 + 2H_2O\ .$$

Here the waste products are carbon dioxide and water, a gas and a liquid at "normal" temperatures.

Now consider an analogous reaction, namely burning monosilane, SH_4. We have

$$2O_2 + SH_4 \Rightarrow SO_2 + 2H_2O\ .$$

The waste products are water and silica. But at "normal" temperatures silica is a solid. As was somewhat humorously pointed out in [1], the "excrement" could be sand bricks or chunks of quartz. The choice of burning methane to obtain energy in the above was to keep the example simple. While certain extremophiles (see below) do burn methane, plants use carbon dioxide, and animals use carbohydrates (molecules of hydrogen, oxygen, and carbon), proteins, and fats. What complex silicon-based molecules an alien lifeform would burn is anyone's guess. Monosilane was chosen to keep it simple.

All life on Earth is carbon-based and water-based. This includes extremophiles, which are microorganisms, especially archaea, living in conditions of extreme temperature, acidity, alkalinity, or chemical concentration. Some of these organisms do not derive energy from the Sun, but from chemicals released from hydrothermal vents at the bottom of the oceans.

Speculating about possible life forms elsewhere, we must be careful not to limit our thinking as to what might be considered alive.

In the stories "The Sign of Four" and "The Adventure of the Blanched Soldier" Sherlock Holmes says

> ...when you have eliminated the impossible,
> whatever remains, however improbable, must be the truth.

This logic, however, is problematic. There is the implicit presumption one knows all the possibilities that need to be considered. Anyone who has spent time solving chess puzzles knows that sometimes when one fails to find a solution and looks at the answer, it is discovered no mistake was made in any move sequence that was considered, but rather the correct sequence is one that was never considered.

The British astronomer Fred Hoyle wrote a science-fiction book *The Black Cloud*. In it a large and essentially immortal cloud of gas that lived in outer space approached our Sun to gain energy. It happened to stop in a position that blocked the Sun from being visible from Earth. Humans managed to communicate with it and asked it to move. It was astonished that living creatures could exist on these small objects called planets. It did move on. Here we have an imagined life form at total variance from the familiar.

If there is life on Saturn's moon Titan that is methane-based, then we will have found a second genesis in the solar system. If life is found on Mars, it might be related to life on Earth. Life could have arisen on Mars and been transported to Earth in meteorites. If life is discovered in Europa's or Enceladus' ocean, or in an under-the-surface ocean of another moon in our solar system, it still might not constitute a second genesis. We now know that comets from other star systems pass through our solar system and leave behind dust. Comet 2I/Borisov (the I is for interstellar) was first observed in 2019 before it approached the Sun. Traveling at 93,000 miles per hour, far faster that any solar-system objects, it left a trail of dust. It received less media attention than the object Oumuamua, which in 2017 was noticed while on its way out of the solar system. In the panspermia theory of life, life started in our solar system from life that was brought in on objects such as 2I/Borisov. If this is true, then life in lunar subsurface oceans might also not constitute a second genesis. How could we tell?

Assuming life in a lunar ocean also uses amino acids, one possibility is to examine the chirality—handedness—of the amino acids. (Chirality relates to a molecule that is not superimposable on its mirror image by any combination of rotations.) Every amino acid molecule is either left-handed or right-handed, like a shoe. All life on Earth has left-handed amino acids. If life in, for example, Europa's ocean is found to use right-handed amino acids, then we could conclude that life on Europa constitutes a second genesis.

Speculation on intelligent life in the Milky Way galaxy.

Then there is the question "How many instances are there of *intelligent* life?" Here I arbitrarily restrict the question to (current or past) intelligent life in the Milky Way galaxy, and by "intelligent" I mean that the life form's civilization is capable of building a radio. Of course this is overly restrictive, and building a radio also requires appendages and an environment that would allow the building of a radio. So dolphins would be excluded. (And, yes, this unfortunately may exclude Fred Hoyle's black cloud.) I do allow for non-biological life forms, including sentient robots. Here I will not define what sentient means.

In thinking about this question, we must remember that we cannot generalize from an instance of one. Probably most people believe there are many instances satisfying this criterion. But, at the current time, it is definitely possible that we are the only such example in the galaxy. Either way, an answer would be profound.

I think one argument increasing the probability of intelligent life eventually arising on a planet is ironically the occurrence of mass extinctions on that planet. If a species doesn't gain intelligence or have the curiosity to study the environment in which its planet exists, i.e., its solar system, then it will have no space program, and will not discover and be able to divert the asteroid or comet headed for a collision with the planet. The species will disappear, allowing some other life form to arise which might develop a space program. Without such extinction events, dominant species that are either dumb or incurious will live on, blocking another species from becoming dominant. After all, the dinosaurs were dominant for about 150 million years. The homo species has only been around for about three million years, with homo sapiens existing for approximately 250,000 years, a blip in time.

The question of longevity arises. If the creatures are robots, they could live forever. I assume any advanced, biological creatures would have solved the aging process—either by freezing or slowing it down to allow for a life span of thousands of years. Even on present-day Earth instances occur of extended longevity in animals.

Greenland sharks (*Somniosus microcephalus*) can live to be at least 400 years old, with females sexually maturing only after about 150 years.[2]

Lobsters do not show any signs of senescence, although they will die when they grow too large for their shells. Their longevity is due to their being able to endlessly repair their DNA.[3]

A jellyfish, *Turritopsis dohrnii*, shows no senescence. This organism appears to be biologically immortal for an unusual reason: facing some environmental stress, it can revert to an earlier form, the equivalent of a butterfly returning to its caterpillar form.[3]

Hydrae—any of several small, freshwater polyps having a simple cylindrical body with a mouth-like opening surrounded by tentacles—may also avoid senescence. The hydra's stem cells appear to have an overabundance of FoxO genes, which scientists believe play an important role in how long cells live.[3]

Where are they? The Fermi Paradox.
This paradox, named after the Italian-American physicist Enrico Fermi, is an attempted rebuttal of the argument that there have been and are many advanced, intelligent civilizations in the Milky Way galaxy. The argument is as follows:

The galaxy has existed for over 10 billion years. Hundreds of billions of planets have existed during this period, allowing for a great number (tens of millions?) of space-faring civilizations. Even at the modest speeds currently envisioned for interstellar travel, over a few million years a civilization could have spread across the galaxy. From the sheer numbers it is argued that the odds are high that this should have happened. Yet there is no evidence that Earth has ever been visited by alien beings or their probes. So where are they?

Even though many opinions have been given on Fermi's Paradox, I will proffer my own (with no claims of originality). I argue that any beings would want a return on any big investment they made. And getting the return in the distant future (say in several thousand years or a million years) would not be appealing. This is even if these beings had solved the aging problem and would still be around—even less so if otherwise. Sending spaceships to distant star systems carrying some of these beings, or artificial life forms that they created, would just take too long in the benefits-versus-rewards category. It could be argued that they might wish to populate the galaxy with their own kind or their creations, with no intention of ever hearing from them again. But I think other projects, such as establishing bases on nearby worlds, would rank higher.

One could also argue that the beings or artificial life-forms on the ship would evolve and after thousands of years might not resemble life on the home planet anymore. Colonists from Britain to the American colonies soon developed their own distinct identity. I suspect if humans terraform Mars and set up colonies, that after the passage of time, the humans on Mars would have their own distinct perspectives, and probably would also undergo physical changes from living in a reduced gravitational field. I believe any beings who sent their own kind, or their creations, in ships to distant star systems would, after the passage of time, find those beings do not resemble them and perhaps do not even identify with the home inhabitants.

Finally, in addition to the long physical travel times involved in interstellar travel, one must also consider the time necessary to send a message. Though the plot of a best-selling and imaginative sci-fi trilogy features instantaneous message transmission (quantum mechanics now permits this through the concept of entanglement, it was claimed), in reality this cannot occur. Current physics does *not* allow message transmission faster than the speed of light. Quantum entanglement does allow the *secure* transmission of a cryptographic key, but the *time* to transmit the key (or any other information) remains limited by the speed of light.

In the classic science-fiction book *Dune*, interstellar travel itself is made possible by "folding space." An interesting idea. This would allow a hand-carried message to reach its destination faster than the speed of light. But of course no claims were made that current physics allows this. Oh, if it were only so.

The current reality is that any civilization that spread out across the galaxy would have to wait 100,000 years for its message to be sent across the galaxy, and then patiently stand by another 100,000 years for a reply. This is another reason an alien House of Representatives would decide to fund only local space travel projects rather than expansion throughout the entire galaxy.

Thoughts on Space travel.

In the near future humanity will again return to the Moon. This time to establish a base. I hope that soon afterward a manned Mars mission will follow.

For over half a century NASA has conducted research on how to protect the human body from the effects of weightlessness. Yet problems still remain, one being loss of bone mass. Astronauts spend considerable time on exercise machines to combat the deleterious effects. We don't want to land human "vegetables" on Mars. Millions have been spent on trying to find a medical solution to a biological problem. I have often wondered why money has not been spent on finding an engineering solution to this biological problem. Here I refer to creating artificial gravity using the centrifugal force created by rotating (at a slow rate) a torus-shaped cabin on a spacecraft. The human body wants to have up and down directions. Solving engineering problems in the creation of artificial gravity could prove simpler than finding medical solutions. And of course, having a combination of medical assistance and artificial gravity would be ideal. We want astronauts to be in good physical shape when they reach Mars.

Another thought: Using a form of nuclear propulsion, in addition to a chemical rocket take off, would allow for continuous propulsion and thus get astronauts to Mars much quicker than using a chemical rocket takeoff alone.

Final thoughts.

Future human space travel and the search for life in the universe require that humans themselves survive and that the Earth's environment be protected.

It seems that so much of humanity's wealth and the Earth's resources are devoted to the production of weapons—which, if ever used, would destroy human civilization as we know it. Centuries after the fall of the Roman Empire, concrete structures built by the Romans could be seen. Yet people had no idea how concrete was produced. It was clear that centuries earlier civilization was more advanced. Regression can occur again.

Many who have spent time on the international space station mention that their most significant impression was how thin Earth's atmosphere really is. Looking toward the edge of the Earth the thickness of the atmosphere is barely noticeable. An astronaut once commented[4] that if you could walk straight up it would be a short, four-mile walk—21,000 feet—where you would run out of the oxygen you need to breathe. When next you take a four-mile walk on land, think about this.

References

1. Trefil, James, & Summers, Michael (2019). *Imagined Life.* Smithsonian Books. p 183-185, 190-191. Interesting book.
2. Milius, Susan (September 17, 2016). Shark takes aim at longevity record. *Science News,* p 13.
3. Berthold, Emma (n.d.). *The animals that can live forever*. Australian Academy of Science. https://www.science.org.au/curious/earth-environment/animals-can-live-forever.
4. Aronofsky, Darren (March 26, 2018). *One Strange Rock.* An American television documentary series, premiered on National Geographic. On July 25, 2018, National Geographic renewed the series for a second season.
5. McKay, Chris (April 21, 2021). *The Dragonfly mission to Titan, and the search for life.* San Francisco Amateur Astronomers meeting. San Francisco, CA.
6. NASA Solar System Exploration. (n.d.). *Enceladus: Ocean moon*. solarsystem.nasa/gov. https://solarsystem.nasa.gov/missions/cassini/science/enceladus/
7. Stolte, Daniel (July 6, 2021). Methane in the plumes of Saturn's moon Enceladus: Possible signs of life? University of Arizona Communications.
https://news.arizona.edu/story/methane-plumes-saturns-moon-enceladus-possible-signs-life
8. Neveu, Marc, et al. (August 6, 2020). *Returning samples from Enceladus for life detection*. Frontiers in astronomy and space sciences. https://www.frontiersin.org/articles/10.3389/fspas.2020.00026/full
9. Seti Institute. https://www.seti.org/drake-equation-index
10. Dobbins, Thomas (May 2022). Do the plumes from Saturn's icy moon reach down to its ocean? *Sky & Telescope,* p 10.

APPENDICES

APPENDIX TOPIC 1

1.1. We show that from the surface of Miranda, Uranus would span about 22.5 degrees, thus appearing about 45 times the size (in diameter) of our Moon.

The radius of Miranda is about 147 miles. The average distance of Miranda from Uranus is 80,642 miles (center to center). So the center of Uranus is about d = 80,642 - 147 = 80,495 miles from the surface of Miranda (at its equator). The angle θ subtended by Uranus at the surface of Miranda is given (in radians) by

$$\theta = s/d \sim D/d. \quad (1)$$

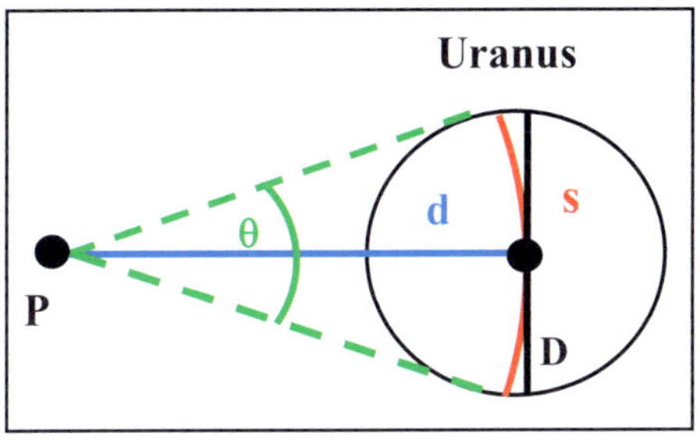

Figure 1. s is the length of the red arc on a circle whose center is at point P on the surface of Miranda closest to Uranus and whose radius d is the length of the blue line from P to Uranus' center. d is so large that s is virtually identical to Uranus' diameter D.

Since the diameter of Uranus is 31,518 miles we have

$$\theta = s/d = 31{,}518/80{,}495 = .392 \text{ radians.}$$

In degrees this is .392 x 57.296 = 22.5 degrees as there are 57.296 degrees in one radian. Our Moon subtends an angle of about a half a degree. So Uranus would appear as a disk about 45 times as wide as our Moon.

1.2. We here show that jumping off Miranda's cliff Verona Rupes results in a fall lasting about 12 minutes with a final velocity of about 125 mph.

In Appendix 50.1, two equations are derived—see equations (6) and (7):

$$v = -gt + v_o. \quad (1)$$

and

$$x = -gt^2/2 + v_o t + x_o. \quad (2)$$

The first gives the vertical velocity v at time t with v_o being the initial vertical velocity and g being the acceleration of gravity at the surface. The second gives the vertical distance x at time t with x_o being the initial vertical distance above the surface.

In the current situation, x = 0 kilometers, $v_o = 0$, $x_o = 20$ kilometers, and g = 0.079 meters/second^2.

Solving for t using (2) gives $t^2 = 2x_o/g$. This results in t = 712 seconds or about 12 minutes.

From (1), v = -gt = -(0.079)(712) = -56 meters/second which is about 202 kilometers/hour or approximately 125 mph. The negative sign indicates that the velocity is downward.

APPENDIX TOPIC 2

2.1. For those interested in how the measurement "seven Earth diameters" given in Topic 2 is calculated, the mathematics is given here: We assume an observer on the surface of Pluto sees Charon directly overhead, i.e., the observer on Pluto is at Pluto's closest point to Charon (although if not, the answer is virtually the same). If d is the distance between the observer on Pluto and the center of Charon, S is the diameter of Charon, and θ is the angular size of Charon (in radians), then (see equation 1, *Appendix 1.1*)

$$\theta \sim S/d\ .$$

S = 750 miles and d = 12,160 - (1,477/2) = 11,422 miles, 1,477 miles being the diameter of Pluto. Thus θ = 750/11,422 = .0657 radians which is 3.76 degrees as there are 57.2948 degrees in one radian. Since our Moon has an angular diameter of about .52 degrees, we have that Charon would appear to be about seven Earth Moon diameters in size. This means that, seen from Pluto, Charon would appear to have about 49 times the area as our Moon as seen from Earth.

APPENDIX TOPIC 3

3.1. Here are a few more details on why the Earth's axis is rotating in a circle with respect to the fixed stars. Consider a gyroscope or a perfect sphere on the Earth that is spinning with its axis of rotation vertical and with the bottom of the axis in a fixed position, say on a table. If one pushes with one's finger sideways on the top of the axis, the axis itself will tilt away from being vertical and the axis will start rotating in a horizontal circle (precession begins). After moving your finger away, the axis will continue to precess in a circle. But continuous circular movement, being acceleration (here the direction but not the magnitude of the velocity changes), implies by Newton that some external force must be acting on the sphere. What is it?

With the center of gravity of the tilted, spinning sphere not over the spot where the axis is fixed on the table, this external force is, of course, the downward gravitational pull of the Earth on the sphere. If the rotating sphere were in space, far away from any gravitational source, then the precession would stop.

Now consider the Earth as that rotating sphere. What is the constant force that keeps the Earth's axis precessing? It is the gravitational pull of the Moon (and to a lesser extent that of the Sun). But if the Earth were a perfect (say uniformly dense) sphere there would be no precession. The extra ingredient needed to have the Moon's gravity torque the Earth's axis into precession is that the Earth is not a perfect sphere. The Earth has an equatorial budge, i.e., the diameter of the Earth from pole to pole is less than its equatorial diameter. The bulge is due to the centrifugal force caused by the Earth spinning on its axis. This is the same force you would feel tugging on a rope that you held in your hand while swinging the rope in a circle with a heavy weight on the other end. The Moon's continuous tugging on this bulge is the external force needed to keep precession going.

APPENDIX TOPIC 10

10.1. We easily show that Saturn's perihelion precession period T is 66,462 years. As stated in this topic, the apsidal (or perihelion) precession rate of Saturn is 19.5 arc seconds per (Earth) year. Since there are 3,600 arc seconds in one degree, we have

$$19.5T = 360^{o} \text{ x } 3{,}600\ .$$

Hence T = 66,462 Earth years.

APPENDIX TOPIC 12

12.1. The radius r of Venus is 3,760.4 miles. So its circumference is 2πr = 23,627 miles. The synodic day is 117 Earth days. So in traveling around Venus' equator, to keep up with the Sun one would have to travel at a rate of 23,627/117 = 202 miles per Earth day or 24 Earth hours which is 8.4 miles an (Earth) hour. One could easily do it riding a bike. In case you are thinking that the equatorial circumference of Venus might be significantly greater than the polar circumference, it isn't. Venus has a very slow axial rotation period. There is not enough centrifugal force in the equatorial region of Venus to cause a bulging out (unlike for the Earth).

12.2. Here we give a proof that for superior planets like Jupiter

$$1/\Sigma = 1/E - 1/P$$

and for inferior planets like Venus

$$1/\Sigma = 1/P - 1/E \, .$$

E is the orbital period of Earth, P is the orbital period of Jupiter, and Σ is the synodic period of Jupiter. Our proof doesn't assume the orbital planes of the Earth and the superior planet are the same. Indeed, if the superior planet were the dwarf planet Pluto, then they would be quite different—Pluto's orbital plane is inclined about 17 degrees to the plane of the ecliptic.

We start with the Sun, Earth, and Jupiter lined up in a straight line, i.e., Jupiter is in opposition. Σ is the time (say in Earth days) for the Earth to make one complete trip around the Sun and then subsequently catch up with Jupiter so that all three are again in a straight line. Let us say that the Earth revolves through an angle of 360 + θ degrees (as measured at the Sun) in going from being in a line with Jupiter to again lapping Jupiter.

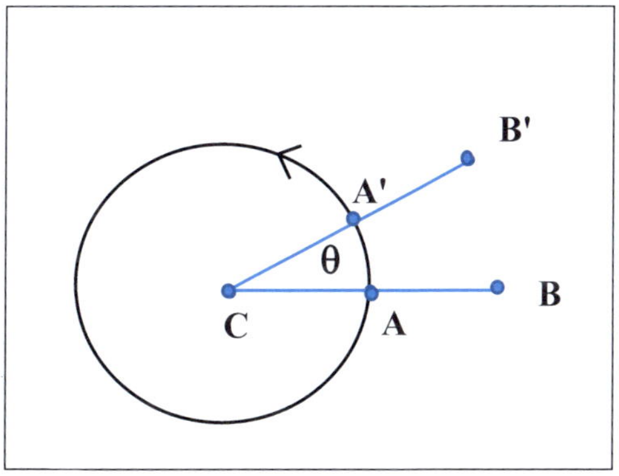

Figure 1. Successive oppositions of Jupiter.

The Sun is at C. We start with the Sun, Earth, and Jupiter in a line, at C, A, B respectively. After time Σ (the synodic period of Jupiter), the Earth has traveled CCW from A, around its orbit back to A, and then to A'. Meanwhile, in time Σ, Jupiter has moved CCW from B to B'. Then the Sun, Earth, and Jupiter are again in a line, namely at points C, A', B' respectively. While the figure depicts the orbital planes of Earth and Jupiter as the same on this two-dimensional flat paper page, and the orbits as circular, the proof doesn't assume this—in three-dimensional space the orbital planes can be at an angle with respect to each other, and the orbits need not be circular.

We have that

360/E is the average angular speed (in degrees per day) of the Earth during one orbit of the Sun.

$360/P$ is the average angular speed of Jupiter during one orbit of Jupiter around the Sun.
$(360 + \theta)/\Sigma$ is the average angular speed of the Earth between successive oppositions with Jupiter.

Finally, θ/Σ is the average angular speed of Jupiter between successive oppositions of the Earth with Jupiter as, during time Σ, Jupiter has revolved by θ degrees around the Sun. We can write that

$$(360 + \theta)/\Sigma = 360/E$$

assuming Earth's average angular speed over time Σ is the same as over time E. This is a reasonable assumption as Σ is greater than E and in fact the Earth's angular speed around the Sun doesn't vary by much as its orbit is close to being a circle.

We can also write that

$$\theta/\Sigma = 360/P$$

assuming Jupiter's average angular speed over time Σ is the same as over time P. Here Σ is much less than P, but it seems to be a good approximation. In fact, this holds even if we substitute Pluto for Jupiter. Pluto's orbit, much less circular than Jupiter's, leads Pluto's average angular speed to vary more than Jupiter's.

Now, with a little algebra, we get our result:

$$360/E = (360 + \theta)/\Sigma = 360/\Sigma + \theta/\Sigma = 360/\Sigma + 360/P\ .$$

Then dividing both sides by 360 we obtain

$$1/\Sigma = 1/E - 1/P\ .$$

In the case where we have an inferior planet, say Venus, then the inner planet is no longer Earth, but Venus. So in the proof, the Earth and the other planet switch roles and we get the same equation but with P and E switched, whence

$$1/\Sigma = 1/P - 1/E\ .$$

APPENDIX TOPIC 13

13.1. The following is a calculation for how long it takes the Sun's disk to clear the horizon on Venus from its equator. From any planet's equator the Sun will rise/set perpendicular to the horizon.

θ_e is the angle (in radians) subtended at the Earth by the Sun. θ_v is the angle (in radians) subtended at Venus by the Sun.

r is the distance to the Sun of the planet in question. For Venus, $r = .723$ astronomical units and for Earth $r = 1$. Let s be the diameter of the Sun in astronomical units.

In general

$$\theta = s/r\ .$$

Let t be the time for the Sun's disk to clear the horizon on Venus. The synodic day or length of one Venusian day is 116.75 Earth days. (See Topic 12.)

$$\theta_e = s/1.\ \ \theta_v = s/.723.\ \ \theta_e = .53^o \sim .0093 \text{ radians.}$$

Thus s ~ .0093 astronomical units. Therefore $\theta_v \sim .0093/.723 = .0128$ rad = .733°. So .733°/360° = t/116.75.

Thus t = .238 days = 5.7 hours, which is the value given in the table for Venus.

APPENDIX TOPIC 14

14.1.

I. Here we derive the equation for c using Figure 2 in Topic 14.

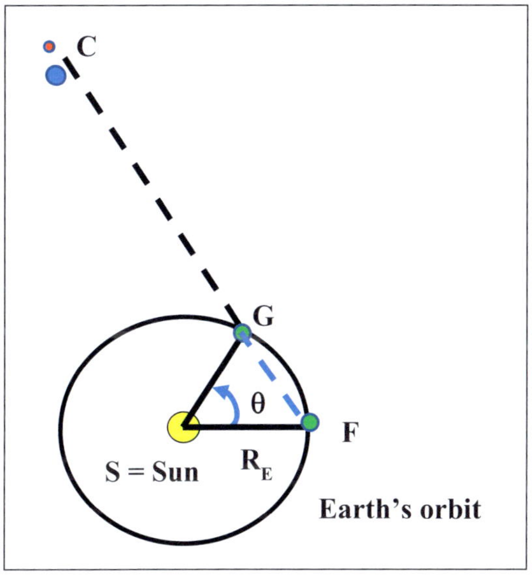

Figure 2 in Topic 14 is repeated here for reference. The Earth is the green dot, Jupiter the blue dot, and Io the red dot. As seen from Earth, Jupiter and Io are close to each other and along the same line of sight. So consider the dashed line as ending at the Jupiter "system."

We wish to prove that

$$c = 2R_E\sin(NP\omega_E/2)/\Delta t\ . \tag{1}$$

First let us review the law of cosines, which is true for any triangle. This simply says that the square of the length of any side of a triangle equals the sum of the squares of the lengths of the other two sides minus two times the product of those other two lengths times the cosine of the angle opposite the first side. In symbols, for the triangle

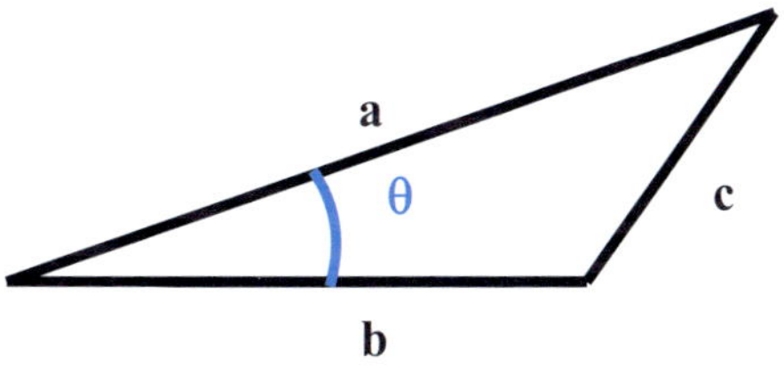

we can write (choosing the first side to be c) that

$$c^2 = a^2 + b^2 - 2ab\cos(\theta) .$$

Let us denote the length of the line segment from F to G by x, and apply the law of cosines to the triangle SFG in Figure 2 above to obtain

$$x^2 = 2R_E^2 - 2R_E^2\cos(\theta) .$$

This gives

$$x = \sqrt{2}R_E\sqrt{1-\cos(\theta)} = 2R_E\sqrt{\frac{1-\cos(\theta)}{2}} = 2R_E\sin(\frac{\theta}{2}) . \quad (2)$$

The last step uses the half-angle formula from trigonometry, namely, $\sin(\frac{\theta}{2}) = \sqrt{\frac{1-\cos(\theta)}{2}}$.

$\Delta t = T_F - T_G$, the light travel time difference. Since distance equals speed times time, we have

$$c\Delta t = \text{distance between C and F minus distance between C and G} = x. \quad (3)$$

From (2) and (3),

$$c = 2R_E\sin(\theta/2)/\Delta t . \quad (4)$$

How much time does it take the Earth to go from F to G? It must be the synodic period P of Io times the number N of orbits around Jupiter that Io makes between the two eclipses. Remembering that ω_E is the angular velocity of the Earth in its orbit, we thus have

$$\theta = NP\omega_E , \quad (5)$$

where $\omega_E = 360/(365 \times 24 \times 60^2)$ degrees per second.

Finally, combining (4) and (5) gives us (1) and we are done.

It is worth pointing out that eclipses of Io into/out of Jupiter's shadow must be used instead of occultations of Io by Jupiter as seen from Earth. The angle between the Sun and Io as seen from Jupiter is relevant when using eclipses, and the repetition of eclipses is independent of the motion of Earth. Using occultations, the important angle is that between Jupiter and Io as seen from Earth in which case the repetition of occultations involves Earth's motion, a complicating factor. I wish to acknowledge my discussion with Akarsh Simha which helped reinforce this point.

II. We now will consider a different situation. Here we will not assume that the two observations are necessarily along the same line (FGC). Instead we will use Figure 3 below in which the beginnings of Jupiter's eclipses of Io are observed at F and G. Here we make one simplifying assumption to allow a less complex derivation of the equation. Namely, we will assume that the second observation of Io entering Jupiter's shadow is made at the time that Jupiter and the Earth are in conjunction, i.e., the Sun, the Earth, and Jupiter are in a straight line.

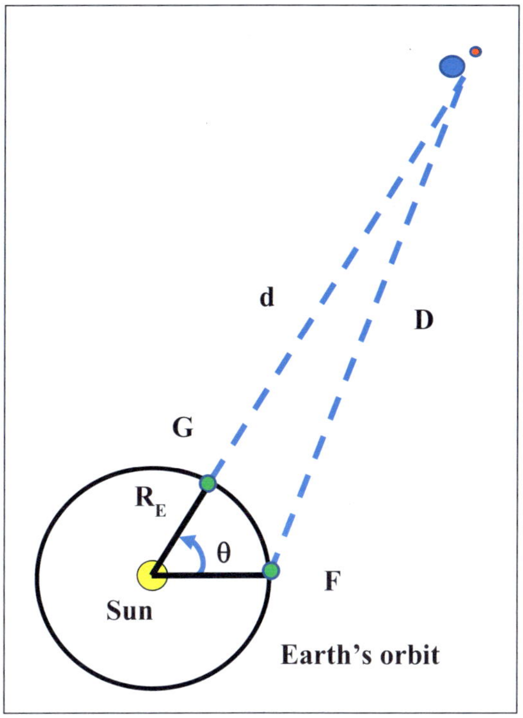

Figure 3. Light takes longer to go from Jupiter's moon Io (red dot) to F than to G, two points on the orbit of the Earth around the Sun. Jupiter is the blue dot and Earth is the green dot. We assume that position G is where Jupiter/Io would be in opposition, i.e., when the Sun, the Earth, and Jupiter are in a straight line. R_J = the radius of Jupiter's orbit around the Sun. $d = R_J - R_E$ (where we assume the orbits of Jupiter and the Earth are almost circles). As before, Jupiter and Io are very far from Earth and so close to each other that the dashed lines from postions F and G should be considered to end at the Jupiter "system."

As in Case I, $\Delta t = T_F - T_G$, the light travel time difference, and $\theta = NP\omega_E$.

We will prove that

$$c = \frac{1}{\Delta t}[(R_E^2 + R_J^2 - 2R_E R_J \cos(\omega_E NP))^{1/2} - d] \ , \qquad (6)$$

where $d = R_J - R_E$, c is the speed of light, and $\Delta t = T_F - T_G$ is the light travel time difference.

In Figure 3, consider the triangle formed by the Sun, Io, and F, whose sides are of lengths R_J, R_E, and D. Remember that the Sun, G, and Jupiter/Io are assumed to be in a straight line. Applying the law of cosines to this triangle we obtain

$$D^2 = R_J^2 + R_E^2 - 2R_J R_E \cos(\theta) \ . \qquad (7)$$

Since distance equals speed times time, we have

$$c\Delta t = D - d \ . \qquad (8)$$

As in Case I,

$$\theta = NP\omega_E \ . \qquad (9)$$

From (8) we have $c = (1/\Delta t)[D - d]$. Our result then follows from (7) and (9).

III. Here we consider the general situation, where we don't assume either of the more restricted situations in Cases I and II.

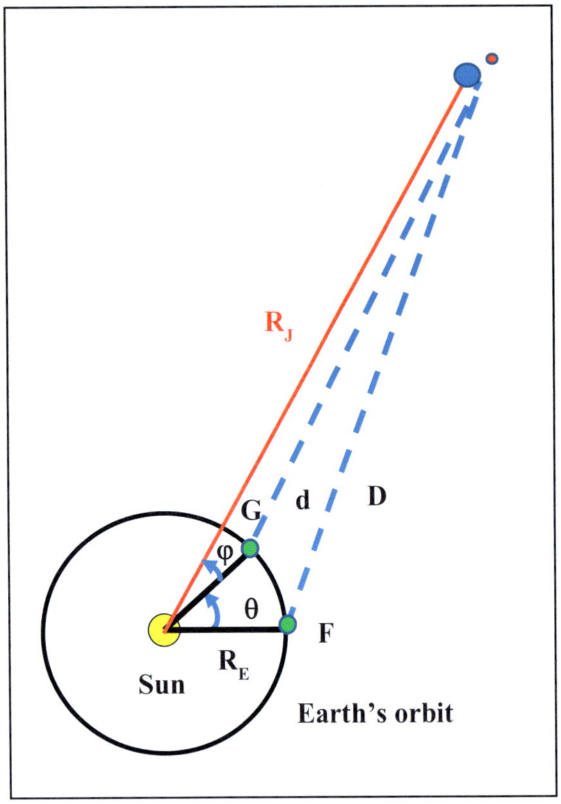

Figure 4. Light takes longer to go from Jupiter's moon Io (red dot) to F than to G, two points on the orbit of the Earth around the Sun. Jupiter is the blue dot and Earth is the green dot. R_J = the radius of Jupiter's orbit around the Sun. Here d is the distance between G and Jupiter/Io. φ is the angle between the line from the Sun to G and the red line from the Sun to Jupiter/Io. If the Earth in its orbit were in conjunction with Jupiter, then G would be on this red line.

We will derive the following equation for c:

$$c = \frac{1}{\Delta t}[(R_E^2 + R_J^2 - 2R_E R_J \cos(\omega_E NP + \varphi))^{1/2} - (R_E^2 + R_J^2 - 2R_E R_J \cos(\varphi))^{1/2}] \qquad (10)$$

Notice that (10) reduces to (6) when $\varphi = 0$. Also, if T is the time for the Earth to move from G to where Jupiter is in opposition (the red line), then $\varphi = T\omega_E$.

We can use the ideas in II. Here d is the distance between G and Jupiter/Io. φ is the angle between the line from the Sun to G and the red line from the Sun to Jupiter/Io. If the Earth in its orbit were in conjunction with Jupiter, then G would be on this red line.

Applying the law of cosines to the triangle formed by the Sun, Jupiter/Io, and G we have

$$d^2 = R_J^2 + R_E^2 - 2R_E R_J \cos(\varphi) . \qquad (11)$$

Applying the law of cosines to the triangle formed by the Sun, Jupiter/Io, and F we have

$$D^2 = R_J^2 + R_E^2 - 2R_J R_E \cos(\varphi + \theta) . \qquad (12)$$

Again

$$c\Delta t = D - d \text{ and } \theta = NP\omega_E . \tag{13}$$

So as before,

$$c = (1/\Delta t)[D - d] .$$

This gives us the equation

$$c = \frac{1}{\Delta t}[(R_E^2 + R_J^2 - 2R_E R_J \cos(\omega_E NP + \varphi))^{1/2} - (R_E^2 + R_J^2 - 2R_E R_J \cos(\varphi))^{1/2}] \text{, which is (10).}$$

APPENDIX TOPIC 18

18.1. We calculate the angular size of the two Martian moons as seen from the surface of Mars. Phobos is about 3,700 miles from the surface of Mars; Deimos is 14,576 miles distant. Phobos' mean diameter is 14 miles; Deimos' is 7.7 miles. The Martian radius is about 2,100 miles. Using

$$\theta = s/d ,$$

where s is the diameter of the object, d is its distance, θ is the angle in radians subtended by the object, and knowing that there are 57.2958 degrees in one radian and 60 arc minutes in a degree, we obtain for Phobos

$\theta = (14/3700) \times 57.2958 \times 60 = 13$ arc minutes when it is overhead;
$\theta = (14/D) \times 57.2958 \times 60 = 8.9$ arc minutes when it is on the horizon,
where $D = [(3700 + 2100)^2 - 2100^2]^{1/2}$ is the distance to Phobos when it is on the horizon.

(The Pythagorean theorem was used to obtain the above expression for D.) For Deimos

$$\theta = (7.7/14{,}576) \times 57.2958 \times 60 = 1.8 \text{ arc minutes.}$$

APPENDIX TOPIC 19

19.1. Here we present an equation that gives the angle I between the ecliptic and the horizon in terms of an observer's latitude L, sidereal time T, and the mean angle θ of inclination (23° 26') of Earth's axis to the ecliptic:

$$\cos(I) = \cos(\theta)\sin(L) - \sin(\theta)\cos(L)\sin(T) . \tag{1}$$

Equation (1) shows that, for a fixed value of L, on *every* day of the year, I runs through the *same* range of angular values (as T runs through almost 24 hours in time). The extreme values for that range are

$$90^\circ - L + \theta \text{ and } 90^\circ - L - \theta .$$

First, one should note that I is *not* the angle at which the Sun rises/sets, i.e., is not the angle between the daily path of the Sun (called the diurnal path) and the horizon, which is discussed in Topic 56. In particular, in that section we see that the angle at which the Sun rises/sets for an observer on the equator is *always* 90° (i.e., vertical). On the equator, L = 0. Thus the range of values for I when on the equator is from 23.5° on one side of vertical to 23.5° on the other side of vertical.

Example: Let L = +51°, $T = 5^h 00^m = 75^\circ$, and θ = 23° 26'. The formula gives I = 62°.

Equation (1) and the example are contained in reference [7] of Topic 19.

19.2. Here we give the basic reason for the surprising fact that a very slight change in elevation (a few feet) can change the time of the green flash by several seconds when the observer is at or near sea level. On the other hand, well above sea level a similar change in elevation doesn't have the same effect. The basic reason for this result is that the time difference between a green flash occurring at sea level and at elevation e (say in meters) above sea level will be seen to be directly proportional to the square root of e and not to e itself. To simplify the mathematics we will ignore the effect of refraction. The following is based on the argument contained in reference [3] of Topic 19.

Let us look at the geometry of the situation and also calculate the angular rate at which the Sun is setting (descending at sunset).

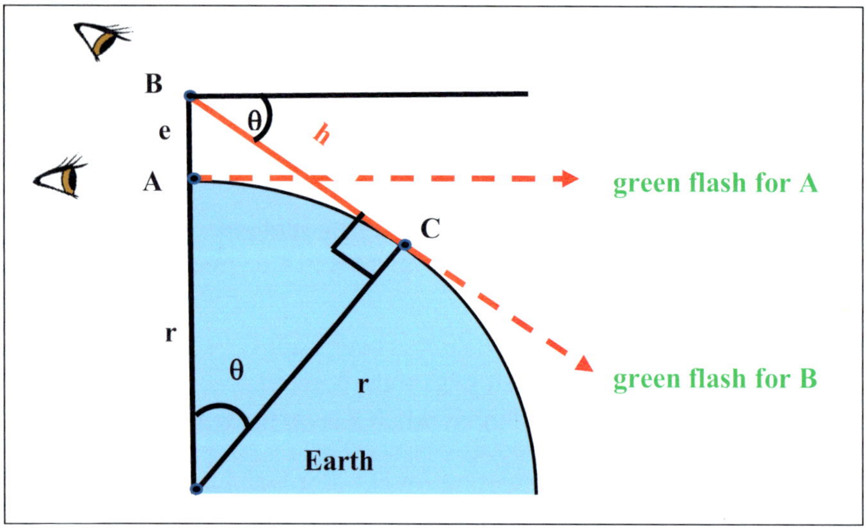

Figure 1. Views of the green flash for observer A (Alice) at sea level and observer B (Bob) e meters above Alice. θ is the angle of depression of the horizon for Bob due to his elevation. r is the radius of the Earth and h is the length of the solid red line (the distance between Bob and point C on his horizon).

Consider the right triangle T formed by points B, C, and the center of the Earth. Since the angle formed by the lines BC and BA is 90° - θ, the angle of T at the center of the Earth must also be θ (as the sum of the internal angles of any triangle is 180°).

By the Pythagorean theorem,

$$h = \sqrt{(r+e)^2 - r^2} = \sqrt{2er + e^2} \sim \sqrt{2er}, \tag{1}$$

since e^2 is much smaller than er when e is much smaller than r.

If θ is a very small angle and in radians, then $\theta \sim \tan\theta$. Thus

$$\theta \sim \tan\theta = h/r \sim \sqrt{2e/r}\,. \tag{2}$$

$r = 6.38 \times 10^6$ meters. Thus if e is in meters,

$$\theta \sim (5.6 \times 10^{-4})\sqrt{e} \text{ radians.} \tag{3}$$

The geometry of the situation has resulted in equation (3).

Now we calculate the rate of descent of the Sun. Since there are 2π radians in a complete circle and 24 hours in a day, we obtain

$$\text{the Sun's rate of descent} = 2\pi/(24 \times 3600) = 7.27 \times 10^{-5} \text{ radians/second.} \quad (4)$$

But that assumes the Sun is descending perpendicular to the horizon, as it does every day at the equator (and never anywhere else)—see Topic 56. Temporarily assuming the observer is on the equator, we have that the number of seconds for the Sun to descend θ radians is

$$(5.6 \times 10^{-4} / 7.27 \times 10^{-5})\sqrt{e} = 7.7\sqrt{e} \text{ seconds.} \quad (5)$$

This would be the fastest possible rate of descent. In general, the angle at which the Sun sets on the horizon is determined by the observer's latitude and the Sun's declination (i.e., the Sun's celestial latitude). An equation for this is given in Topic 56. From this angle the actual rate of the Sun's descent can be determined. The smaller the angle the less the rate of the Sun's descent. For our purpose we will not consider all possible rates of descent. Instead we will (as is done in reference [3] Topic 19) choose a rate that is somewhat slower than the maximum possible. We will use the rate of descent that Garriott experimentally determined to be the case on July 22, 2002, at his latitude. He measured the time between when the Sun first touched the horizon to when it disappeared. It was almost exactly three minutes. The Sun's apparent diameter is 0.533° which is 0.0093 radians. Thus

$$\text{the Sun's rate of descent} = .0093/(3 \times 60) = 5.17 \times 10^{-5} \text{ radians/second.} \quad (6)$$

Compare this to (4). Hence the number of seconds for the Sun to descend θ radians is

$$(5.6 \times 10^{-4}/5.17 \times 10^{-5})\sqrt{e} = 10.8\sqrt{e} \text{ seconds.} \quad (7)$$

Because we obtained (3) from (2) by using the value of r in meters, in (7) e must also be given in meters. Since there are .3048 meters in one foot, we obtain the number of seconds for the Sun to descend θ radians is

$$\theta = 10.8\sqrt{.3048e} \text{ seconds,} \quad (8)$$

where now e is in feet.

Let us now compare Bob's green flash time to that of Alice's, assuming Alice is a sea level. The values of e that we consider are two feet, six feet, 100 feet, and 104 feet.

e in feet	delay = $10.8\sqrt{.3048e}$ seconds
2	8.4
6	14.6
100	59.6
104	60.8

Notice that the time difference between two and six feet is 6.2 seconds and that between 100 and 104 feet is 1.2 seconds. Thus if an observer were *sitting* at two feet above sea level and *stood up*, so as to be four feet higher above sea level, the delay in seeing the green flash would be 6.2 seconds. But the delay from sitting at 100 feet above sea level and standing up so as to be four feet higher is only 1.2

seconds. These values are close to the values of four seconds and less than one second given in reference [2] Topic 19. In calculating delay times, remember that we ignored refraction.

APPENDIX TOPIC 20

20.1. As stated in Topic 20 (see Speed's effect on time) a reader familiar with special relativity might see a discrepancy in the discussion, namely that a satellite in orbit around the Earth is *not* traveling at a *constant* velocity around the Earth. The speed would be a constant, but because of the continually changing direction of the satellite, the velocity is not a constant. Velocity is a vector having both magnitude and direction. The satellite is undergoing centripetal acceleration. Also, Alice, being on Earth's surface, is in a rotating coordinate system. Alice is rotating with a period of 24 hours about the Earth's axis; the satellite has a period of 12 hours around the Earth's center. So the velocity of the satellite is constantly changing as seen by Alice. The equations of special relativity generally apply to the situation when the relative velocity between the two (here Alice and the satellite) is a constant, in other words when there is no acceleration. But special relativity can also handle acceleration of an object.

Here's how. One breaks the path of the satellite as seen by Alice into segments of tiny lengths, assumes a constant velocity over each segment, which allows the equations of special relativity to be used over any segment, and then, using calculus, integrates over the path to obtain the elapsed time for the satellite clock as seen by Alice. So special relativity *can* be applied to accelerating objects.

20.2. It was said that a GPS satellite clock would be ticking faster by about 45 microseconds per day than a clock on the Earth's surface. Here's the calculation.

Considering only gravitation (assume the clocks don't move with respect to each other), what is the time interval Δt_2 of a clock at distance r_2 from the center of the Earth in terms of the time interval Δt_1 at distance r_1 from the center of the Earth? Answer:

$$\Delta t_2 \sim \Delta t_1[1 + (GM/r_1c^2)] / [1 + (GM/r_2c^2)] .$$

G is the gravitational constant, M is the Earth's mass, and c is the speed of light in a vacuum. Let us plug in the relevant distances for the GPS. r_1 is the radius of the Earth, approximately 4,000 miles or 4,000 x 1.61 = 6,440 kilometers, and $r_2 = r_1 + 20{,}000 = 26{,}440$ kilometers.

$G = 6.67259 \times 10^{-20}$ km^3/s^{2}kg, $M = 5.98 \times 10^{24}$ kg, $\Delta t_1 = 24$ hours or $24 \times 3{,}600 \times 10^6$ microseconds, and c = 299,792 km/second. Plugging in these numbers and using a calculator we get that Δt_2 is about 44.9 microseconds greater than Δt_1.

20.3. Here we give an argument that *at the event horizon* the surface gravity of a black hole is inversely proportional to its mass.

We will first use *Newtonian* physics to try to get a formula for the surface gravity on an object at the event horizon of a black hole of mass M.

The equation giving the radius R of the event horizon of the black hole in terms of its mass is

$$R = 2GM/c^2 , \qquad (1)$$

where c is the velocity of light and G is the gravitational constant.[5] For an object of mass m, the force of gravity on m due to the black hole is given by Newton's law as

$$F = GMm/R^2 . \qquad (2)$$

Substituting the right side of the first equation in place of R in the second equation gives

$$F = [mc^4 /4G]/M . \qquad (3)$$

Using F = ma, the surface gravity a is given as

$$a = [c^4 /4G]/M . \qquad (4)$$

Since the term within the brackets is a constant, a is inversely proportional to M.

The above derivation of (4) (which uses Newtonian physics, specifically equations (2) and (3)) was given to help give a clue as to why (4) might be true. But the derivation is actually not a proof of (4) since Newtonian physics is modified in general relativity. But, using general relativity, it turns out that (4) is actually still correct.

From (4) it is clear that the more massive the black hole (i.e., the larger the M), the lower is the surface gravity a.

Finally, we note that equation (1) for the Schwarzschild radius of a black hole of mass M, which can be proved in general relativity, also has a suggestive non-relativistic derivation**:** The escape velocity v of a body of mass M is the velocity needed to allow an object to escape from the surface of the body, never to return. This means that the kinetic energy $(1/2)mv^2$ of an object of mass m at distance r from the center of the body must equal the *magnitude* of the object's potential energy (the potential energy being -GmM / r). So we need

$$(1/2)mv^2 = GmM / r . \qquad (5)$$

Setting v = c, r = R = the Schwarzschild radius of the body, and solving for R gives (1).
This has been derived for a particle with mass m, but can also be shown to be valid for a photon (which has no mass).

<u>APPENDIX TOPIC 21</u>

21.1. Three relevant distances:

(a) The distance d_{emit} of a distant object, e.g., a quasar, from Earth at the time the light left the quasar.
(b) The distance d_{LT} that the light travels before reaching us. (light travel distance).
(c) The distance d_{now} of the quasar from Earth at the time the light reaches us.

It must be remembered that space is expanding even as the light travels toward us. Thus

$$d_{emit} < d_{LT} < d_{now} . \qquad (1)$$

We give formulae for these three distances. See lecture 8 in the appendix of the transcript book mentioned in reference [3] Topic 21 for a further discussion of the three distances.

Space and wavelength are stretched by the same factor rsf, which is called the *redshift stretch factor*.

If λ_{emit} and λ_{now} are respectively the wavelength of the light when it was emitted and now, then

$$rsf = d_{now}/d_{emit} = \lambda_{now}/\lambda_{emit} . \quad (2)$$

It is to be noted that astronomers prefer to use

$$z = rsf - 1 . \quad (3)$$

By spectroscopically examining the light from the object astronomers can determine rsf (or equivalently z).

$$d_{LT} = c(t_{now} - t_{emit}) \text{ and } d_{emit} \text{ x } rsf = d_{now} \quad (4)$$

where t_{now} and t_{emit} are respectively the times when the light arrived at Earth and when the light was emitted. The look-back time, i.e., the light travel time is $t_{now} - t_{emit}$. Equation (4) allows d_{emit} to be determined if d_{now} is known.

Obtaining d_{now} is more involved. Suppose D is the current distance between two points in the universe. S(t)D is their distance at time t in the past. This defines the *scale factor* function S(t). S(t) gives the separation between any two points at time t relative to their separation today.

$$S(t) = 0 \text{ at the Big Bang and } S(t_{now}) = 1 . \quad (5)$$

Also,

$$S(t_{emit}) = 1/rsf = 1/(z+1) . \quad (6)$$

Consider a small segment of distance c(dt) between us and an object at time t. At time t_{now} this small distance has become stretched to length (1/S(t))cdt. To obtain d_{now} we must add together all these differently sized infinitesimal distance segments from start to finish. This means we must use integral calculus, obtaining

$$d_{now} = \int_{t_{emit}}^{t_{now}} [c/S(t)]dt . \quad (7)$$

The expansion history of the universe, given by S(t), depends on the value H(t) of the Hubble constant and the universe's density parameters for radiation, matter, and the vacuum. S(t) is discussed extensively in the cosmology lectures referenced above in [3] Topic 21.

We note that using (4) to determine d_{LT}, and (7) to determine d_{now} requires knowing t_{emit}. But (6) can be solved for t_{emit} in terms of rsf or z.

On page 193 in the appendix of the transcript book mentioned in Topic 21 reference [3], the equation

$$S(t) = [(3/2)(t/t_H)]^{2/3} \quad (8)$$

is derived, where $t_H = 1/H_0$, with H_0 being the value of the Hubble parameter for today. The derivation assumes the *Einstein-de Sitter (EdS) model* of the universe where there is only matter and the universe is flat. From (8) we easily obtain

$$t_{emit}(z) = (2/3)(1/H_0)(1+z)^{-3/2} . \quad (9)$$

The EdS model is the simplest model for the universe even though our universe does not follow this model. Being mathematically simple it is often used to obtain a rough estimate of $t_{emit}(z)$. Compare this with (13) below.

Here we give *the general equation for $t_{emit}(z)$* that involves not only the Hubble constant H_0, but parameters Ω_M, Ω_K, and Ω_Λ, respectively for matter density, curvature of space, and vacuum density.[9] It is assumed that the radiation parameter Ω_R is small enough to be ignored.

$$t_{emit}(z_1) = t_{now} - H_0^{-1}\int_0^{z_1} \frac{dz}{(1+z)[\Omega_M(1+z)^3 + \Omega_K(1+z)^2 + \Omega_\Lambda]^{1/2}} \tag{10}$$

where $\Omega_M + \Omega_K + \Omega_\Lambda = 1$. For high values of z, (10) gives us the approximation

$$t_{emit}(z) - t_{now} \sim (2/3)(1/H_0)\Omega_M^{-1/2}[(1+z)^{-3/2} - 1]\,. \tag{11}$$

The beginning of the universe is when $t_{emit} = 0$, i.e., $z = \infty$. Thus from (11) we have

$$\textit{the age of the universe} = t_{now} \sim (2/3)(1/H_0)\Omega_M^{-1/2}\,. \tag{12}$$

Combining (11) and (12) gives

$$t_{emit}(z) \sim (2/3)(1/H_0)\Omega_M^{-1/2}(1+z)^{-3/2} \tag{13}$$

for high values of z. Here $t_{emit}(z)$ is the *age of the universe at red shift z*. Compare (13) with the situation for the EdS model given by (9).

Finally, in many articles on distant celestial objects only the value of z is stated for an object; no values for any of the distances are given. Alternatively, two sources may give the same value for z, but give conflicting values for one of the distances. Why? While different sources may agree on z, they may give slightly different values for t_{emit}, and hence for d_{LT}, d_{emit}, and d_{now}. The reason is that while the value of z may be accurately known, the value of t_{emit} depends not only on z but on the constants in (10). The exact values of these constants (within a small range of values) are still subject to debate, and some sources may use different values than others.

APPENDIX TOPIC 22

22.1. Here we give a derivation of the formula for γ, the fraction of the speed of light that appears to be the transverse speed of an object as seen from Earth. We assume the object's actual speed (which will have radial and transverse components) is v and that the object is traveling at an angle of θ to our line of sight to the object.

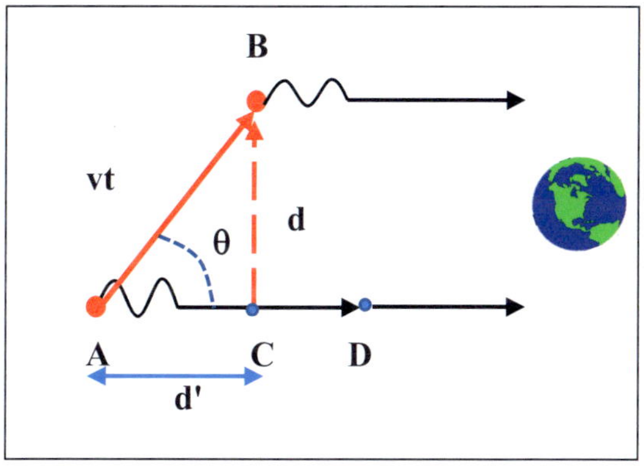

Figure 1. Superluminal speed for an arbitrary angle θ.

For convenience we define some variables.

t = the time for the object to go from A to B, which is also the time for a photon emitted at A to go to D.
t' = the time for a photon to go from A to C.
d = the distance from B to C, d' = the distance from A to C.

We have that $d' = ct'$.

Let $\beta = v/c$ = the speed of the object as a fraction of the speed of light. We have that

$$\sin(\theta) = d/vt \text{ and } \cos(\theta) = d'/vt = ct'/vt . \qquad (1)$$

Let v_a = the apparent velocity of the object as seen from Earth, where it appears to move transversely from C to B a distance d. Then $\gamma = v_a/c$.

We then have the key equation

$$d = v_a (t - t') . \qquad (2)$$

To prove (2) we argue as follows: A photon that starts at A and heads toward Earth is at D at time t. At the same time t, a photon starts from B and heads toward Earth. The difference in their arrival times at Earth is the time for a photon to go from C to D, which is $t - t'$. For someone on Earth this is the time it takes to see the blob apparently move from C to B. Equation (2) follows as distance equals rate times time.

Now, with a little algebra, we will obtain our result. Using the second equation in (1) we have that

$$t'/t = v\cos(\theta)/c = c\beta\cos(\theta)/c = \beta\cos(\theta) . \qquad (3)$$

Thus using (2) and the first equation in (1),

$$v_a = d/(t - t') = vt\sin(\theta) / (t - t') = v\sin(\theta) / (1 - t'/t) = c\beta\sin(\theta) / (1 - \beta\cos(\theta)) .$$

Hence the desired result:

$$\gamma = v_a/c = \beta\sin(\theta) / [1 - \beta\cos(\theta)] .$$

APPENDIX TOPIC 23

23.1. We prove that M31, as seen from M33, would be 3.6 times M31's length as seen from Earth, be 12.76 times brighter, and have magnitude 0.68 (first magnitude).

As seen from M33, M31 would have 2.5 million/700,000 = 3.6 times its length as seen from Earth. The brightness of an object of luminosity L at distance d from the viewer is

$$b = L/4\pi d^2 .$$

Since $4\pi d^2$ is the surface area of a sphere of radius d and centered on a star/galaxy, b is the luminosity of the star/galaxy per unit area of this sphere.

If b_e is the brightness of M31 as seen from Earth, b_{33} is the brightness of M31 as seen from M33, d_e is the distance of M31 from Earth, and d_{33} is the distance of M31 from M33, then

$$b_{33}/b_e = (d_e/d_{33})^2 = (2{,}500{,}000/700{,}000)^2 = 12.76.$$

The equation for the absolute magnitude M of an object that is distance d away and has apparent magnitude m is

$$M = m - 5[\log_{10}d - 1] .$$

Let M be the absolute magnitude of M31, m_{33} the apparent magnitude of M31 as seen from M33, and m_e the apparent magnitude of M31 as seen from Earth. Then

$$m_e = M + 5[\log_{10}d_e - 1] \text{ and}$$

$$m_{33} = M + 5[\log_{10}d_{33} - 1] .$$

Thus

$$m_{33} - m_e = 5\log_{10}(d_{33}/d_e) = 5\log_{10}(.28) = -2.76 .$$

So $m_{33} = 3.44 - 2.76 = 0.68$, as $m_e = 3.44$.

APPENDIX TOPIC 24

24.1. Here we give the defining mathematical relationship between apparent magnitude m and brightness b, and between luminosity L and brightness.

$$m = -2.5\log(b)$$

defines m, where the log is to base 10. Thus $m \propto \log(b)$.

Thus if m_1 and m_2 are the apparent magnitudes of two different stars, and b_1 and b_2 are their respective brightnesses, then

$$m_2 - m_1 = 2.5 \times \log(b_1/b_2) .$$

So

$$m_2 - m_1 = 5 \text{ if and only if } b_1/b_2 = 100 ,$$

and

$$m_2 - m_1 = 1 \text{ if and only if } b_1/b_2 = \text{the } 5^{th} \text{ root of 100 which is} \sim 2.512 .$$

In words, this says that

the magnitude difference is 5 if and only if one star is 100 times as bright as the other,

and

the magnitude difference is 1 if and only if one star is the 5th root of 100 (~2.512) times as bright as the other.

As for the relationship between the brightness of an object of luminosity L at distance d from the viewer, we repeat what was said in *Appendix 23.1*:

$$b = L/4\pi d^2 .$$

Since $4\pi d^2$ is the surface area of a sphere of radius d and centered on the star, b is the luminosity of the star per unit area of this sphere.

24.2. A discussion of parallax and the definition of parsec (a unit of distance).

One can use parallax to determine the distance to relatively nearby stars. Hold a finger at arm's length in front of your face. Close your left eye and look with your right eye at your finger and see what it lines up with in the background, say a tree. Then close your right eye and with your left eye line your finger up with another background tree. By measuring the angular distance in arc seconds between the two background trees and knowing the distance between your eyes, the distance from your face to your finger can be calculated.

Now consider the Earth in its orbit around the Sun and a foreground star F. We wish to determine the distance D from the Earth to the star. We assume the distances from Earth, from any two points in its orbit around the Sun, to star F are essentially the same. This can be assumed as the distance to any star is vastly greater than the diameter of the Earth's orbit.

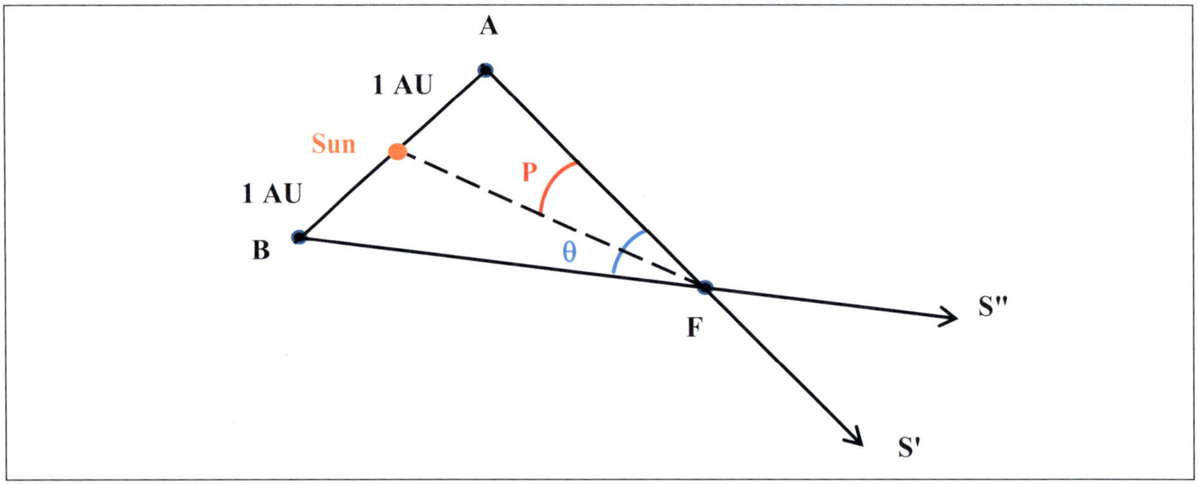

Figure 1. Determining the distance from the Earth to a star F using the star's parallax P (not to scale). A and B are opposite points on the Earth's orbit around the Sun. Each is one AU from the Sun. S' is a background star in a line with foreground star F as seen from point A. S" is a background star in a line with F as seen from point B. θ is the angle in arc seconds measured between S' and S". The indicated angle P is the parallax of star F. P = θ/2 as any star F is so far from the Sun compared to the diameter of the Earth's orbit that the two points A and B on Earth's orbit are virtually the same distance from F.

Because of the vastly different distances between the diameter of the Earth's orbit and the distance to a star F, an illustration more to scale would be the following:

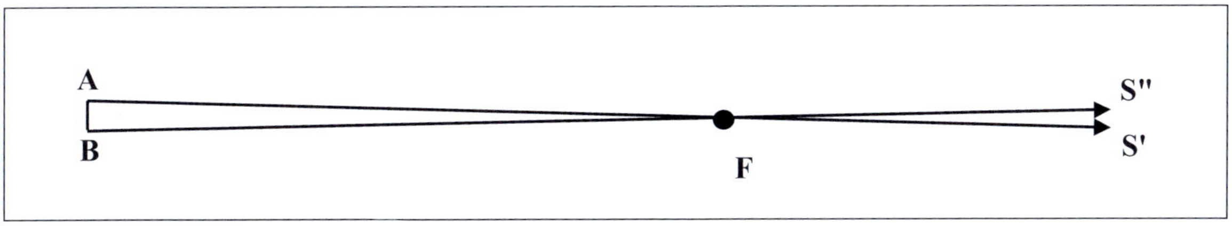

Figure 2. A diagram more to scale.

When the Earth is at some point A in its orbit, we look at F and line it up with some (very distant) background star S'. Six months later, when the Earth is on the opposite side of its orbit, at some point B, we again look at F and line it up with a (different) background star S". We measure the angular distance θ (it will be a fraction of an arc second) between the background stars S' and S". Knowing that the distance between the Sun and the Earth is one AU we can calculate the distance D. In effect, points A and B correspond to our two eyes, the foreground star F corresponds to our finger, and stars S' and S" correspond to the background trees.

Now we show in detail how D is calculated. In Figure 1, θ is the angle measured between the two background stars. Because the distance of A to the Sun is the same as the distance of B from the Sun (namely one AU), and because of our assumption that the two distances AF and BF from the Earth to F are the same, the indicated angle P equals $\theta/2$. (This follows as the triangle ABF is an isosceles triangle with the line from F to the Sun bisecting the base AB.) Angle P is called the *parallax of star F*.

A *parsec* (a unit of distance) is defined as the radius of a circle such that an arc S of length one AU on the circumference of the circle subtends an angle of one arc second at the center of the circle. One parsec is about 3.26 light years.

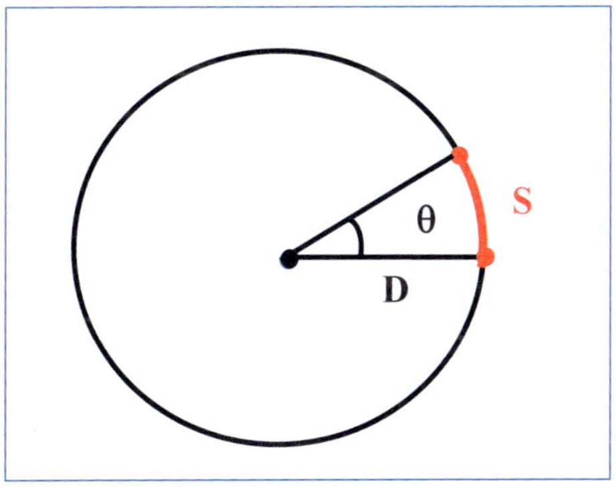

Figure 3. Definition of when a distance D is 1 parsec: D = 1 parsec if S = 1 AU and θ = 1 arc second. (not to scale)

We have

$$\theta = 1/D$$

where θ is the parallax in arc seconds and D is in parsecs. Knowing θ we obtain D.

An example: Proxima Centauri, the closest star to Earth other than the Sun, is a member of the three-star Alpha Centauri system. Its parallax is $\theta = 0.76813$ arc seconds, giving a distance of $1/\theta = 1/0.76813 = 1.302$ parsecs or about 4.24 light years.

APPENDIX TOPIC 25

25.1. The diameter of Sun is 865,370 miles. The average distance of the Earth from the Sun is about 93 million miles. And the Earth's orbit is almost a circle, with the minimum and maximum distances for Earth from the Sun being 91 million and 94.5 million miles. Mercury's distance from the Sun varies from 28.5 to 43 million miles. Quite a variation.

The angular size of the Sun's diameter from Mercury would be $\theta = 865{,}370/28{,}500{,}000$ radians or 1.74° and $865{,}370/43{,}000{,}000$ radians or 1.15° at the minimum and maximum distances respectively.

The angular size of the Sun as seen from Earth is $\theta = 865{,}370/93{,}000{,}000$ radians or $.53^\circ$.

Thus the ratio of the angular size of the Sun as seen from Mercury to that as seen from Earth is $1.74/.53 = 3.3$ when Mercury is at perihelion and $1.15/.53 = 2.2$ when Mercury is at aphelion.

The ratio of the solar intensity at Mercury to that at Earth varies from $(93/43)^2 = 4.7$ to $(93/28.5)^2 = 10.6$. Here we have used the fact that intensity is inversely proportional to the square of the distance from the source of the radiation.

APPENDIX TOPIC 30

30.1. Here we show how the size of each planet as seen from a given planet is calculated in terms of the size of our Moon.

In Figure 1 each planet's distance from the star is given in AU (astronomical units) and each planet's size (radius/diameter) d is given in terms of the Earth's size. We illustrate the process by calculating the size of planet g as seen from planet e. The basic formula behind the process is

$$\theta = d/D \qquad (1)$$

with d being planet g's diameter in miles, D being g's distance from planet e in miles, and θ being the angle subtended by g's diameter at (the center of) e in radians. Let d' be d in Earth diameters, and D' be D in AU. d' and D' are given in Figure 1 of Topic 30.

We have that the ratio of planet g's angular size (as seen from planet e) to the angular size of our Moon is

$$[(d/D)\,(57.2957^\circ)]/0.5^\circ = [(8000d' / 93{,}000{,}000D')\,(57.2957^\circ)]/0.5^\circ = (d'/D')\,(.00986). \qquad (2)$$

Here 8,000 miles is Earth's diameter, 93,000,000 miles is Earth's distance from the Sun, 57.2957 is the number of degrees in one radian, and 0.5 is our Moon's diameter in degrees. From Figure 1 Topic 30, $D' = 0.0469 - 0.0293$ AU and $d' = 1.15$ Earth diameters. Equation (2) gives the ratio as .64.

30.2. Here we answer the question "What is the magnitude of Alpha Centauri A as seen from Proxima?"

Proxima Centauri is about 0.2 light years distance from Alpha Centauri.

A few abbreviations will be helpful:

ACA = Alpha Cent A, ACB = Alpha Cent B, P = Proxima Cent.

Absolute magnitude is defined as the brightness of an object at a distance of 10 parsecs, i.e., at 32.6 light years.

The absolute magnitudes of ACA and ACB are 4.38 and 5.71 respectively. Both are 4.37 light years from Earth. P is 4.24 light years from Earth.

Let b be the brightness of ACA as seen from 32.6 light years.
Let B be the brightness of ACA as seen from 0.2 light years.
Let m be the magnitude of ACA as seen from 32.6 light years.
Let M be the magnitude of ACA as seen from 0.2 light years.

As $m = -2.5\log_{10}b$ and $M = -2.5\log_{10}B$, the relationship between the four variables is

$$B/b = 100^{(m-M)/5} . \qquad (1)$$

Solving for M we have

$$M = m - (5/2)\log_{10}(B/b) . \qquad (2)$$

The absolute magnitude of ACA is m = 4.38.
$B/b = (32.6/0.2)^2$ since brightness drops as the square of the distance.

Plugging this value and that of m into (2) gives

M = -6.68 as the magnitude of ACA as seen from P.

The variation in the magnitude of Venus as seen from Earth is -3.7 to -4.5.

We can use (1) to compare the brightness b_A of ACA as seen from P to the maximum brightness b_V of Venus as seen from Earth to obtain

$$b_A/b_V = 100^{[-4.5 - (-6.68)]/5} = 7.45 . \qquad (3)$$

ACA as seen from P is thus about 7.45 times as bright as Venus can be from Earth.

APPENDIX TOPIC 31

31.1. Here is the calculation that at 3,050 feet below the ocean the pressure is 90 times that at the surface of Earth.

The pressure in kg/m^2 at the bottom of a column of water that is H meters high and one meter square at its base is the weight of the column of water, and this equals the volume of the column (which is H cubic meters) times the density of water. The density of water (at 4° C) is 997 kg/m^3.

Two other units of pressure are Pa (1 Pascal) and ATM (1 atmosphere). 1 ATM = $10^6/9.9$ Pa and 1 Pa = $1/9.808\ kg/m^2$. So 1 ATM = $10^6 / (9.9 \times 9.808)\ kg/m^2$.

Thus to find H (in meters) when the pressure is 90 ATM (the pressure at the surface of Venus) we write

$$997H = 90\ [10^6 / (9.9 \times 9.808)\ ]\ .$$

This gives H = 929.7 ~ 930 meters, which is 930 x 3.28 = 3,050 feet.

APPENDIX TOPIC 34

34.1. We show that the tidal force of the Sun is about 46 percent that of the Moon. We also derive equation (2) in Topic 34.

Let M_M, M_S, and M_E respectively be the masses of the Moon, Sun, and Earth.
Let R_M and R_S respectively be the distances of the Moon and Sun from the Earth.

At first one might think the answer is given by looking at the ratio of the force of gravity of the Sun at the Earth divided by the force of gravity of the Moon at the Earth, namely

$$GM_SM_E / R_S^2 \text{ divided by } GM_MM_E / R_M^2\text{, which is}$$

$$(M_S / M_M) / (R_S / R_M)^2\ . \qquad (1)$$

But remember from the discussion on tides in Topic 34 that it is the *differential* force vectors, not the force vectors, that are important for tidal effects. So we must first determine an equation for the differential force vector. This can be done without using calculus, but it is immediately obtained using calculus, which we will do.

The gravitational force vector is $F = -GMm/R^2$ for masses m and M at a distance R from each other. Differentiate this with respect to R to obtain

$$dF/dR = 2GMm / R^3\ . \qquad (2)$$

Thus with a small change $R' = \Delta R$ in R, there is a change $F' = \Delta F$ in F given by

$$F' = \Delta F = (2GMm / R^3)\Delta R = (2GMm / R^3)R'\ . \qquad (3)$$

F' is the (magnitude of the) differential force. In our situation, with M being the mass of the Earth, m the mass of the Moon, and R the distance between (the centers of) the Earth and Moon, R' is the radius of the Earth. A similar statement holds when replacing the Moon with the Sun. Note that we have now proved equation (2) in Topic 34, as that equation is the same as (3) above.

So let us now consider the ratio of the *differential* force of gravity of the Sun at the Earth divided by the *differential* force of gravity of the Moon at the Earth, namely

$$2GM_SM_ER' / R_S^3 \text{ divided by } 2GM_MM_ER' / R_M^3\text{, which is}$$

$$(M_S / M_M) / (R_S / R_M)^3\ . \qquad (4)$$

(4) differs from (1) in that the denominator now involves a power of 3 instead of 2.

Get out your calculator and plug into (4) the values:
$M_S = 1.989 \times 10^{30}$ kg, $M_M = 7.347 \times 10^{22}$ kg, $R_M = 2.389 \times 10^5$ miles, $R_S = 9.296 \times 10^7$ miles. We obtain the ratio of .46, i.e., the Sun's tidal effect is about 46 percent of the Moon's.

Finally, it must be noted that equation (4) was derived for the point A on the Earth's surface that is exactly opposite the Moon/Sun, or to its antipode A'. For convenience let us just refer to point A. Here the gravitational force *vector* at the center of the Earth is conveniently parallel to the force vector at A.

In the general case the force vector at the center of the Earth is not parallel to the force vector at other points on the surface of the Earth. But the tidal vector at any point is still the force vector at that point minus the force vector at the center of the Earth. Also, the height by which the tide rises is proportional not to the (magnitude of the) tidal force at the point, but to the vertical component of that tidal force.

In the derivation regarding A, the vertical component of the tidal force is conveniently the tidal force itself (with a tangential component of zero). So the ratio, given by (4), of the tidal force for the Sun divided by the tidal force of the Moon at A is the same as the ratio of the height of the tide at A due to the Sun divided by the height due to the Moon.

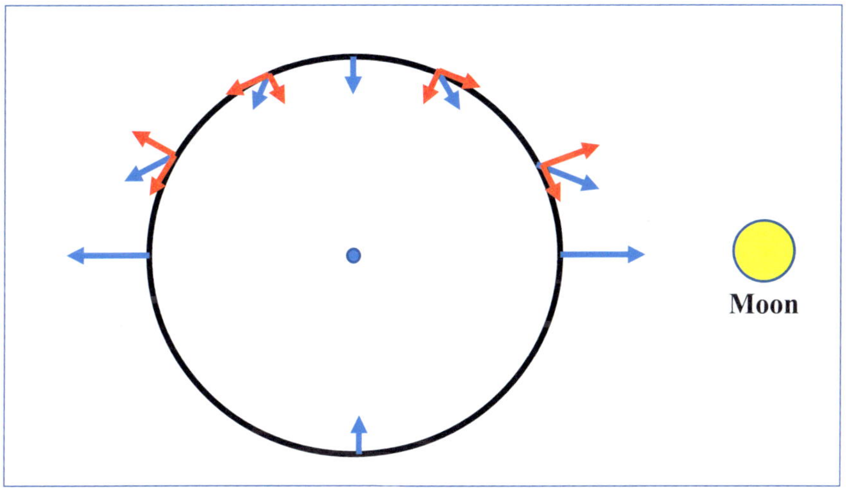

Figure 5 from Topic 34 is repeated here for easy reference. Tidal forces (blue arrows) are indicated at a few points with their vertical and tangential components indicated with red arrows.

What about at a point, call it B, on the Earth's surface where the gravitational force vector at B is not parallel to that at the center C and where the tidal force vector at B (which is the gravitational force vector at B minus the gravitational force vector at C) is not perpendicular to the Earth's surface at B? It turns out that (4) also holds at B, that is, (4) still gives the *ratio* of the height of the tide at B due to the Sun divided by the height at B due to the Moon. See Topic 34 reference [2] for a derivation for the general case.

34.2. We show that, in Figure 6, point A (at latitude 40° N) is 12.5° from the tidal bulge point D while B (where a person at A will be about 12 hours later) is 67.5° from the bulge point D'. Thus it certainly is the case that the tidal bulge at A is higher than at B.

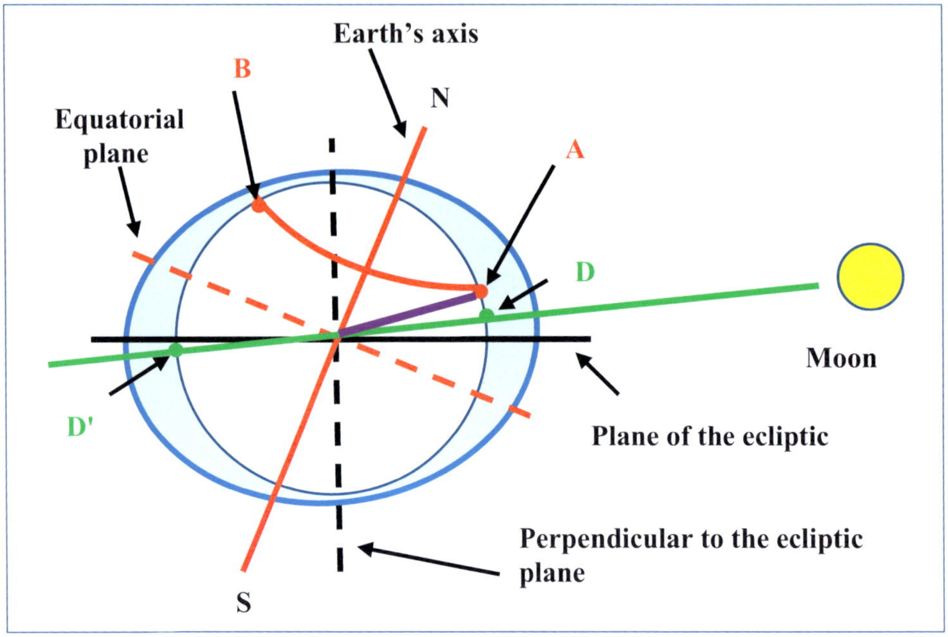

Figure 6. The high tide for a person at location A is higher than the high tide for that same person approximately 12 hours later when he/she has rotated to location B.

Figure 6 from Topic 34 is repeated here for easy reference. The tilt of the Earth's axis to the ecliptic, which is the angle between the dashed black line and the solid red line, is 23.5°. We assume the Moon is 4° above the ecliptic plane (it varies between -5.1° to +5.1° from the plane during a lunar month), i.e., 4° is the angle between the green line and the horizontal black line.

We assume that a person is at point A, which is at latitude 40° N, and 12 hours later that person is now at B due to the Earth's rotation.

We wish to show that point A is 12.5° from the tidal bulge at D and B is 67.5° from its tidal bulge at D'.

Because the solid and dashed black lines are perpendicular to each other, and the same is true for the solid and dashed red lines, the angle between the solid black line and the dashed red line is 23.5°.

The angle between the green line and the sold black line is 4°. Since the latitude of point A is assumed to be 40°, this means that the purple line (from the center of Earth to A), makes an angle of 40° with the dashed red line. Note that (where angle(x, y) is the angle between lines x and y)

angle(purple line, green line) =
angle(purple line, dashed red line) *minus* angle(green line, solid black line)
minus angle(solid black line, dashed red line).

This means that point A is 40 - 4 - 23.5 = 12.5° from point D. Similarly, B is 40 + 23.5 + 4 = 67.5° from D'. (Or you might notice the angle = 80° - 12.5° as the antipode of A is 80° from B.)

34.3. Here we show that in 12 hours the Moon moves about 6.6° in its orbit around the Earth, and thus it takes about 12.44 hours and not 12 hours from one high tide to the next.

Suppose the Moon moves d degrees in its orbit around the Earth in 12 hours. It takes about 27.3 x 24 hours for the Moon to orbit the Earth. (Here sidereal time is used.) We thus have that

$$12/(27.3 \times 24) = d/360^\circ .$$

This gives d = 6.6°. Suppose it takes T hours for the Earth to rotate 6.6° on its axis. Since it takes 24 hours to make a complete rotation, we have that

$$6.6^\circ/360^\circ = T/24 .$$

Thus T = .44 hours. So the Earth must rotate an additional .44 hours after rotating 12 hours to reach the second high tide.

APPENDIX TOPIC 38

38.1. Here we prove that the tangential velocity of the Moon in its orbit about the Earth is inversely proportional to the square root of its distance from the (center of the) Earth.

Assume the Moon orbits in a circle, m = Moon's mass, r = distance of the Moon from the Earth's center, P = period of the Moon, v = orbital velocity of the Moon, and ω = the angular velocity (cycles/second) of the Moon. Then (using $x \propto y$ to mean x is *directly proportional* to y, i.e., there is a constant k such that x = ky)

the Moon's angular momentum is $L = mrv$,

$$v = \omega r ,$$

$P^2 \propto r^3$ (from Kepler's 3rd law), and

$$1 \text{ cycle} = \omega P .$$

Thus

$$L \propto r^{1/2} .$$

Since $mrv = L \propto r^{1/2}$, we have that

$$v \propto r^{-1/2} .$$

From a more general point of view, where the Moon is in an elliptical orbit about the Earth, one can show that

$$v = \sqrt{[(2/r)-(1/a)]gR_e}$$

where a = semi-major axis of the Moon's orbit, g is the acceleration of gravity at the Earth's surface, and R_e is the radius of the Earth. Thus, setting a = r, we again obtain $v \propto r^{-1/2}$ for a circular orbit.

38.2. We show that, given the assumptions in the text, the Moon's size (i.e., its diameter) as seen from Earth initially was 12 times its size today. Let θ be the angular size (in radians) of the Moon as seen from Earth when the Moon was first formed. Let θ' be the angular size as seen today (which is 1/2 degree). Let d be the diameter of the Moon, r the initial distance of the Moon from the Earth, and r' the current distance. Then

$$\theta = d/r \text{ and } \theta' = d/r' .$$

So $\theta/\theta' = r'/r = 238{,}000/20{,}000 = \sim 12$. Thus $\theta = 12(.5) = 6$ degrees.

APPENDIX TOPIC 40

40.1. We assume the angular size of the Tarantula is 20 arc minutes, its distance is 160,000 light years, and the distance of the Orion Nebula is 1,500 light years. We show that the diameter of the Tarantula is on the order of 1,000 light years, and if it were at the distance of the Orion Nebula it would cover an area of the sky with a diameter of about 36 degrees. This is all very easy if we remember the formula

$$\theta = S/R ,$$

where θ is the angular size of an object in radians, S is the diameter of the object, and R is its distance.

Let S be the diameter of the Tarantula. Let θ be the angular size of the Tarantula at the distance of $R = 160{,}000$ light years. Let θ' be the angular diameter of the Tarantula at a distance of $R' = 1{,}500$ light years.

First, 20 arc minutes is $20 / (60 \times 57.3) = 0.00582$ radians. So $.00582 = \theta = S/160{,}000$, which gives

$$S = 931 \text{ light years, or about } 1{,}000 \text{ light years.}$$

We have $\theta = S/R$ and $\theta' = S/R'$. Algebra gives us $\theta' = \theta R/R'$. Since we have angles on both sides of the equation, we don't have to use radians. We obtain

$$\theta' = 20 \times 160{,}000 / 1{,}500 = 2{,}133 \text{ arc minutes or } 35.6 \text{ degrees.}$$

APPENDIX TOPIC 50

50.1. We give proofs for the following five equations:

$$h = (g_e/g)h_e \qquad (1)$$

$$v_{esc} = \sqrt{2GM/R} \qquad (2)$$

$$v_o^2 = 2g_e h_e \qquad (3)$$

$$R^2 = R_e h_e(\rho_e/\rho) \qquad (4)$$

The ratio of a person's weight on the surface of a solar system body to a person's Earth weight is

$$g/g_e = (R/R_e)(\rho/\rho_e) . \qquad (5)$$

For convenience let the solar system object in question be Mars, but it could be any celestial body.

Remember that g_e, R_e, ρ_e, and M_e are respectively the surface acceleration of gravity, the radius, the (average) density, and the mass of the Earth. g, R, ρ, and M are the same quantities, but for Mars. h_e and h are the heights at which Alice can spring up respectively from Earth and from Mars.

We first notice that (3) implies (1):

The initial velocity v_o of Alice is the same for *all* bodies. So from (3),

$$v_o^2 = 2g_eh_e \text{ on Earth and } v_o^2 = 2gh \text{ on Mars.}$$

Thus $2g_eh_e = 2gh$, whence $h = (g_e/g)h_e$, which is (1).

We directly prove (5):

Let m be Alice's mass. (It will drop out of the equation, but we need it for the proof.)

From Newton's universal law of gravitation the force F_e between the Earth and Alice (on the surface of the Earth) is

$$F_e = GM_em/R_e^2 .$$

From Newton's Second Law,

$$F_e = mg_e .$$

Equating the right sides of the two equations we have that

$$GM_e/R_e^2 = g_e .$$

Now replace the Earth by Mars and have Alice on the surface of Mars, with F being the gravitational force between Alice and Mars. Similarly we have $F = GMm/R^2 = mg$, and so

$$GM/R^2 = g .$$

Dividing the left side of the last equation by the left side of the previous equation, ditto for the right sides, and equating the two resulting numbers, one obtains

$$g/g_e = (M/M_e)(R_e^2/R^2) .$$

Since density is mass/volume, $\rho_e = M_e/(4/3)\pi R_e^3$ and $\rho = M/(4/3)\pi R^3$. So

$$M/M_e = (\rho/\rho_e)(R^3/R_e^3) .$$

Inserting this into the previous equation, we obtain

$$g/g_e = (\rho/\rho_e)(R/R_e) \text{ which is (5).}$$

Next we show that (2) together with (3) implies (4):

In the proof we will use (5), which we already know to be true.

(2) says $v_{esc}^2 = 2GM/R$. (3) says $v_o^2 = 2g_eh_e$.

We now set $v_o = v_{esc}$. In doing so, note that we are here assuming that v_{esc} (the escape velocity of a *moon/asteroid*) is much less than the escape velocity of *Earth*. This assumption is needed in order that h_e be finite (it would be infinite if $v_o \geq v_{esc}$ the escape velocity of the Earth) and also that h_e not be so large that the acceleration of gravity at height h_e above the Earth's surface is significantly less than at the Earth's surface.

With m as Alice's mass, $mg = GMm/R^2$ by Newton's law of gravitation and his second law. Thus $g = GM/R^2$. Using this and (2) we have that

$$v_{esc}^2 = 2gR .$$

Setting $v_{esc} = v_o$, and using (3) we thus have that

$$g/g_e = h_e/R .$$

Now we use (5) to obtain

$$(\rho/\rho_e)(R/R_e) = h_e/R .$$

Solving for R we have

$$R^2 = R_eh_e(\rho_e/\rho) ,$$

which is (4).

All five of the equations will have been proved if we can directly prove that both (2) and (3) are true. Both proofs use some elementary calculus. If you have not studied calculus, you can stop here. At least some of the "magic" has been taken out of the equations, as you have seen how equations (1), (4), and (5) are derived.

Proof that equation (3) $v_o^2 = 2g_eh_e$ **is true:**

Since we are only concerned with the Earth, let us drop the use of the subscript e.
m is the mass of Alice.

The force of gravity is

$$F = -mg .$$

We use the negative sign to indicate that the force is directed downward. For us the positive direction is up. x will denote Alice's distance above the Earth's surface, with v_o being her initial upward velocity. We will say that at time t Alice is at height h with velocity v. It is convenient to say that she jumps up at time t = 0. So x_o (her initial height) is zero.

$$F = -mg = m(d^2x/dt^2) .$$

So

$$d^2x/dt^2 = -g .$$

Integrate to get

$$v = dx/dt = -gt + C_1 \text{ for a constant } C_1 .$$

Let t = 0 to obtain $v_o = C_1$. Thus

$$v = dx/dt = -gt + v_o . \tag{6}$$

This is the general equation giving the velocity of Alice at any time t.

Integrate again to get

$$x = -gt^2/2 + v_ot + C_2 \text{ for a constant } C_2 .$$

Let t = 0 to obtain $x_o = C_2$. So

$$x = -gt^2/2 + v_ot + x_o . \tag{7}$$

This is the general equation giving the height of Alice at any time t.

For us, $x_o = 0$. Now Alice will be at her highest distance h above the surface when her velocity v is zero. Set v = 0 in equation (6) to obtain

$$t = v_o /g .$$

Then replace t in equation (7) with v_o/g and x with h to get

$$h = v_o^2/2g .$$

So

$$v_o^2 = 2gh ,$$

which is (3) if we restore the subscript e to the g and h.

Proof that equation (2) $v_{esc} = \sqrt{2GM / R}$ is true:

m is the mass of Alice. She is at distance r from the center of a solar system body of mass M and radius R. The gravitational force between Alice and the solar system body is

$$F = -GMm/r^2 = m(dv/dt) .$$

So (using the chain rule)

$$-GM/r^2 = (dv/dr)(dr/dt) = (dv/dr)v .$$

When r = R, Alice is on the surface of the body and has initial velocity v_{esc}. We set the final velocity, when Alice is at infinity, as zero. Then, integrating both sides of the above equation, we obtain

$$\int_R^{\infty} (-GM / r^2) dr = \int_{v_{esc}}^{0} v dv .$$

Evaluating both sides we have

$$-GM/R = -v_{esc}^2/2 .$$

Thus $v_{esc}^2 = 2GM/R$, which is (2).

APPENDIX TOPIC 51

51.1.

II. The determination of the circumference of the Earth using two cities that are on the same meridian. Shadow measurements are made simultaneously at both cities.

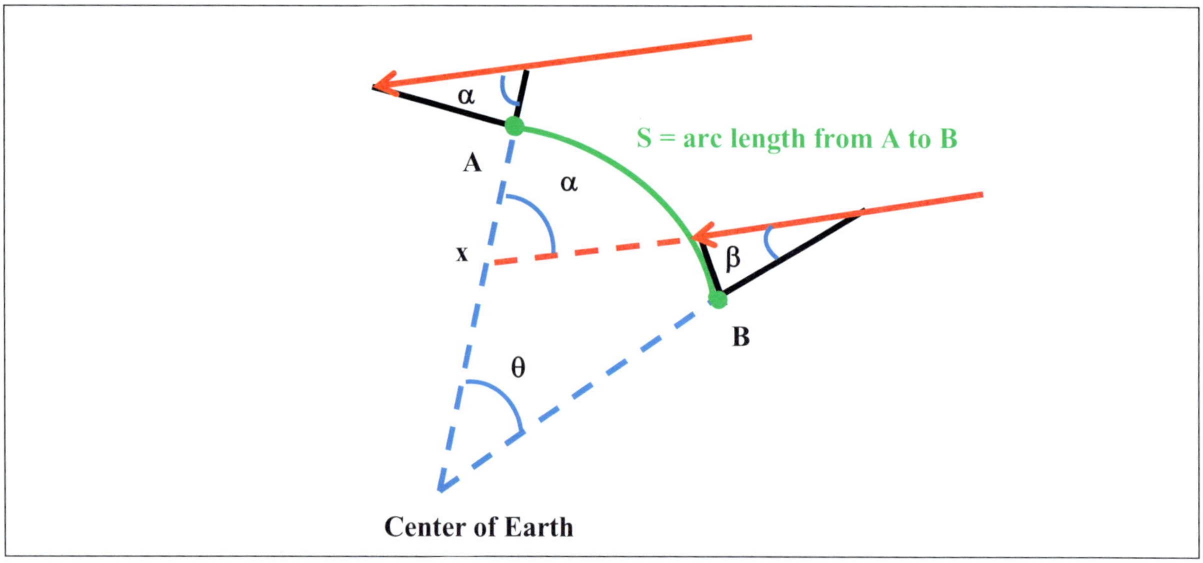

Figure 1. A and B are cities where the angle α is larger than β (which, given the direction of the Sun's rays in the diagram, places A north of B if A and B are in the northern hemisphere). Parallel Sun rays (in red) strike the two cities. At noon at A an upright pole casts a shadow whose length can be measured. The same holds at B. θ is the angle subtended at the center of the Earth by the arc of length S, which is the distance between the two cities. The angle between the dashed red and blue lines equals α because the two red rays are parallel. For clarity, the two poles and their shadows (at A and B), both in black, are greatly out of scale.

As in Case 1,

$$C = 360 \times S/\theta .$$

The angles α and β are measured the same as in Case I, using the lengths of the poles and the shadows cast by these poles. It is easy to see that θ is given by the equation

$$\theta = \alpha - \beta .$$

To see this, look at the triangle formed by the center of the Earth, x, and the top of the pole at B. The sum of the interior angles of any triangle is 180°. The three angles for this triangle are θ, β, and $180 - \alpha$. Thus $\theta + \beta + 180 - \alpha = 180$, whence $\theta = \alpha - \beta$.

Note:
1. If $\beta = 0$, then Case II reduces to Case I (with $\theta = \alpha$).
2. In Figure 1, the shadows cast by both poles are to the north. If the Sun's rays cast both shadows to

the south, we still have $\theta = \alpha - \beta$ with the convention that we adopted, namely that α is larger than β, which means that A now denotes the southern city, instead of the northern city (as in Figure 1).
3. The only other possibility is that the Sun's rays at the northern city cast a shadow of its pole to the north and cast a shadow of the southern city's pole to the south. We show that in this case we obtain

$$\theta = \alpha + \beta \,.$$

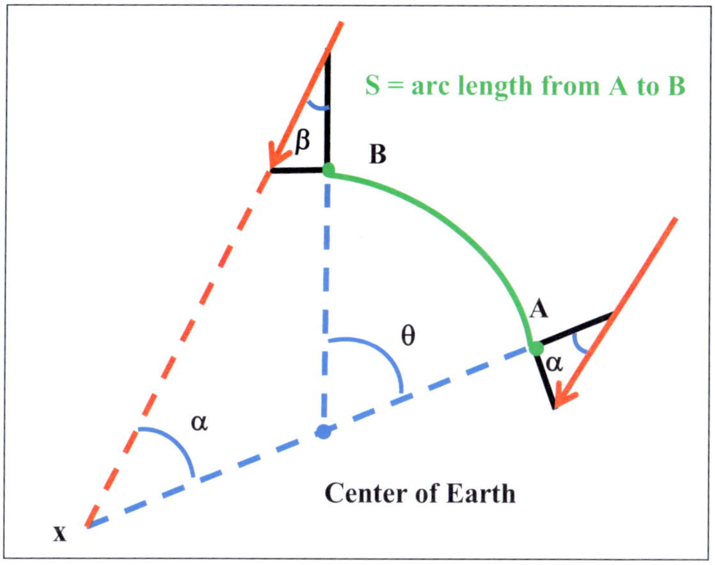

Figure 2. A modification of Figure 1 for the situation where the shadows cast by the poles at cities A and B are now in opposite directions. Again, lengths of the poles and the shadows are greatly exaggerated for clarity.

Because the Sun's rays (in red) are parallel, the angle at x between the dashed red line and the dashed blue line is α. The interior angles of the triangle formed by x, the top of the pole at city B, and the center of the Earth are α, β, and $180 - \theta$ respectively. Since the sum of the interior angles of any triangle is 180°, we have

$$\alpha + \beta + (180 - \theta) = 180 \,,$$

whence $\theta = \alpha + \beta$.

This covers all possible scenarios when the two cities are on the same meridian.

III. Determining the Earth's circumference using two cities that are not necessarily on the same meridian.

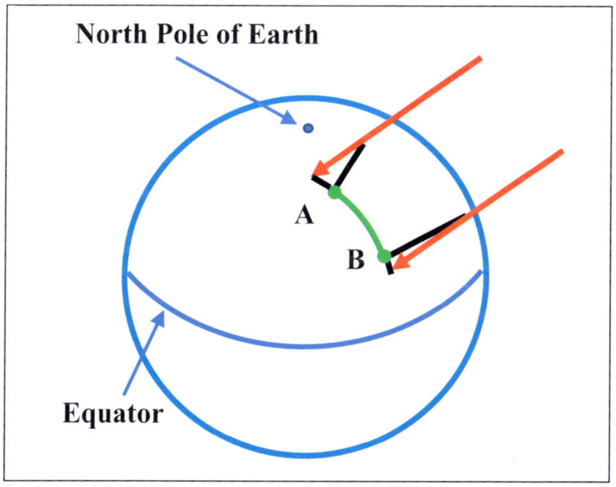

Figure 3. Cities A and B are connected by an arc (in green) that is part of a great circle running through the two cities on the Earth's surface. The arc need not lie on a longitudinal meridian. The Sun's rays are in red.

This is simply Case II in disguise. The green arc joining cities A and B in Figure 3 is part of the great circle on the Earth passing through A and B—call this the "green great circle." (Recall that a great circle on a sphere is one whose center is at the center of the sphere.) The green great circle takes the place of the meridian of longitude (another great circle) in Case II. In Case II we assumed the Sun was at noon, i.e., the Sun was on the meridian. Here the counterpart to this situation is to require the Sun to be on the green great circle.

While it is easy to tell when the Sun is on the meridian (i.e., when it is noon), one might ask how one can tell when the Sun is on the green great circle? But that is easy. A person at either city is presumed to know the direction toward the other city. The direction to the Sun must be in alignment with the arc between the cities. (More technically, the Sun must be in the plane that passes through the green great circle.) Put another way, the shadow cast by the pole at either city must be in alignment with the direction to the other city. In Case II, the arc between the cities is north-south. So the Sun has to be on the meridian (i.e., it is noon). Just as in Cases I and II, in Case III we have either θ equal to 0, $\alpha - \beta$, or $\alpha + \beta$. For Figure 3, $\theta = \alpha + \beta$. And of course $C = 360 \times S/\theta$.

APPENDIX TOPIC 52

52.1. We prove equations (9) and (10) in Topic 52.

(3) $\Rightarrow e(S + U) = Us \Rightarrow U = eS/(s - e)$.
(4) $\Rightarrow uU = (U - M)e \Rightarrow U = Me/(e - u)$.

Thus $Me/(e - u) = eS/(s - e) \Rightarrow M/S = (e - u)/(s - e)$.

Using (7), this implies $m/s = (e - u)/(s - e) \Rightarrow (s - e)/s = (e - u)/m \Rightarrow 1 - e/s = e/m - u/m \Rightarrow$

$$e/m + e/s = 1 + u/m = 1 + B\,. \qquad (*)$$

Thus $(e/m)(1 + m/s) = 1 + B$. By (7), $A = s/m$. Therefore $(e/m)(1 + 1/A) = 1 + B$, from which we obtain $m/e = (A + 1)/A(B + 1)$, the second equation in (9).

So $(e/s)(A + 1) = (e/s)(s/m + 1) = e/m + e/s = 1 + B$, using equation (*) again. From this we obtain $s/e = (A+1)/(B+1)$, which is the first equation in (9).

We are almost done. By (6), $m/M = s/S = \alpha$, if α is in radians.
Thus $S/e = (s/e)/\alpha$ and $M/e = (m/e)/\alpha$. Since $1^\circ = \pi/180$ radians,

$$S/e = (s/e)/(\pi\alpha/180), \text{ and } M/e = (m/e)/(\pi\alpha/180) \text{ if } \alpha \text{ is in degrees.}$$

These are the two equations in (10), completing the derivations.

APPENDIX TOPIC 54

54.1. We give a more general proof of Topic 54 Case 2, where now we don't assume an angle of 50°.

Here we assume that

At time $t = 0$, θ is not close to being either 0° or 180°.

The more general proof continues as follows:

Equivalently,

> *At time $t = 0$, θ is not within an angular distance of d degrees from 0° or 180°, for some d much larger than 0.*

This implies that neither θ nor $180^o + \theta$ is within d degrees of being 0 at time $t = 0$. But after n orbits by Pluto (and thus at a later time), Neptune will be $[180n + \theta](\text{mod } 360)$ degrees from C_N (with Pluto at P). (XmodY is the remainder when X is divided by Y. It is "clock arithmetic." E.g., 29mod24 is 5. So 29 hours is 5 a.m., assuming midnight is the zero hour.) Thus, for n = 0, 1, 2, etc., we have Neptune at angles θ, $180 + \theta$, θ, *repeating ad infinitum.*

So Neptune will never be closer than d degrees from C_N when Pluto is at point P in its orbit, which is close to C_P. Since the point P is any point that is close to C_P, if d is large enough, resonance will keep Neptune from ever getting close enough to Pluto to have Pluto's orbit altered.

APPENDIX TOPIC 56

56.1. Further discussion for Topic 56:
The arctic circle is that circle of latitude where the Sun is on the horizon at midnight at summer solstice (when the Sun is directly above the Tropic of Cancer). At all other times during that day the Sun is above the horizon. Equivalently, the arctic circle is that circle of latitude where the Sun is on the horizon at noon at winter solstice (when the Sun is directly above the Tropic of Capricorn). At all other times during that day the Sun is below the horizon.

The latitude of the arctic circle is $90^o - 23.5^o = 66.5^o$ where 23.5^o is the latitude of the Tropic of Cancer.

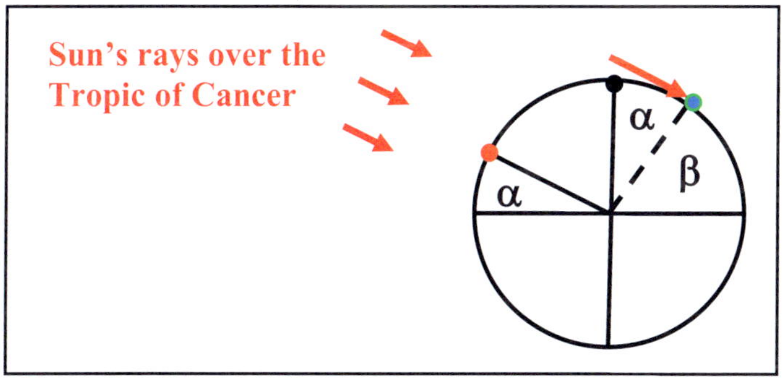

Figure 1. Shows the arctic circle is at latitude 66.5°. The red, black, and green dots are respectively the Tropic of Cancer, the north pole, and the point on the arctic circle where the Sun is on the horizon at midnight. α (23.5°) is the latitude of the Tropic of Cancer. β is the latitude of the arctic circle.

Consider equation (1) in Topic 56: $\cos(J) = \sin(L)/\cos(D)$. Of course, above the arctic circle there are days when the Sun doesn't set, in which case the equation is invalid (there is no angle J).

What happens at the north pole? Here $L = 90^o$, so $\cos(J) = 1/\cos(D)$. Since $\cos(D)$ is less than 1, unless $D = 0$ (for an acute angle D), and since $\cos(J)$ cannot be greater than 1, it must be the case that $D = 0$ and $J = 0$, i.e., the Sun is on the equator and thus is at an equinox. So at the north pole the Sun can be seen to set/rise only when the Sun is at an equinox and the angle of the Sun with respect to the horizon is zero. When this is the case, the Sun circles the *horizon* during one 24-hour period. On all other days the Sun is either always above or below the horizon (and moving in a circle parallel to the horizon). We have used the equation to deduce this, but thinking about the geometry between an observer at the north pole and the location of the Sun easily shows the same thing.

INDEX OF NAMES

Made in the USA
Las Vegas, NV
04 November 2026

57428327R00176